sustainability

Sustainability Transition Towards a Bio-Based Economy

New Technologies, New Products, New Policies

Edited by
Piergiuseppe Morone

Printed Edition of the Special Issue Published in *Sustainability*

www.mdpi.com/journal/sustainability

MDPI

Sustainability Transition Towards a Bio-Based Economy

Sustainability Transition Towards a Bio-Based Economy

New Technologies, New Products, New Policies

Special Issue Editor

Piergiuseppe Morone

MDPI • Basel • Beijing • Wuhan • Barcelona • Belgrade

MDPI

Special Issue Editor
Piergiuseppe Morone
Unitelma Sapienza University of Rome
Italy

Editorial Office
MDPI
St. Alban-Anlage 66
4052 Basel, Switzerland

This is a reprint of articles from the Special Issue published online in the open access journal *Sustainability* (ISSN 2071-1050) in 2018 (available at: https://www.mdpi.com/journal/sustainability/special_issues/Sustainability_Transition)

For citation purposes, cite each article independently as indicated on the article page online and as indicated below:

LastName, A.A.; LastName, B.B.; LastName, C.C. Article Title. *Journal Name* **Year**, *Article Number*, Page Range.

ISBN 978-3-03897-380-5 (Pbk)
ISBN 978-3-03897-381-2 (PDF)

Cover image courtesy of pexels.com user McKylan Mullins.

Contents

About the Special Issue Editor

Piergiuseppe Morone is Full Professor of Economic Policy at Unitelma Sapienza with a strong interest in green innovation and sustainable development pushing his research at the interface between innovation economics and sustainability transitions, an area of enquiry that has attracted growing attention over the last decade. His work regularly appears in prestigious innovation and environmental economics journals. He is the coordinator of STAR-ProBio, an H2020 research project aiming at developing sustainability assessment tools for bio-based products and promoting their market penetration. Piergiuseppe is head of the Bioeconomy in Transition at Unitelma Sapienza, a research group studying the emergence of a circular and bio-based economy. Piergiuseppe is member of the Editorial Boards of Current Opinion in Green and Sustainable Chemistry (ELSEVIER) and Open Agriculture (De Gruyter Open) and acts as Guest Editor for various journal including the *Journal of Cleaner Production* (ELSEVIER) and *Sustainability* (MDPI).

sustainability

MDPI

Editorial

Sustainability Transition towards a Biobased Economy: Defining, Measuring and Assessing

Piergiuseppe Morone

Bioeconomy in Transition Research Group (BiT-RG), Unitelma Sapienza University of Rome, 00161 Rome, Italy; piergiuseppe.morone@unitelmasapienza.it

Received: 24 July 2018; Accepted: 25 July 2018; Published: 26 July 2018

Abstract: This Special Issue proposes an array of 11 key papers aimed at investigating the complex and multifaceted nature of the biobased economy, focusing both on a conceptual understanding of the transition and on the measurement issues associated to it. More specifically, collected papers can be broadly divided in two groups: (1) those aiming at adding to our understanding of the transition process towards a sustainable biobased economy; and (2) those aiming at adding to the definition and measurement of the emerging sustainable biobased economy. In the guest editor view, papers collected in this Special Issue offer valuable and complementary insights to our understanding of the ongoing transition towards a biobased economy, providing a logical framework to understand the transitions, as well as an overview of existing tools to assess and measure it. Ideally, policy makers will benefit from the papers included in this Special Issue and, hopefully, it will contribute to make a further step to the much-needed transition towards sustainability.

Keywords: biobased economy; sustainability transition; standards; indicators

1. Introduction

It is widely acknowledged that, currently, two economic models co-exist, side-by-side, i.e., the dominant fossil-economy and the emerging biobased economy. The rise of a new biobased economic model underlines the need to undergo a paradigm shift towards sustainability in order to meet society's long-term goals and emerging challenges, which include the following: decoupling economic growth from environmental pressure, managing natural resources in a sustainable way, improving food security, reducing poverty, etc. Although sustainability has become a core theme of innovation economics, strategies for achieving this goal—and hence, for supporting the paradigm shift—remain under-investigated, mainly due to the complexity related to the manifold nature of the concepts involved.

Among others, key drivers of this paradigm shift (or transition towards sustainability, we might say) involve the following: (1) development and diffusion of new green technologies (eco-innovations) for a biobased economy; (2) development of a holistic approach for sustainability assessment of biobased products (e.g., sustainability schemes, standards, and eco-labelling initiatives); (3) policy measures for promoting market uptake of biobased products and creating a level playing field among biobased products and conventional (fossil-based) alternatives.

However, before looking closely at these leveraging points we need to better define the biobased economy first, and properly measure it, subsequently. The lack of agreed-upon definitions, as well as measurement tools, can be explained by the complex nature of the phenomenon under investigation. Indeed, the biobased economy sits at the intersection of many overlapping concepts, including sustainable development, circular economy, and green technologies, which are complex notions on their own.

Hence, initiating a transition towards such a new socio-economic paradigm (i.e., based on biomasses and circularity principles) is not just a matter of scaling-up an innovative technology that has emerged in a niche [1] (p. 63), but rather it involves the emergence of a new and complex set of relations among stakeholders acting at the production level, as much as the consumption level. These intersect with institutional actors playing a fundamental role in steering the transition process altogether.

In order to account for these complex relations, the literature on sustainability transitions (see, among many others) [2], has argued in favor of adopting a multi-level perspective (MLP), where socio-technical changes are seen as the outcome of a combined pressure exerted upon the incumbent dominant technological regime (i.e., the meso-level), operated simultaneously from the landscape level (i.e., the macro-level consisting of a set of deep structural trends) and the technological niche level (i.e., the micro-level where new technologies are developed in a protected environment).

Stemming from these considerations, this Special Issue proposes an array of 11 papers aimed at investigating the complex and multifaceted nature of the biobased economy, focusing first on a conceptual understanding of the *transition* and, subsequently, on the measurement issues associated with it.

Papers included in this Special Issue provide contributions coming from researchers trained in hard sciences (chemistry, agronomy, engineering) as well as in social sciences (economics, management, policy analysis). This broad spectrum of academic knowledge is well complemented by more applied perspectives coming from international institutions, NGOs, consultants, and independent analysts. This plurality of voices gives the reader an overall picture of the transition towards a biobased economy, highlighting the mounting complexity characterizing the system under investigation. Papers collected in this Special Issue can be broadly divided in two groups: (1) those aiming at adding to our understanding of the transition process towards a sustainable biobased economy; and (2) those aiming at adding to the definition and measurement of the emerging sustainable biobased economy. In the first case, authors often take a country-specific or a sector-specific perspective to resolve the complexity of the challenge. In the second case, a European-wide (or even worldwide) outlook is often privileged as well as a life-cycle perspective that spans from feedstocks' procurement to end-of-life options.

2. Understanding Sustainability Transition Pathways to a Biobased Economy

Four papers of this Special Issue broadly refer to the transition process, with a sectoral focus on transition pathways.

Bennich et al. [3] focus their attention on the Swedish agricultural sector, attempting to identify "high order leverage points", which could best initiate a systemic change. Authors provide valuable insights on the social and ecological processes contributing to or hindering the transition in this sector. The analysis is performed using a combined methodology that relies on systems analysis and expert interviews, and concludes that the assumption that a transition process would necessarily entail an expansion of agricultural production for the purpose of growing crops for non-food biomass applications is challenged. As shown by the authors, different pathways have different implications in terms of biomass demand and supply, and, depending on the objectives set by the policymaker, alternative pathways should be supported.

In a subsequent paper, Bennich et al. [4] use a similar twofold methodology for the forestry sector in Sweden, assessing potential future transition pathways. Desired change processes identified include a transition to diversified forest management, a structural change in the forestry industry to enable high-value added production, and increased political support for the biobased economy concept. Hindrances identified include the difficulty in demonstrating the added value for end users of novel biomass applications, and the uncertainty linked to a perceived high level of polarization in the forestry debate.

Alaerts et al. [5] focus on biobased plastics and assess how the introduction of these new polymers in a circular economic model could disturb the current recycling of fossil-based plastics, inhibiting the closure of plastic cycles. This is a rather relevant research question to address in order to assess how the co-existence of the two economic systems mentioned in the introduction (the fossil-economy and the emerging biobased economy) might produce unexpected side effects, hindering the transition pathway. As claimed by the authors, the co-living with the recycled fossil-based plastics presents no risks for biobased plastics as a group. However, several potential sources of contamination arise when considering separately different bio-polymers. For instance, PLA (polylactic acid) shows a severe incompatibility with PET (polyethylene terephthalate); hence, future risks are assessed by measuring the amounts of PLA ending up in PET waste streams. For PHA (polyhydroxy alkanoate), there is currently no risk, but it will be crucial to monitor future application development. Bearing this in mind, the authors stressed that any introduction of novel plastics must be well guided from a system perspective, properly pondering incompatibilities with current and upcoming practices in the recycling of plastics.

Urmetzer et al. [6] underline the knowledge dimension associated with the biobased economy transition. Taking an innovation systems (IS) perspective, the authors claim that, to successfully contribute to a sustainability transition, a knowledge-based bioeconomy should broaden its scope beyond the techno-economic dimension. Along this line of reasoning, the authors propose to include systems knowledge, normative knowledge, and transformative knowledge in research and policy frameworks for a sustainable knowledge-based bioeconomy (SKBBE). In order for policy makers to be able to effectively steer the bioeconomy transformation onto a sustainable path, the authors claim that a stronger focus on the characteristics of *dedicated* knowledge (including stickiness, locality, context specificity, dispersal, and path dependence) and its creation, diffusion, and use are necessary for the knowledge-based bioeconomy to become truly sustainable.

3. Measuring the Biobased Economy

Measurement issues associated with the biobased economy have been addressed both at the macro and micro levels. Specifically, two papers focus on indicators to measure the biobased economy as a whole, whereas the remaining five focus on value-chains and specific biobased products.

When looking at macro data, Ronzon and M'Barek [7] observe how the monitoring of the European biobased economy is hampered by a lack of statistics on emergent and partially biobased sectors (i.e., those sector where biobased products are produced along non-biobased products). The authors, to tackle this issue, propose a simplified socioeconomic indicator framework for the bioeconomy in the EU. Subsequently, they use the proposed indicators to assess economic performance through a labor productivity measure. This exercise provides insights related to the growth potential of specific bioeconomy sectors in individual EU Member States. The authors first position Member States on a transition path to higher productivity and, subsequently group them looking specifically at the East-West bioeconomy disparities within Europe; these leads to the suggestions of a set of measures to promote the development of the EU biobased economies.

Bracco et al. [8] move from the observation that, worldwide, most countries focus on the contribution of the bioeconomy sectors to gross domestic product (GDP), turnover, and employment. However, this approach offers an incomplete picture as environmental and social aspects, which are unanimously considered as fundamental pillars of the biobased economy, are dropped from the analysis. Bearing this in mind, the authors provide a critical assessment of the national methods used for the measurement, monitoring, and reporting of the bioeconomy contribution to the total economy. The analysis, based on research and surveys conducted on six countries (Argentina, Germany, Malaysia, The Netherlands, South Africa, and the United States), shows the lack of a homogenous definition of bioeconomy across the six considered countries—a fact which hinders any straightforward comparison of the relevance of bioeconomy in the different economies. Moreover, as observed by the authors, the bioeconomy targets set by nation-wide strategies often reflect the country's priorities and comparative advantages linked, for instance, the availability of natural resources, traditional industries, labor productivity, and past investments in R&D.

In order to improve the measurement and monitoring of the bioeconomy, the authors call upon national governments to enhance and coordinate communication among domestic agencies, establishing protocols for sharing data, formalizing biobased industry measurement standards, developing a comprehensive survey for the biobased industry and commodity usage, and improving industry classification systems.

Moving down to the micro level of analysis, Lokesh et al. [9] present a mapping exercise aiming, first, at identifying the most relevant biobased value chains to attain a fully functional biobased circular economy, and subsequently, at visualizing/foreseeing the strengths, weaknesses, opportunities, and challenges associated with it. Value chains selection was done by means of a two-step methodology based on multi-criteria decision analysis. The selection process led to the identification of five key value chains, for each of which specific maps were developed. As claimed by the authors, this exercise demonstrates the highly informative nature of this tool and its crucial role in understanding the complex interactions among the various processes and stakeholders within complex bioeconomy value chains.

Both Falcone and Imbert [10] and Martin et al. [11] look at a specific dimension of sustainability, i.e., the social one. As mentioned earlier, this is an area often neglected in sustainability assessment and, indeed, deserves greater attention. Falcone and Imbert [10] provide an overall assessment of social impact categories and indicators that should be included in a social sustainability assessment of biobased products. The study is performed following a three-step methodology. First, the authors carry out a literature review on existing social life cycle studies, identifying most relevant social categories and indicators. Subsequently, the categories' list and indicators are validated with the help of a focus group bringing together industrial experts and academics. Finally, a zoom-in into consumers' perceptions of social indicators was obtained by conducting semi-structured interviews with consumer representatives. Results showed the need to better exploit consumers' role in the ongoing process of market uptake of biobased products. More specifically, this need entails the effective inclusion of some social indicators (i.e., end users' health and safety, feedback mechanisms, transparency, and end-of-life responsibility) in the social life cycle assessment scheme for biobased products.

On a similar ground, Martin et al. [11] review the scientifically published life cycle studies on biobased products, investigating the extent to which they include important sustainability indicators. Results suggest that there is a discrepancy between the indicators considered important in established frameworks for sustainability assessments, and the indicators that are frequently included in published scientific studies. The authors suggest that greater attention should be paid to categories such as workers conditions, water depletion, indirect land use change, and impacts on ecosystem quality and biological diversity—all elements noted as very relevant also by the United Nations (UN) Sustainable Development Goals.

The last two papers of this Special Issue are dedicated to certification and standards for the emerging biobased economy. Majer et al. [12] performed a rather impressive gap analysis in order to assess the current status of sustainability certification and standardization in the biobased economy. The methodology used was twofold, involving a comprehensive desk analysis complemented by expert interviews. The analysis revealed an impressive amount of existing certification frameworks, criteria, indicators and applicable standards. However, the authors identified major gaps in: (1) existing criteria sets; (2) the practical implementation of criteria in certification processes; (3) the legislative framework; (4) end-of-life processes; as well as (5) necessary standardization activities.

Fonseca and Domingues [13] take a narrower view, focusing on a specific standard (ISO 14001:2015) and assessing the transition towards the revised version of the voluntary environmental management systems (EMS) certification scheme among Portuguese organizations. By means of an on-line survey, the authors collected data on 108 organizations. Respondents viewed *Determination of risks and opportunities* and the *Life cycle perspective* as both the most useful concepts of ISO 14001:2015 but also as the hardest difficulties to overcome by the organizations when implementing or transitioning to the 2015 edition of ISO 14001. Indeed, those organizations that successfully completed the transition reported benefits, with an enhanced environmental performance of the organization and an improvement in the compatibility with other management system standards, such as ISO 9001. Authors also observed that the perception of the benefits achieved with ISO 14001:2015 certification varies with the size of the organization, whereas the motivation to proceed with certification is independent of organization profile.

4. Concluding Remarks and Further Issues on the Research Agenda

This special issue has succeeded to collect eleven key papers addressing, from different angles, the topic of a sustainability transition towards a biobased economy. Contributions include research by European academia, research centers as well as international organizations and institutions. The plurality of voices shows the growing interest around this topic in Europe. Major efforts have been invested to foster the transition both at the single Member State level and at the European Commission level. Since 2012, Europe has adopted a strategy for the bioeconomy that is currently undergoing a revision process to ensure that Europe focuses its efforts in the right direction. As stated by the Commission, "*Bioeconomy is Europe's response to key environmental challenges the world is facing already today. It is meant to reduce the dependence on natural resources, transform manufacturing, promote sustainable production of renewable resources from land, fisheries and aquaculture and their conversion into food, feed, fibre, biobased products and bio-energy, while growing new jobs and industries*". This statement shows a strong commitment to reduce the impact on the environment without compromising on job creation and "healthy" economic growth.

For these objectives to be simultaneously achieved, major efforts are needed in defining properly the target and setting the right policy to reach it. Yet, technological uncertainty makes the bioeconomy a moving target, hence continuous adjustments are required to hit it. In this regard, papers collected in this Special Issue make a useful contribution by providing a logical framework to understand the transitions, as well as existing tools to assess and measure it. Ideally, policy makers will benefit from the papers included in this Special Issue and, hopefully, it will contribute to making a further step to the much-needed transition towards sustainability.

Funding: This research received no external funding.

Conflicts of Interest: The authors declare no conflicts of interest.

References

1. Diaz, M.; Darnhofer, I.; Darrot, C.; Beuret, J.E. Green tides in Brittany: What can we learn about niche-regime interactions? *Environ. Innov. Soc. Transit.* **2013**, *8*, 62–75. [CrossRef]
2. Geels, F.W. Technological transitions as evolutionary reconfiguration processes: A multi-level perspective and a case-study. *Res. Policy* **2002**, *31*, 1257–1274. [CrossRef]
3. Bennich, T.; Belyazid, S.; Kopainsky, B.; Diemer, A. Understanding the Transition to a Bio-Based Economy: Exploring Dynamics Linked to the Agricultural Sector in Sweden. *Sustainability* **2018**, *10*, 1504. [CrossRef]
4. Bennich, T.; Belyazid, S.; Kopainsky, B.; Diemer, A. The Bio-Based Economy: Dynamics Governing Transition Pathways in the Swedish Forestry Sector. *Sustainability* **2018**, *10*, 976. [CrossRef]
5. Alaerts, L.; Augustinus, M.; Van Acker, K. Impact of Bio-Based Plastics on Current Recycling of Plastics. *Sustainability* **2018**, *10*, 1487. [CrossRef]
6. Urmetzer, S.; Schlaile, M.P.; Bogner, K.B.; Mueller, M.; Pyka, A. Exploring the Dedicated Knowledge Base of a Transformation towards a Sustainable Bioeconomy. *Sustainability* **2018**, *10*, 1694. [CrossRef]
7. Ronzon, T.; M'Barek, R. Socioeconomic Indicators to Monitor the EU's Bioeconomy in Transition. *Sustainability* **2018**, *10*, 1745. [CrossRef]
8. Bracco, S.; Calicioglu, O.; Gomez San Juan, M.; Flammini, A. Assessing the Contribution of Bioeconomy to the Total Economy: A Review of National Frameworks. *Sustainability* **2018**, *10*, 1698. [CrossRef]
9. Lokesh, K.; Ladu, L.; Summerton, L. Bridging the Gaps for a 'Circular' Bioeconomy: Selection Criteria, Bio-Based Value Chain and Stakeholder Mapping. *Sustainability* **2018**, *10*, 1695. [CrossRef]
10. Falcone, P.M.; Imbert, E. Social Life Cycle Approach as a Tool for Promoting the Market Uptake of Bio-Based Products from a Consumer Perspective. *Sustainability* **2018**, *10*, 1031. [CrossRef]

11. Martin, M.; Røyne, F.; Ekvall, T.; Moberg, Å. Life Cycle Sustainability Evaluations of Bio-based Value Chains: Reviewing the Indicators from a Swedish Perspective. *Sustainability* **2018**, *10*, 547. [CrossRef]

12. Majer, S.; Wurster, S.; Moosmann, D.; Ladu, L.; Sumfleth, B.; Thrän, D. Gaps and Research Demand for Sustainability Certification and Standardisation in a Sustainable Bio-Based Economy in the EU. *Sustainability* **2018**, *10*, 2455. [CrossRef]

13. Fonseca, L.M.; Domingues, J.P. Exploratory Research of ISO 14001:2015 Transition among Portuguese Organizations. *Sustainability* **2018**, *10*, 781. [CrossRef]

sustainability

MDPI

Article

Understanding the Transition to a Bio-Based Economy: Exploring Dynamics Linked to the Agricultural Sector in Sweden

Therese Bennich [1],*, Salim Belyazid [1], Birgit Kopainsky [2] and Arnaud Diemer [3]

[1]   Department of Physical Geography, Stockholm University, SE-106 91 Stockholm, Sweden;
      salim.belyazid@natgeo.su.se
[2]   System Dynamics Group, Department of Geography, University of Bergen, Postboks 7802, 5020 Bergen,
      Norway; birgit.kopainsky@uib.no
[3]   Center for Studies and Research on Internal Development (CERDI), University of Clermont Auvergne,
      FR-320, 63009 Clermont-Ferrand, France; arnaud.diemer@uca.fr
*   Correspondence: therese.bennich@natgeo.su.se; Tel.: +46-(0)72-554-9592

Received: 5 March 2018; Accepted: 3 May 2018; Published: 10 May 2018

Abstract: There is a growing interest in the bio-based economy, evident in the policy domain as well as in the academic literature. Its proponents consider it an opportunity to address multiple societal challenges, and the concept has broad reach across different sectors of society. However, a potential transition process is also linked to areas of risk and uncertainty, and the need for interdisciplinary research and for the identification of potential trade-offs and synergies between parallel visions of the bio-based economy have been emphasized. The aim of this paper is to contribute to addressing this gap by using an approach combining tools for systems analysis with expert interviews. Focusing specifically on dynamics in the agricultural sector in Sweden, an integrated understanding of the social and ecological processes contributing to or hindering a transition in this area is developed, high order leverage points are identified, and potential impacts of proposed interventions explored. The paper also considers cross-sectoral linkages between the forestry and agricultural sectors.

Keywords: sustainability transitions; systems analysis; causal loop diagrams; bio-based economy; bio-economy; agriculture

1. Introduction

Efforts aimed at understanding and supporting the transition to a bio-based economy have been increasing over the past two decades, to a large extent linked to developments in the policy sphere. Strategies on regional, national, and local levels have been brought forward, and now at least 45 countries have developed their own bio-based economy agendas [1]. Partly as a consequence of the broad reach of the concept, these strategies adopt different visions and definitions of the bio-based economy, often reflecting the context and preconditions of a specific nation or actor [2–4]. For example, the bio-based economy is variously seen as an opportunity to achieve climate change mitigation, competitive advantage linked to knowledge generation and novel biomass applications, improved global governance of biological resources, decentralized modes of production, rural development, and a lower dependency on finite, fossil-based resources [5–8]. One example of how a transition to a bio-based economy is defined is as a shift away from an economy dependent on fossil-based resources, to an economy utilizing renewable, biological resources to provide the products and services demanded in society [9,10]. A parallel understanding does instead stress the role of biotechnology, perceiving it as central to the bio-based economy, and aiming for its use and share of economic output to increase [7,11].

In Sweden, factors such as biomass availability, a long history of traditional industries, a skilled labor force, and high access to infrastructure and markets have been identified as beneficial in the context of facilitating a transition process [12–14]. Specific initiatives addressing developments toward a bio-based economy include a research and innovation strategy, commissioned by the Swedish Government and published in 2012. It outlined overarching aims of the Swedish bio-based economy, such as replacing fossil-based resources with bio-based resources, developing smarter products, increasing resource use efficiency, and to change consumption patterns and behaviors. The strategy further stressed the need for collaboration, knowledge generation, and formation of new partnerships in order to enable a transition [9]. Moreover, the bio-based economy could directly, or indirectly, affect the attainment of several of the Swedish Environmental Quality Objectives [15], in addition to climate-related targets such as achieving zero net emissions of greenhouse gases by 2045 and a fossil-free transport fleet [16–18]. Furthermore, the bio-based economy is an integral part of the current Innovation Partnership Program initiated by the Swedish Government to address critical societal challenges [19]. Thus, also in Sweden the bio-based economy is considered a mean to achieve multiple objectives, and its premises to a large extent seen as based on the broad and cross-sectoral reach of the concept.

Nevertheless, a large number of the road maps and initiatives promoting the Swedish bio-based economy have been centered around single sectors and technological pathways. The forest sector plays a dominant role, reflected in many of the ongoing efforts to facilitate a transition process [20–22]. Additionally, in the broader debate surrounding the concept, aside from the opportunities being highlighted, uncertainty and potential risk factors have also been brought forward. Questions about the viability and sustainability of the emerging bio-based economy have been raised, stressing potential conflicts and adverse environmental and social impacts arising from the multitude of competing and growing claims on biological resources [23–25], a lack of focus on multi-functionality, public goods, local knowledge and social innovation in the bio-based economy [26], and that ecological sustainability is often overlooked in official strategies and the broader bio-based economy discourse [23,27,28]. The discourse surrounding the concept has further been criticized for being promissory [29], and a lack of clear meaning, indicators for sustainability, or measures of success have been underlined [2,30,31]. Additionally, the need for interdisciplinary approaches, the identification of synergies and trade-offs between parallel visions of the bio-based economy, and insight into the larger societal impacts of a transition process have been highlighted [4,9,32–34].

The aim of our work is to contribute to an integrated understanding of the multiple processes and proposals suggested to underpin and facilitate a transition to a bio-based economy in Sweden. The focus of the study is on terrestrial sources of biomass from forestry and agriculture. For the sake of space, the present paper outlines the results focused on the agricultural sector. Results linked to dynamics in the forestry sector, also emphasizing a political dimension of a transition process, are presented in Bennich et al. (2018) [35]. This paper is organized as follows: The first section outlines the methodological framework. Thereafter, the results are presented, divided into the change processes identified as key by the actors participating in the study, followed by an exploration of interconnectedness, the impact of proposed interventions, and of the potential future transition pathways to which they might contribute. Our conclusions are drawn in the final section, together with suggestions for future work.

2. Methodological Framework and Research Process

The study was structured around the following guiding research questions: What are the social-ecological dynamics currently enabling or hindering a transition to a bio-based economy? What actions are proposed to facilitate a transition to a bio-based economy, and what pathways do they form? The research design was based on the use of qualitative tools for systems analysis, specifically Causal Loop Diagrams (CLDs). The empirical basis for the CLDs was expert interviews. This section gives an overview of the methodological framework and research process, while a more in-depth presentation can be found in Bennich et al. (2018) [35].

CLDs are diagramming tools, originating from the fields of system dynamics and cybernetics [36]. They consist of variables connected by arrows, where the links between the variables represent hypotheses about causal relationships. CLDs are commonly used to conceptualize and communicate system structure, for clarification of assumptions, and for creating new knowledge and learning in situations where problems are perceived as complex or unstructured [37,38]. Lane (2008) provides an overview of the emergence of the use of CLDs, as well as a critical reflection on their strengths and weaknesses [39]. In relation to the aims of this study, the strength of CLDs lies in their ability to simultaneously map social and ecological systems, as well as different actor perspectives, in an integrated way. CLDs are in addition meant to document both chains of cause and proposed effects, but also any possible system feedbacks, thereby moving from linear to non-linear conceptualization. The limitations of CLDs include potential trade-offs between precision and simplification. The specific focus on depicting and communicating key elements and feedbacks of the system under study might make the underlying rationale of the causal relationships less evident. Moreover, while CLDs can provide structural insight, they do not allow for rigorous inference about the dynamic behavior arising from the system structure [39]. For more information on the process of developing CLDs, as well as on their use in this research, see Bennich et al. (2018) [35]. Expert interviews are often referred to as efficient means to gain exploratory insight about a specific topic and as an entry points to fields where it might otherwise be difficult to gain access [40]. In the context of this study, expert interviews were used to identify key drivers, hindrances, and interventions linked to the transition to a bio-based economy in Sweden, as perceived by different actor groups. The interview data were then used as a basis for elicitation of the causal relationships included in the CLDs.

Thus, our starting proposition is that systems analysis and the use of CLDs as an analytical tool can contribute to developing and documenting a qualitative understanding of the structural components governing a transition to a bio-based economy, by mapping expert knowledge in integrated conceptual maps. Based on this understanding, effective interventions may be identified, in contrast to technical, end-of-pipe solutions, and the ability to avoid policy resistance can be strengthened. Policy resistance refers to the failure of interventions to generate anticipated results, caused by unexpected internal responses in the system, and partly attributed to linear cause and effect thinking [39,41].

To support reasoning about potential leverage points and to qualitatively assess the potential impact of suggested proposals to facilitate change, we build on the Leverage Points Framework [42]. Leverage points can be understood as places in a system where a small disturbance or initial change may ultimately lead to large-scale system change, i.e., places where well-directed interventions may lead to substantial and lasting improvements [43]. The framework may serve as a basis for analyzing the effectiveness of interventions, starting from leverage points at the lower end of efficiency (e.g., changing the value of parameters, constants, the physical structure of a system, or the relative length of delays), to higher impact leverage points (changing the relative strength of balancing and reinforcing feedback loops, adding information feedbacks, or changing the rules guiding the behavior of the system) [42].

As also outlined in Bennich et al. (2018) [35], the first step of the research process consisted of actor analysis and outreach. The process was informed by an initial literature review, and was structured by the guidelines for actor analysis as presented by Lelea, et al. (2014) [44]. Fourteen experts were selected for semi-structured interviews. The experts were selected based on (1) their knowledge and experience of sectors relevant to the Swedish bio-based economy; (2) their ability to represent larger actor groups in these sectors; and (3) their ability to provide diverse standpoints. The process made use of snowball sampling, specifically aiming to identify actor perspectives that might currently be overlooked in the bio-based economy debate. Semi-structured interviews were selected for data collection, as they allow for an in-depth exploration of perceptions and views in situations where issues are complex, information seemingly conflicting, and where there is diversity in the personal history, profession, and educational background of the study group [45,46]. In the second step, the interviews were carried out, either physically in Stockholm or via Skype. The interviews, lasting from 60 to 120 min, were based on a number of open ended questions, formulated to enable the identification of key variables, causal relationships, and reinforcing or balancing feedbacks. Measures to ensure validity and reliability in the data collection phase included the development and testing of an interview guide in preparation for the interviews, as well as digitally recording and transcribing the interviews. In the third step, the data was analyzed and used as a basis for the development of the CLDs, following the method presented by Kim and Andersen (2012) [47]. For each interview an initial CLD was developed. These were subsequently integrated by aggregating similarities while maintaining differences, thereby ensuring that the multiple perspectives of the interviewees remained visible. The process was documented, to create visible connections between the data segments and the integrated CLDs. The CLDs were then sent to the interviewed experts for confirmation, along with an explanatory text and a set of complementary questions. The questions were aimed at generating additional insight, clarification, and completeness of coverage in terms of proposed system structure. The CLDs were subsequently revised based on the feedback from the participants, and then used as a basis to perform a qualitative exploration of the impact of proposed interventions and the transition pathways they formed. Two researchers were engaged in the process of developing and analyzing the CLDs, and the resulting output was compared and discussed in an iterative process, aiming to reduce researcher biases. For an interviewee overview and interview guide, see Bennich et al. (2018) [35].

3. Results

The results, in the form of CLDs, are presented under different sub-headings. Firstly, the four dynamic change processes in the agricultural sector identified as important in the context of the broader transition to a bio-based economy are outlined. More specifically, the first section presents dynamics governing the overarching objective of expanding and maintaining farming activities in Sweden. This is followed by three more disaggregated CLDs focusing on the desired objectives of employing more environmentally friendly practices in conventional agriculture, of securing biomass supply for both food and non-food purposes, and of facilitating a shift to regenerative production, respectively. The next section presents an integrated CLD, highlighting cross-scale interactions. This section also introduces the leverage points and interventions that were identified during the interviews. The final section elaborates on interconnectedness between the agricultural and forestry sectors. Variable names are indicated by quotation marks.

3.1. Identification of Dynamics Governing the Expansion and Maintenance of Farming Activities

The first change process identified as important during the interviews relates to the underlying ability to expand and maintain activities within the agricultural sector, perceived as a pre-condition to achieve other objectives of the bio-based economy. It was emphasized that change in the agricultural sector is path dependent, and that it takes time and effort to reverse a potential decline in the size of the sector. The variable "Farming activities" (capitalized, Figure 1) represents the combinations of activities directly enabling primary production in the agricultural sector (e.g., tilling, planting, fertilizing, irrigation, pathogen mitigation and harvesting), thus referring to farming in a general sense. During the interviews, several reinforcing feedbacks either supporting an expansion or triggering a loss of "Farming activities" were identified. The more "Farming activities" carried out, the higher the "Learning-by-doing" among the practitioners, making the "Agricultural capacity" expand. The expansion of "Agricultural capacity" drives the "Conversion of fallow land to actively cultivated land", enabling further "Farming activities" (reinforcing feedback, R1, Figure 1). Additionally, more "Farming activities" make the "Available support functions for farmers" increase, including advisory services and research centers. Having such support functions in place contributes to the buildup of "Agricultural capacity", a higher "Conversion of fallow land to actively cultivated land", and ultimately to more "Farming activities" (reinforcing feedback, R2, Figure 1). An expansion of "Farming activities" could also contribute to a higher "Attractiveness of profession", through increasing visibility and awareness in society, in turns supporting the "Education of new farmers" (reinforcing feedback, R3, Figure 1).

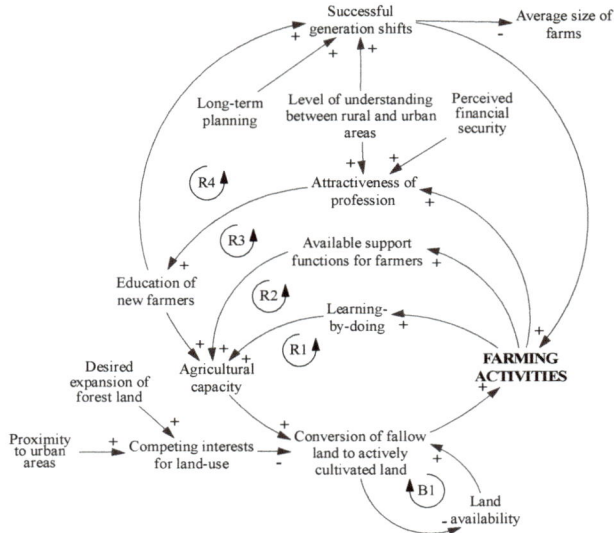

Figure 1. Causal Loop Diagram (CLD) displaying the dynamics suggested to govern the expansion and maintenance of "Farming activities". CLDs consist of variables connected by arrows, representing hypotheses about causal relationships among the variables. Each link is assigned a polarity, either positive (+) or negative (−). A positive link indicates that the dependent and independent variable move in the same direction (if the independent variable increases, so does the dependent variable, and if the independent variable decreases, so does the dependent variable). A negative link indicates that the variables move in opposite directions (if the independent variable increases, the dependent variable decreases, and if the independent variable decreases, the dependent variable increases). Important feedbacks are highlighted in the diagram. These could be either reinforcing (denoted by a R) or balancing (indicated by a B). For a further introduction to the use of CLDs in this research, see Bennich et al. (2018) [35].

The difficulty of generation shifts was identified as a key issue to overcome in the agricultural sector. The larger the number of people educated to become farmers, the higher the number of "Successful generation shifts", resulting in more "Farming activities" than what would otherwise have been (reinforcing feedback, R4, Figure 1).

Some variables affect, but are not part of, the reinforcing feedbacks R1–4 (Figure 1). A lack of "Long-term planning" was identified as a factor hindering the process of handing over farms from one generation to the next. The higher the degree of "Long-term planning", the more "Successful generation shifts". Another variable potentially having an impact on the amount of "Successful generation shifts" is the "Level of understanding between rural and urban areas". A higher "Level of understanding between rural and urban areas" could make the pool of potential new farmers grow, the level of trust between the current and potential future generation of farmers increase, and additionally ensure that the needs of the part of the population living in rural areas are recognized to a larger extent. In terms of the latter, it was suggested that the social welfare system is poorly adapted to the living conditions and work situation of farmers, where potential improvements include better pensions and financial support in the case of sickness. Aside from access to social welfare, enabled by a high "Level of understanding between rural and urban areas", also the "Perceived financial security" in the agricultural sector was identified as having an impact on the "Attractiveness of profession". The variable "Perceived financial security" refers specifically to the views and understandings of the financial situation of farmers among those not themselves active in the profession.

As emphasized by the balancing feedback B1 (Figure 1), land is a limited resource. The more land that is converted to cultivated land, the lower the potential to further expand cultivation. In addition, the "Conversion of fallow land to actively cultivated land" is affected by "Competing interest for land-use". Factors such as "Proximity to urban areas" and a "Desired expansion of forest land" could increase competition and pressure on the land, drive prices up, and thereby make it more difficult to allocate land to farming activities. Another proposed relationship with respect to land is that between "Successful generation shifts" and the "Average size of farms". If the farm is not handed over from one generation to the next, it is likely that the land will be transferred to the neighboring farm, thereby making the average farm size increase.

3.2. Employment of More Environmentally Friendly Practices in Conventional Agriculture

A second change process identified as important in the context of the transition to a bio-based economy is the "Employment of more environmentally friendly practices" in conventional agriculture (capitalized, Figure 2). The need to adopt more environmentally friendly practices relates to a number of overarching challenges in the agricultural sector, where examples highlighted during the interviews include a need to reduce greenhouse gas emissions from primary production and processing, to lower the use of mineral fertilizers and pesticides, and to overcome the dependency on fossil-based sources of energy. Three drivers explaining the employment of more environmentally friendly practices were suggested: A financial logic (it makes sense from a cost saving perspective), an emotional rationale (arising from factors such as personal experience of the negative impacts of intensive use of production inputs), and the availability of options (i.e., the farmer would employ more environmentally friendly practices if the option to do so was available).

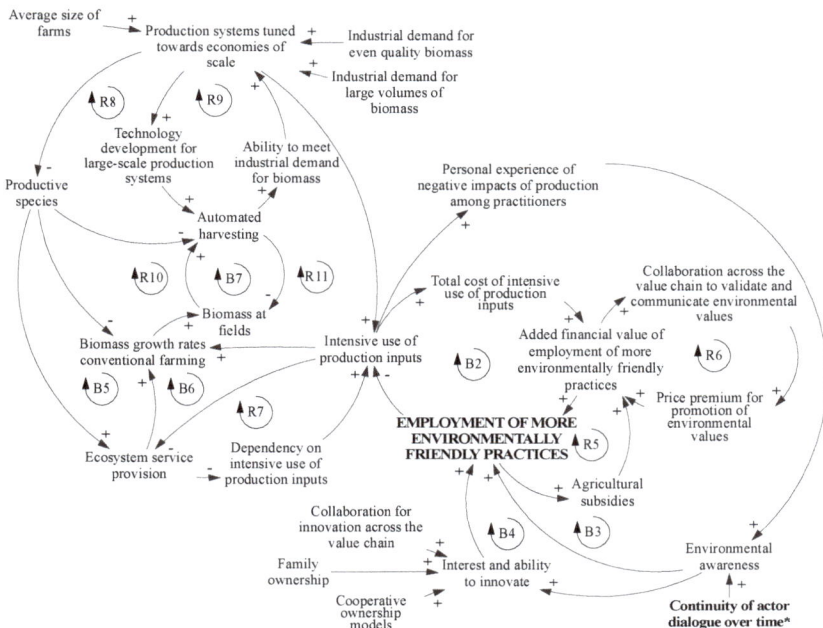

Figure 2. Dynamics hypothesized to govern the employment of more environmentally friendly practices in conventional agriculture. * Underlying dynamics of variable related to the political dimension of the bio-based economy concept, explained in further detail in Section 3.6.

The higher the "Intensive use of production inputs", the larger the "Total cost of intensive use of production inputs". Consequently, the "Added financial value of employment of more environmentally friendly practices" increases, supporting the "Employment of more environmentally friendly practices", thereby reducing the "Intensive use of production inputs" (balancing feedback, B2, Figure 2). Adhering to sustainability criteria opens up the possibility of receiving "Agricultural subsidies" for environmentally sound practices, adding to the financial incentives for promoting environmental values (reinforcing feedback, R5, Figure 2). The financial viability of environmentally friendly practices is, aside from being affected by potential cost reductions or subsidies, to a certain extent dependent on the ability of the farmer to receive a price premium. With "Collaboration across the value chain to validate and communicate environmental values", the willingness of the end consumer to pay a price premium is expected to increase. The price premium, in turns, contributes to the added financial value of employing environmentally friendly practices, consequently also increasing the ability to engage more actors in the work to promote environmental values (reinforcing feedback, R6, Figure 2).

The emotional rationale for employing more environmentally friendly practices is partly explained by the "Personal experience of negative impacts of production among practitioners", creating a stronger "Environmental awareness", in turns reducing the "Intensive use of production inputs" (balancing feedback, B3, Figure 2). Additionally, with higher "Environmental awareness", a greater "Interest and ability to innovate" for sustainability may be created. Innovation is suggested to be important as it can allow for the development of less harmful production inputs and more resource efficient methods, thereby further contributing to the "Employment of more environmentally friendly practices" (balancing feedback, B4, Figure 2).

While a high environmental awareness could support innovation for sustainability, a number of additional variables explaining relative innovation capabilities for sustainability in the bio-based economy were identified during the interviews. One factor to consider is the role of ownership, where it was suggested that cooperatives and family owned businesses might be better able to support innovation for sustainability. This may partly be explained by these structures allowing for longer planning horizons, and in terms of cooperatives, the possibility to spread risk among the owners. An additional factor thought of as supporting innovation is "Collaboration for innovation across the value chain", the underlying assumption being that novel ideas and solutions are created when actors with different perspectives meet and work together. Examples of areas of innovation brought up during the interviews include digitalization, precision agriculture, and new means of collecting and analyzing data to inform decision making (e.g., the internet of things).

Another factor suggested to affect the ability to employ more environmentally friendly practices is the structure of the agricultural system. In the current industrial system, a strive to maximize "Biomass growth rates in conventional farming" requires a low number of "Productive species" and an "Intensive use of production inputs" such as mineral fertilizer and pesticides. Intensive use of production inputs harms the "Ecosystem service provision", thereby creating a self-reinforcing dependency on these same inputs (reinforcing feedback, R7, Figure 2). This reinforcing feedback may create a lock-in effect, where it becomes increasingly difficult shift to more environmentally friendly modes of production.

An additional number of reinforcing feedbacks linked to industrial agricultural production that may create lock-in effects, and that have an impact on the ability to employ more environmentally friendly practices, were identified during the interviews. "Industrial demand for large volumes of biomass", "Industrial demand for even quality biomass", and an increasing "Average size of farms", drive the expansion of "Production systems tuned towards economies of scale". In these systems, the fewer the "Productive species" and the more extensive the use of "Automated harvesting", the larger the "Ability to meet industrial demand for biomass", further generating "Production systems tuned towards scale" (reinforcing feedback, R8, Figure 2). Additionally, "Production systems tuned towards economies of scale" require "Technology development for large-scale production systems". This technology further supports "Automated harvesting", and therefore leads to a larger "Ability to meet industrial demand for biomass", again reinforcing the development of production systems based on economies of scale (reinforcing feedback, R9, Figure 2). Aside from technology, the selection of specialized species and the intensive use of production inputs in these systems boost biomass growth rates, thereby also supporting the "Ability to meet industrial demand for biomass" (reinforcing feedbacks, R10–11, Figure 2). In the long run, however, two balancing feedbacks counteract this development. Few productive species and the intensive use of production inputs negatively affect the provision of ecosystem services (such as natural pest control), thereby having an adverse impact on biomass growth rates and consequently also on the ability to meet industrial demand for biomass (balancing feedbacks, B5 and B6, Figure 2).

3.3. Identification of Dynamics Governing Biomass Availability

The sources and quantities of biomass available in a transition process were brought forward as central during the interviews. These sources include the biomass derived from "Domestic food production" as well as the production contributing to the "Total biomass supply for non-food applications". Examples of non-food applications are biomass being converted into bioenergy, chemicals, industrial products, or manufactured goods. During the interviews, it was stressed that national food self-sufficiency and domestic production capacity are increasingly perceived as key issues for the future. In addition, the biomass demand for non-food applications is expected to grow, for instance with an emerging "Interest and ability to innovate". With innovation, novel biomass applications are developed and demonstrated, and at the point when maturation is reached and markets materialize the "Non-food biomass applications" in the agricultural sector would start to increase, as long as the "Total biomass supply for non-food applications" is able to support this development.

Three ways of meeting future biomass demand were identified during the interviews, all with different implications. First, harvest residues and waste products from the food industry could be utilized to a further extent. The larger the "Domestic food production", the more "Litter" and "Food waste" generated and potentially available to the bio-based economy, contributing to the "Total biomass supply for non-food applications". The "Outtake of harvest residues" is hindered by a lack of "Perceived market potential for residues", described as a match-making problem where farmers are not connected with potential industrial customers. Nevertheless, when the "Ability and interest to innovate" increase, solutions such as more data driven and automated means of connecting actors across the value chain are expected to be developed and employed, thereby increasing the "Perceived market potential for residues".

A second way to increase biomass supply in the bio-based economy is to increase the "Conversion of fallow land to actively cultivated land", thereby giving rise to more "Land available for expanding food or non-food crop production". The choice between producing food and non-food crops is represented by the balancing feedbacks B8 and B9 (Figure 3). The more land allocated to the production of food crops, the less land is available for the production of non-food crops, and vice versa. Third, biomass supply for the bio-based economy can increase by utilizing "Novel crop rotation schemes", alternating between the production of crops for food consumption and crops for other purposes. The balancing feedbacks B10–B13 (Figure 3) stress the limits to biomass supply, in the sense that any allocation of biomass or area of use is restricted by the biomass availability. The reinforcing feedbacks R12 and R13 (Figure 3) highlight that the "Biomass growth rate" is affected by the soil quality, which is enhanced by the amount of "Litter" returned to the soil. The effect of increasing agricultural production may thereby be reinforced by an improvement in soil quality caused by more litter and harvest residues on the farm land and hampered by a larger "Outtake of harvest residues". Crop rotation schemes hold the potential to improve "Biomass growth rates" through an improvement in soil quality, thereby creating synergies between productivity related objectives and novel uses of biomass in the bio-based economy.

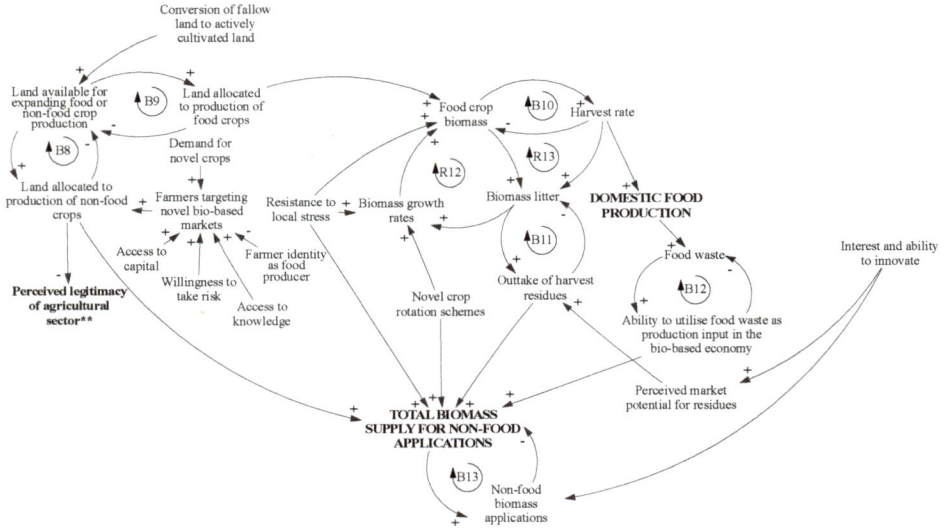

Figure 3. Sources of biomass in the bio-based economy. ** Variable influencing the political dimension of the bio-based economy, for further explanation see Section 3.6.

3.4. Introduction to Diversified Farming and Regenerative Production

An additional perspective brought forward during the interviews stresses the need for a shift towards diversified farming and regenerative production. While the variable "Employment of more environmentally friendly practices" (Figure 2) represents a shift in conventional farming toward more sustainable modes of production (e.g., toward practices enabling a higher resource use efficiency), the variable "Regenerative production" (capitalized, Figure 4) refers to a more fundamental shift in the way agricultural production is carried out. The creation of value in regenerative production is based on ecological improvement, thereby placing ecosystem functioning at the core of the agricultural practice. Diversity is a fundamental characteristic of regenerative production, as it is assumed to increase both resilience and productivity. Thus, the term diversified farming is in this context used to refer to the practices allowing for the production to be regenerative. Diversified farming is based on a number of design principles, striving to maximize the ability of plants to photosynthesize, to optimize the use of surface areas, and to utilize the qualitative components of biomass, as well as the synergetic relationships between species. Energy use should be optimized in every step of the process, aiming to realize the potential of agricultural production to have a positive energy balance. Another aspect highlighted as important during the interviews is the need to achieve self-sufficiency within each farm or cluster of farms, through utilizing combinations of animal husbandry, fodder production, and food crop production.

3.4.1. The Transition to Diversified Farming and Regenerative Production

The practices enabling regenerative production were identified as knowledge intensive, requiring both specific practical knowledge and commitment. When "Regenerative production" is carried out, practical knowledge is generated through learning-by-doing, and feelings of confidence and will-power among the practitioners grow. As a result, the overall "Ability to carry out diversified farming" increases, which in turns facilitates more "Regenerative production" (reinforcing feedback R14, Figure 4). Additionally, diversified farming requires a shift in mind-set among farmers. "Successful generation shifts" (see Figure 1) are perceived as an opportunity in this regard, as they may create an "Openness to new ideas", allowing for changes in the way agricultural production is carried out.

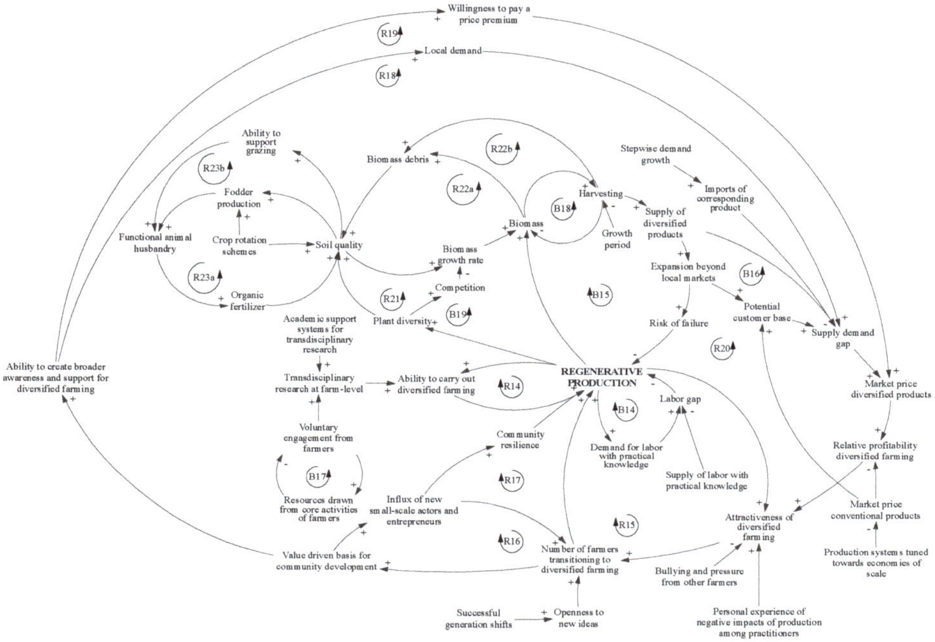

Figure 4. Dynamics suggested to govern the shift to diversified farming and regenerative production.

The variable "Attractiveness of diversified farming" was identified as crucial, as it determines the "Number of farmers transitioning to diversified farming", thereby enabling more "Regenerative production". More "Regenerative production" does, in turn, have a positive impact on the "Attractiveness of diversified farming", creating a reinforcing feedback (R15, Figure 4). Several factors were suggested to explain why an expansion of regenerative production may increase the attractiveness of diversified farming. For instance, diversified farming is expected to contribute positively to the well-being and quality of life of the practitioners. The more "Regenerative production", the larger the awareness about such benefits in the larger society, thereby making the perceived "Attractiveness of diversified farming" increase. Additionally, the more "Regenerative production", the lower the uncertainty linked to the production process, another factor contributing positively to the "Attractiveness of diversified farming".

The "Attractiveness of diversified farming" is also affected by the financial viability of this type of production, as compared to more conventional farming practices. The relative profitability is affected by the "Market price of conventional products". The higher the "Market price of conventional products", the lower the relative profitability of diversified farming, thereby having a negative impact on the attractiveness of diversified farming. A higher "Market price of conventional products" does on the other hand also give rise to a larger "Potential customer base" and demand for diversified products, in this way positively contributing to the attractiveness of diversified farming. The "Attractiveness of diversified farming" is also promoted by the "Personal experience of negative impacts of production among practitioners" (see Figure 2) in conventional farming, while being eroded by "Bullying and pressure from other farmers". "Bullying and pressure from other farmers" could arise from an underlying fear for the spread of pests or diseases among the neighboring farms, as well as from embedded cultural preferences regarding the appearance of the farm land.

Diversified farming and regenerative production are suggested to have spill-over effects not only on the quality of life of the individual farmer, but also on the surrounding community. The larger the "Number of farmers transitioning to diversified farming", the stronger the "Value driven basis for communal development". The value driven basis may refer to a common understanding and appreciation for the environmental and social benefits provided by diversified farming and regenerative production. This, in turn, may attract others with similar values, creating an "Influx of new small-scale actors and entrepreneurs". This would support the transition process, making the number of farmers shifting to diversified practices increase (reinforcing feedback R16, Figure 4) and facilitating cluster development that enhances overall "Community resilience" (reinforcing feedback R17, Figure 4). Additionally, the stronger the "Value driven basis for community development", the greater the "Ability to create broader awareness and support for diversified farming" in the larger community. This awareness is suggested to create "Local demand" as well as a "Willingness to pay a price premium" for the production output. These effects would strengthen the "Relative profitability of diversified farming", thereby further supporting the "Attractiveness of diversified farming" (reinforcing feedbacks R18–19, Figure 4). As more "Regenerative production" is carried out, the "Supply of diversified products" increases, thereby creating a potential for "Expansion beyond local markets". When entering new markets, the "Potential customer base" grows, thereby contributing to the "Relative profitability of diversified farming" through buoying the market price of diversified farming products (reinforcing feedback R20, Figure 4).

3.4.2. The Biophysical Basis for Diversified Farming and Regenerative Production

As "Regenerative production" expands, the output in terms of standing "Biomass" increases. Diversified farming supports a large "Plant diversity", and this diversity in combination with "Debris" in the form of plant litter and harvest residues enhance "Soil quality" over time. Better soil quality increases the "Biomass growth rate" and thereby also the standing volume of "Biomass" (reinforcing feedbacks R21–22b, Figure 4). A higher "Soil quality" also supports biomass production for purposes other than food crop production. Through a higher "Fodder production" and "Ability to support grazing", more "Functional animal husbandry" can be carried out. "Functional animal husbandry" refers to practices where the animals are seen as an integral part of farming, valuing their ability to optimize energy use and support production of food crops rather than holding animals primarily for meat or diary production. With a larger number of animals integrated into the farming, more "Organic fertilizer" becomes available, again enhancing the "Soil quality" (reinforcing feedbacks 23a–b, Figure 4).

Achieving an optimal animal intensity based on the capacity of the land to hold these animals is key to regenerative production. Thus, part of the ability to facilitate a transition to diversified farming and regenerative production is structural, in the sense that certain farms have better preconditions and therefore are relatively easier to convert. For example, they may already have a sufficient number of animals to provide manure for the lands, as well as enough land to grow fodder or allow for grazing, thereby ensuring self-sufficiency on the farm level.

3.4.3. Dynamics Identified as Counteracting a Transition to Diversified Farming and Regenerative Production

A number of factors could potentially counteract growth in diversified farming and regenerative production. One aspect is the ability to find labor with practical skills, considered a bottle-neck in a transition towards diversified farming as the current educational model is perceived to be built on a strong theoretical basis. As regenerative production expands, the "Demand for labor with practical knowledge" increases, thereby making the "Labor gap" grow, and if not addressed, reducing regenerative production in the long run (balancing feedback B14, Figure 4). Another aspect brought forward is the ability to expand to new markets. As the production output increases and actors move outside the local market, there is a greater deal of anonymity and competition, and therefore also a greater "Risk of failure" (balancing feedback 15, Figure 4). Anonymity in this case refers to a shift from selling on a local market where demand is built on reputation and personal contacts, to a market where the producer has no means to directly communicate with the end consumer (but rather is dependent on the communication and marketing carried out by the distributor). Also, the nature of demand growth matters. If the demand growth is step-wise rather than smooth, an immediate shortage of supply might follow. In response, demand might be met with imports rather than by the domestic production capacity adjusting to the higher level of demand, thereby halting the transition to regenerative production on a national scale. Additionally, as the supply of the product increases, the price would normally fall, negatively affecting the "Relative profitability of diversified farming" (balancing feedback B16, Figure 4).

The "Ability to carry out diversified farming" is dependent on the problem-solving ability of the farmer, which is assumed to increase with "Transdisciplinary research at farm-level". "Voluntary engagement" from farmers enables this type of research, but the larger the voluntary engagement the more "Resources drawn from core activities of farmers", ultimately reducing the "Voluntary engagement" (balancing feedback B17, Figure 4). Additionally, to be able to carry out "Transdisciplinary research at farm-level", not only the "Voluntary engagement" from farmers needs to be in place, but also the necessary "Academic support systems for transdisciplinary research".

There are also biophysical limits to "Regenerative production". For example, the more biomass that is harvested and sent to the market, the less standing biomass (balancing feedback B18, Figure 4), the longer the "Growth period" of a specific specie the lower the "Harvesting", and the larger the plant diversity the higher the "Competition" for resources, also limiting "Biomass growth rates" (balancing feedback B19, Figure 4).

3.5. Proposed Leverage Points and Interventions

Aside from the dynamics governing the change processes suggested to underpin a transition to a bio-based economy, a number of leverage points (i.e., places to intervene in the agricultural system to support the transition), as well as specific proposals targeting these leverage points, were suggested during the interviews. Additionally, while the individual CLDs in Figures 1–4 depict dynamics on different scales, also cross-scale interlinkages were identified in the interview process. Figure 5 displays an integrated CLD, highlighting suggested interventions (in italics), as well as the proposed cross-scale interlinkages.

3.5.1. Interventions Linked to Environmentally Friendly Practices and Biomass Availability

Interventions aiming at supporting the "Employment of more environmentally friendly practices" in conventional agriculture include utilizing "Environmental taxes" or other financial instruments to a larger extent, to make the intensive use of production inputs more expensive. The "Added financial value of employment of more environmentally friendly practices" would thereby increase, ultimately lowering the "Intensive use of production inputs" (strengthening the balancing feedback, B2, Figure 5). Another proposal in this area is to facilitate a "Redirection of farming intensity from high to low quality farm land". Aside from lowering the pressure on the most productive lands, this shift would entail intensifying farming activities on lands that are on the verge of becoming overgrown. This could positively affect the "Ecosystem service provision", thereby also holding the potential to reduce the dependency on "Intensive use of production inputs" (directly affecting the reinforcing feedback R7, Figure 5).

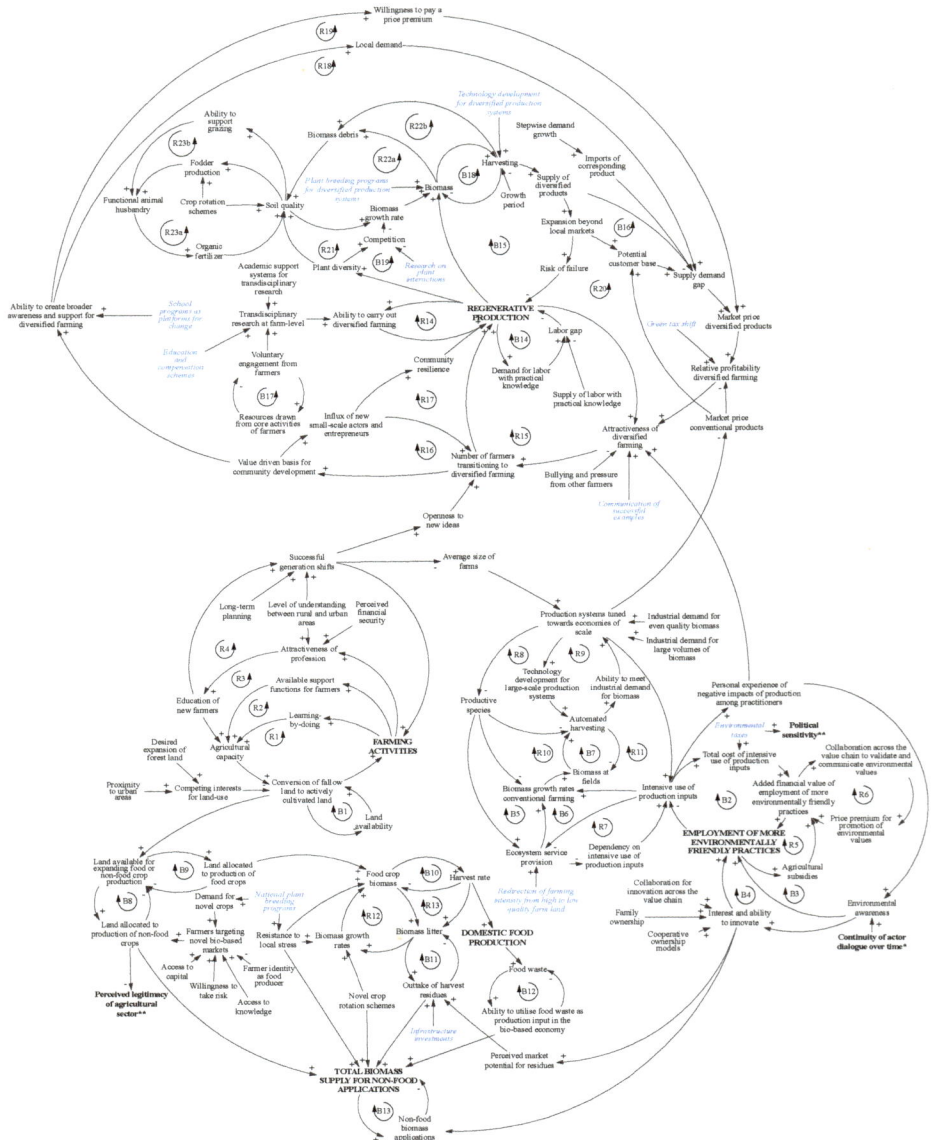

Figure 5. Proposed interventions (variables in italic) to facilitate desired change in key variables (capitalized). The CLD also highlights proposed cross-scale interlinkages, as identified during the interviews. * Impacted by the political dynamics governing a transition; ** Influencing the political dynamics governing a transition, see Section 3.6.

Specific proposals targeting biomass availability in the bio-based economy include efforts addressing a perceived lack of capacity to transport and handle harvest residues from the agricultural sector. By "Infrastructure investments" (bottom, centered, Figure 5), the "Outtake of harvest residues" would increase, contributing to the "Total biomass supply for non-food applications" and thereby enabling more "Non-food biomass applications". Infrastructure may in this context be understood in a broad sense, referring to physical but also organizational, technological, and logistical structures enabling new markets for harvest residue and production side streams to be established. Another suggestion is to develop and implement "National plant breeding programs" (bottom, left, Figure 5), aiming to generate a better fit of breeding techniques to Nordic conditions, and a larger variation in plant characteristics and chemical composition. Expected impacts of this proposal include a better plant "Resistance to local stress", generating higher "Biomass growth rates" and larger biomass supplies for both food and non-food applications. In addition, greater variation in plant characteristics would create a larger "Demand for novel crops", making the number of "Farmers targeting novel bio-based markets" increase, resulting in more "Land allocated to the production of non-food crops", thereby contributing to a higher "Total biomass supply for non-food applications".

3.5.2. Interventions Linked to Diversified Farming and Regenerative Production

Leverage points and proposed interventions affecting the feedbacks governing the transition to diversified farming and regenerative production fall within a range of categories, including knowledge production, the organization of labor and technology in diversified farming systems, and the emergence of new markets.

Developing design principles for diversified farming systems was suggested as crucial to fully utilize the potential of these systems to provide food for human consumption in a resource efficient manner. To enable this development, a need to expand the knowledge base was identified. "Plant breeding programs for diversified systems" could, if successful, increase the output from these systems in terms of edible, nutritious, and tasty yields, and make them adapted to Nordic geographical conditions (strengthening the reinforcing feedbacks R22a–b, Figure 5). Also "Research on plant interactions" could contribute to the development of smart design principles, through the identification of ways to minimize competition for resources while maximizing synergies between species (strengthening the reinforcing feedbacks R21–22b, while reducing the strength of the balancing feedback B19, Figure 5).

Another aspect of knowledge generation for diversified production systems concerns a suggested gap between scientific knowledge and the reality of the farmer. Advice is perceived as unable to account for multiple complexities and the contextual nature of challenges at the farm. Carrying out more research among practitioners was identified as a mean to address this gap, thereby enhancing the "Ability to carry out diversified farming". Transdisciplinary and farm-based research would help practitioners to formulate research questions and design suitable interventions to address them, thereby creating a better problem-solving ability at the farm level. Research processes involving practitioners would also support the development of design principles for diversified systems, integrating knowledge currently tacit among those active in the field. A proposed intervention to support this development is to implement "Education and compensation schemes" to ensure participation among farmers. By strengthening the interest and ability to participate through education and compensation, the need for voluntary commitments from the individual farmer could be reduced, thereby enabling transdisciplinary research projects without being affected by the balancing feedback B17 (Figure 5).

Moreover, a transition to more efficient, diversified farming systems would entail rethinking the organization of technology and labor. The current technology development within the agricultural sector has been described as tailored for production of scale and mono-cultures, while there is a lack of technology for diversified farming. As compared to conventional production, diversified production systems are characterized by larger complexity, holding a greater number of species within multiple habitats. In addition, these systems are smaller in scale, bound by the utilization of renewable resources and ecosystems services such as natural pest control. Acknowledging and addressing the technology development gap, through "Technology development for diversified production systems", would enable larger harvests, thereby contributing positively to the "Supply of diversified products".

Diversified farming is relatively labor intensive. The proposed "Green tax shift" is assumed to benefit this type of production, through reducing taxation of labor, thereby making the "Relative profitability of diversified farming" increase. In addition, it was pointed out that there is an overall need to rethink labor in the emerging bio-based economy. The agricultural sector is suggested to hold the potential to create new jobs, offering an opportunity to address societal challenges such as unemployment, inequality, and integration. To utilize this potential, diversified production systems could be based on an organizational design that combines technology for small-scale, complex systems, with a larger input of human labor. Technology development for diversified farming would then not replace, but complement, human labor, and a "Green tax shift" would be a mean to realize the potential of the agricultural sector to contribute to addressing broader societal challenges.

Uncertainty about the market potential and viability of the production process have been identified as thresholds to overcome in order to facilitate a transition to diversified farming. Market uncertainty could be reduced by targeting the leverage point "Ability to create broader awareness and support for diversified farming", in turns creating "Local demand" and a higher "Willingness to pay a price premium". By social means, a value shift could be created that aligns diets to local and seasonal supply. One proposed intervention is to design and use "Schools programs as platforms for change" (having an impact on the reinforcing feedbacks, R18–19, Figure 5). Successful examples were highlighted during the interviews, including municipalities where schools serve primarily seasonal and locally produced food. Another potential leverage point is the "Attractiveness of diversified farming", targeted by implementing proposals such as "Communication of successful examples" of diversified farming (affecting reinforcing feedbacks R15–19, Figure 5).

3.6. Interconnectedness between the Agricultural and Forestry Sectors

The study serving as a basis for the results outlined in the present paper covered bio-based resources derived both from the agricultural and forestry sector. In Bennich et al. (2018) [35], the dynamics governing transition pathways in the forestry sector are presented. During the interviews, it was highlighted that feedbacks directly related to primary production and processing in the forestry sector are also coupled with a political dimension of the bio-based economy concept. Key linkages were pointed out, such as the perceived legitimacy of the forestry sector having an impact on the political support for the bio-based economy, while political support in turns enable resource mobilization for innovation or investments in green jobs and traineeship programs [35]. Similarly, dynamics in the agricultural sector were identified as having an impact on the political support for the bio-based economy. Firstly, the "Perceived legitimacy of agricultural sector", to a certain extent determined by the adherence to the food-first principle, may influence the public support for the bio-based economy. The level of public support does ultimately either contribute to or erode political support. Secondly, the proposal to increasingly use "Environmental taxes" or other financial instruments (Figure 5) to make the use of fossil-based resources relatively more expensive is thought to create "Political uncertainty" linked to the bio-based economy concept. This sensitivity may reduce the ability to create a shared understanding and definition of the bio-based economy concept and its objectives, thereby halting a transition process, also in the forestry sector [35]. Other cross-sectoral linkages brought forward in the interviews include the connection between the ability of the bio-based economy concept to bring actors together, which is suggested to be partly governed by political factors, and the creation of environmental awareness in the agricultural sector. More specifically, the bio-based economy is perceived as a platform to discuss environmental issues in a non-threatening way. The larger the "Continuity of actor dialogue" over time, the greater the chances of creating "Environmental awareness" (Figures 2 and 5). Another factor that was emphasized is that land is a limited resource, and that developments in the forestry sector might have an impact on the ability to expand cultivated land (a potential "Desired expansion of forest land" contributing to "Competing interests for land-use", thereby lowering the "Conversion of fallow land to actively cultivated land" in the agricultural sector, Figure 1). This linkage was not recognized by interviewees in the forestry sector (i.e., acknowledging that a potential expansion of agricultural land might impede developments in the parts of the bio-based economy linked to the forestry sector). Thus, as also stressed in Bennich et al. (2018) [35], the interconnectedness of developments in the agricultural and forestry sectors, and the dynamics governing public and political support for the bio-based economy, may create synergies supporting a transition, but could also lead to negative spillover effects halting a broader transition process. Moreover, perceptions about these linkages differ among the actors in the bio-based economy.

4. Transition Pathways towards a Bio-Based Economy

4.1. Summary of Proposed Interventions

Transition pathways in the agricultural sector contributing to the emergence of a bio-based economy may be explored by asking "what-if" questions, using the CLDs and proposals identified during the interviews as a basis. Recalling the leverage points framework, not all leverage points are seen as equally efficient in terms of their ability to generate systemic change. However, with respect to the leverage points suggested during the interviews, many of them rank relatively highly in terms of efficiency, as they directly affect the relative strength of feedbacks or the overarching goals of the agricultural system. The specific proposals identified by the interviewees are summarized in Table 1, a majority of which target leverage points linked to the employment of environmentally friendly practices in conventional agriculture, biomass availability, or the shift to diversified farming and regenerative production. Leverage points related to the expansion and maintenance of farming activities were addressed to a lesser extent. The content of Table 1 is restricted to factors specifically brought up during the interviews.

In addition to the focus of the interventions summarized in Table 1, alternative leverage points were identified during the interviews. While no specific proposals to facilitate change in these variables were suggested, they may serve as a starting point for further thinking around potential interventions.

In order to support the expansion and maintenance of "Farming activities", potential leverage points identified include the "Perceived financial security" in the agricultural sector, the "Level of understanding between rural and urban areas", and the amount of "Long-term planning" carried out, all contributing to a higher number of "Successful generation shifts" or a greater "Attractiveness of profession". In terms of the "Employment of more environmentally friendly practices" in conventional agriculture, examples of additional leverage points are the "Collaboration across the value chain to validate and communicate environmental values" and the actors' "Interest and ability to innovate".

The aim of securing biomass supplies for the emerging bio-based economy was suggested to be fulfilled for instance through the implementation of national plant breeding programs, and by expanding farming activities, thereby increasing the conversion of fallow land to actively cultivated land. Additional leverage points in regards to securing biomass supplies, particularly for non-food biomass applications, are the variables linked to the number of "Farmers targeting novel bio-based markets". A development with more "Farmers targeting novel bio-based markets" could be facilitated by a growing market demand for novel crops, but also by interventions increasing farmers "Access to capital", "Access to knowledge", and their "Willingness to take risk". The shift to "Regenerative production" is assumed to be supported by interventions targeting the relative profitability of diversified farming and the ability to create a broader awareness and support for this type of production. Other potential leverage points may be the "Supply of labor with practical knowledge" and "Community resilience". The latter was emphasized based on the observation that isolated practices are vulnerable, and therefore not likely to be sustained over time.

In summary, a broad range of proposed interventions and places to intervene have been suggested, ranging from areas such as technology development, use of financial instruments, and efforts to facilitate change in consumer habits and community values. While certain proposals and places to intervene in the system were specifically highlighted during the interviews, it should be noted that all variables in the hypothesized system structure linked to farming activities, the employment of more environmentally friendly practices, biomass availability, and the shift to regenerative production could serve as a basis for discussing additional points of intervention.

Table 1. Summary of proposed interventions in the agricultural sector.

Targeted Objective	Proposed Intervention	Desired Change	Potential Unintended Consequences, Sources of Policy Resistance, or Systemic Risk	Uncertainty and Examples of Questions Remaining to Explore
Employment of more environmentally friendly practices in conventional agriculture	"What if the intensity of agricultural production is leveled out?" (reducing farming intensity on high quality farm land while intensifying cultivation on low quality farm land at risk of becoming overgrown)	Environmental values and long-term productivity are promoted.		How to determine and achieve optimal land-use and farming practices? What would be the net-effect on biomass availability?
	"What if financial instruments, such as environmental taxes, are used to a larger extent?"	Extracting and using fossil-based, finite resources would become relatively more expensive, promoting the employment of more environmentally friendly production practices.	The implementation, or prospect of implementing, such financial instruments could be perceived as a threat by some actors in the economy. If these interventions are linked to efforts to promote the bio-based economy, it could make the concept politically sensitive, reducing the ability to create a shared understanding of its objectives, and thereby hindering a broader transition to a bio-based economy.	How to "get the prices right" while simultaneously ensuring that the bio-based economy is perceived as an opportunity for a broad group of actors, also those currently not directly linked to bio-based sectors of the economy?
	"What if infrastructure is developed, allowing better transport and handling of harvest residues from the agricultural sector?"	Harvest residues would be utilized as an input to the bio-based economy to a larger extent. Resource use efficiency would increase, while pressure on other sources of biomass would be reduced.	Nutrient loss due to the removal of harvest residues and litter from agricultural land.	How can circularity be achieved, specifically in terms of nutrient recycling? How to solve the perceived match-making problem, where primary producers have biomass they cannot sell while industrial biomass demand is not fulfilled?
Biomass availability (total biomass supply for non-food applications, domestic food production)	"What if national plant breeding programs are further expanded?"	The resistance of plants to local stress, as well as the variability in chemical composition and plant characteristics, would increase. The demand for novel crops would grow, with a consequent increase in the number of farmers targeting new markets.	More land allocated to the production of non-food crops could make the perceived legitimacy of the agricultural sector decrease, lowering the overall ability to facilitate a transition to a bio-based economy.	What pathways make biomass demand and supply increase or decrease, and what are the consequences in terms of ability to meet biomass demand? Demand for novel crops is suggested to make the number of farmers allocating land to crop production for non-food purposes increase. What actors could take lead in this development, those currently active in the agricultural sector or completely new actors?

Table 1. Cont.

Targeted Objective	Proposed Intervention	Desired Change	Potential Unintended Consequences, Sources of Policy Resistance, or Systemic Risk	Uncertainty and Examples of Questions Remaining to Explore
	"What if plant breeding programs for diversified production systems were implemented?"	The ability of diversified production systems to provide edible yields grows.		What are the design principles that would allow these systems to function optimally? How much food are these systems able to provide in a Nordic context? How do consumption patterns need to change to allow production to meet demand?
	"What if the knowledge base on plant interactions was expanded?"	Competition between species in diversified systems would be reduced, and productivity would thereby increase.		
	"What if education and compensation schemes to ensure that farmers can participate in transdisciplinary research programs were employed?"	The interest and ability of farmers to engage in transdisciplinary research would increase, contributing to both theoretical and practical knowledge supporting diversified farming.		How to ensure that the potential of the ideas generated in these research programs is leveraged? How to ensure that transdisciplinary research programs are supported in an academic setting?
Shift to regenerative production	"What if technology for diversified production systems was developed?"	Diversified production systems would become more efficient, leveraging the potential of technology to complement manual labor, and harvests would increase.		Whom should take the lead in this development? What will be the net effect on the labor market?
	"What if taxation of labor was reduced, as part of a green tax shift?"	The relative profitability of the labor-intensive practices in diversified farming would increase, thereby strengthening the attractiveness of diversified farming.	With an increase in regenerative production, the demand for labor with practical skills would grow, increasing the labor gap. Unless measures are taken, this would limit a further expansion of regenerative production.	What are potential effects on the labor market and larger economy of reducing the tax burden on labor?
	"What if school programs were used as platforms for social change?"	Public procurement would support the expansion of diversified farming, through creating broader awareness.		What are suitable diets for a bio-based economy? What attitudinal changes are needed to support this change? How to implement these programs in a way so that they can be sustained over longer time periods?
	"What if successful examples of diversified farming practices were communicated?"	The attractiveness of diversified farming would grow, making more farmers transition and the regenerative production increase.	As regenerative production increases, also an expansion beyond local markets might take place. This could entail larger competition and risk, and unless complementary measures are taken, a larger failure rate.	

4.2. Synergies and Trade-Offs

The results outline multiple and diverse perspectives on the change needed in the agricultural sector for a transition to a bio-based economy to be facilitated. One area of controversy is the perceived need to increase adaptation of environmentally friendly practices in conventional agriculture. Views ranged from this being a necessity for the transition to a bio-based economy, to perspectives considering the Swedish agricultural sector sufficiently sustainable already. Another area where views differed, and where there are trade-offs, concerns the allocation of land. One view promotes an expansion of the cultivated land area, seeing more farm land as a requirement for meeting an anticipated increase in biomass demand in the transition to a bio-based economy. Another view suggests a shift in farming intensity rather than an expansion, with the objective of establishing agricultural practices that enhance ecosystem service provision. Different management approaches will also have different implications in terms of the combinations of labor and technology required, thus having an impact on the overall ability of the bio-based economy to contribute to new employment opportunities and rural development.

Aside from areas where priorities differ and where trade-offs exist, there are also change processes that are seemingly compatible. Interventions with synergistic effects, supporting the attainment of multiple objectives, include the implementation of new national plant breeding programs, contributing to biomass production for both food and non-food purposes. Moreover, while some of the identified objectives are seemingly separated from each other, challenges might still be shared. One example is the strive to achieve a price premium for promoting environmental values in primary production. In diversified farming, this is currently achieved predominantly through building local networks. In conventional farming, a price premium is suggested to be enabled by collaboration across the value chain to increase communication of the environmental values provided by agricultural production. Learning across domains may in this case be beneficial in order to achieve the objective of obtaining a price premium in both diversified and conventional farming. Lastly, timing and the specific order of intervention might be critical. For instance, the ability to expand and maintain farming activities might be seen as a prerequisite for a transition, as it ultimately affects the employment of more environmentally friendly practices in conventional agriculture, the utilization of new sources of biomass for the bio-based economy, as well as the shift to diversified farming and regenerative production.

5. Conclusions and Future Research

While agricultural sources of biomass have been recognized as important for the bio-based economy in Sweden, comparatively little attention has been given to the specific change processes underpinning a transition process in the agricultural sector. This paper begins to address this gap in an integrated way by using an approach combining systems analysis and expert interviews in the development of conceptual maps. These maps, representing causal hypotheses about the interplay of desired change processes and potential hindrances in the agricultural sector, were subsequently used to explore proposed interventions and the transition pathways they form. The study also considered cross-sectoral linkages between the forestry and agricultural sectors.

The results underline the diversity in objectives and views held by different actors. As identified during the interviews, desired change processes include an expansion of farming activities in Sweden, the employment of more environmentally friendly practices in conventional agriculture, securing biomass availability in the bio-based economy, and facilitating a shift to diversified farming and regenerative production. Even though no consensus on the desired change was held among the actors represented in the interviews, the process of making hypothesized system structure explicit allows for the development of a qualitative understanding of the synergies and trade-offs between different objectives. The approach and results in the form of the CLDs also highlight under which conditions certain arguments and areas of controversy apply. The polarization created by the fuel versus food debate may serve as an example. By creating a systemic analysis, the assumption that a transition process would necessarily entail an expansion of agricultural production for the purpose of growing

Sustainability **2018**, *10*, 1504

crops for non-food biomass applications is challenged. Different pathways have different implications in terms of biomass demand, as well as supply, and depending on the objectives either one of these pathways may be supported by action for change.

The results presented in this paper are exploratory, and further research could entail re-examining each proposed causality, probing additional insight about the assumptions and conditions under which it holds. Additionally, the CLDs presented in this paper are based on the accounts of the interviewed experts. A worthwhile next step could entail comparing these results with other sources of information, to identify system structure that may have been overlooked during the interviews, as well as to support the identification of knowledge gaps in the existing literature. Another area of future research may include a further exploration of the implementation phase of the proposed interventions, identifying necessary change in causal structure for these actions to be undertaken, or alternatively to explore additional proposals and their potential impacts. Another direction for future research includes identifying actor perspectives currently being overlooked in the general debate, for example linked to the dynamics governing biomass demand. Furthermore, the aim of the study serving as the basis for the present paper was to provide an integrated understanding of change in the coupled social and ecological systems enabling or hindering a transition process in Sweden, with a specific focus on terrestrial biomass sources. Results linked to the dynamics in the forestry sector are outlined in Bennich et al. (2018) [35] and could be used to put the CLDs linked to the agricultural sector in perspective. The results also exemplify how actors perceive cross-sectoral relationships differently, and a way to further explore transition pathways could entail identifying and mapping additional ways in which the sectors in the bio-based economy are interrelated, how these linkages contribute to or hinder an overall transition process, as well as how different actors in the bio-based economy perceive these interlinkages.

Finally, CLDs do not allow for rigorous inference about the relative strength of the feedbacks identified, or about how the interplay of these feedbacks causes the variables in the system to change over time. Another avenue for further research may therefore be in the direction of simulation-based analysis, to test the hypothesized system structure and enable learning about the behavioral output of the system. The results outlined in the present paper as well as in Bennich et al. (2018) [35] provide a basis for systemic, holistic thinking around future transition pathways towards a bio-based economy. A number of feedbacks with the potential to drive or contribute to a transition were identified during the interviews. Yet, many of these might currently be working in the opposite direction, moving the system away from a desired state or outcome, thereby preventing a transition from happening. Moreover, for change to be sustained, there might be a need to pass critical thresholds. Thus, while the potential to facilitate a transition is perceived as relatively large, there is a need to acknowledge the inherent difficulties in shifting the direction and dominance of central feedbacks and to deepen the understanding of the magnitude of change actually required for a transition to happen.

Author Contributions: All authors were part of designing the study and developing the methodological frame. T.B. conducted the data collection, S.B. and T.B. carried out the data processing and analysis. All authors contributed to the writing of the paper and have read and approved the final version.

Acknowledgments: This project was funded by the European Union's Horizon 2020 research and innovation programme under the Marie Skłodowska-Curie grant agreement No. 675153 (ITN JD AdaptEconII). We are thankful to the participants of this study for contributing their time, and for sharing their knowledge and experiences. We also thank Nuno Videira, Jenneth Parker, and Eline de Jong for useful input and discussions.

Conflicts of Interest: The authors declare no conflict of interest. The founding sponsors had no role in the design of the study; in the collection, analyses, or interpretation of data; in the writing of the manuscript, and in the decision to publish the results.

References

1.   German Bioeconomy Council. *Bioeconomy Policy (Part II) Synopsis of National Strategies around the World*; German Bioeconomy Council: Berlin, Germany, 2015.

2. Dubois, O.; Juan, M.G.S. *How Sustainability Is Addressed in Official Bioeconomy Strategies at International, National and Regional Levels—An Overview*; Food and Agriculture Organization of the United Nations: Rome, Italy, 2016.

3. Priefer, C.; Jörissen, J.; Frör, O. Pathways to Shape the Bioeconomy. *Resources* **2017**, *6*, 10. [CrossRef]

4. Bugge, M.; Hansen, T.; Klitkou, A. What Is the Bioeconomy? A Review of the Literature. *Sustainability* **2016**, *8*, 691. [CrossRef]

5. Lange, L.; Björnsdóttir, B.; Brandt, A.; Hildén, K.; Hreggviðsson, G.Ó.; Jacobsen, B.; Jessen, A.; Karlsson, E.N.; Lindedam, J.; Mäkelä, M.; et al. *Development of the Nordic Bioeconomy*; Nordic Council of Ministers: Copenhagen, Denmark, 2016.

6. EU. Innovating for Sustainable Growth: A Bioeconomy for Europe. 2012. Available online: http://ec.europa.eu/research/bioeconomy/pdf/official-strategy_en.pdf (accessed on 13 January 2017).

7. OECD. *The Bioeconomy to 2030 Designing a Policy Agenda*; OECD Publishing: Paris, France, 2009; Available online: https://www.oecd-ilibrary.org/economics/the-bioeconomy-to-2030_9789264056886-en (accessed on 8 May 2018).

8. El-Chichakli, B.; von Braun, J.; Lang, C.; Barben, D.; Philp, J. Five cornerstones of a global bioeconomy. *Nat. News.* **2016**, *535*, 221–223. [CrossRef] [PubMed]

9. Formas. *Swedish Research and Innovation Strategy for a Bio-Based Economy*; Swedish Research Council for Environment, Agricultural Sciences and Spatial Planning, Formas: Stockholm, Sweden, 2012.

10. McCormick, K.; Kautto, N. The Bioeconomy in Europe: An Overview. *Sustainability* **2013**, *5*, 2589–2608. [CrossRef]

11. The White House. National Bioeconomy Blueprint. *Ind. Biotechnol.* **2012**, *8*, 97–102.

12. Refsgaard, K.; Teräs, J.; Kull, M.; Oddsson, G.; Jóhannesson, T.; Kristensen, I. *The Rapidly Developing Nordic Bioeconomy: Exerpt from State of the Nordic Region 2018*; Nordic Council of Ministers: Copenhagen, Denmark, 2018.

13. Skånberg, K.; Olsson, O.; Hallding, K. *Den Svenska Bioekonomin: Definitioner, Nulägesanalys och Möjliga Framtider*; Stockholm Environment Institute: Stockholm, Sweden, 2016.

14. Teräs, J.; Lindberg, G.; Johnsen, I.; Perjo, L.; Giacometti, A. *Bioeconomy in the Nordic Region: Regional Case Studies*; Nordregio Working Paper; Nordic Centre for Spatial Development: Stockholm, Sweden, 2014.

15. SEPA. *Sweden's Environmental Objectives—An Introduction*; Swedish Environmental Protection Agency: Stockholm, Sweden, 2016.

16. IVL and Fossil Free Sweden. *Hoppfulla Trender för ett Fossilfritt Sverige En Rapport från Initiativet Fossilfritt Sverige*; IVL and Fossil Free Sweden: Stockholm, Sweden, 2017.

17. Government of Sweden. *The Swedish Government's Overall EU Priorities for 2017*; Government of Sweden: Stockholm, Sweden, 2017.

18. Regeringskansliet. *Fact Sheet: Proposal Referred to the Council on Legislation on a Climate Policy Framework for Sweden*; Regeringskansliet: Stockholm, Sweden, 2017.

19. Regeringskansliet. Innovation Partnership Programmes—Mobilising New Ways to Meet Societal Challenges. 2016. Available online: http://www.government.se/articles/2016/07/innovation-partnership-programmes-mobilising-new-ways-to-meet-societal-challenges/ (accessed on 5 April 2017).

20. Swedish Forest Industries Federation. *Skogsnäringen Driver Tillväxt i Världens Bioekonomi—Mål på Vägen mot Visionen*; Swedish Forest Industries Federation: Stockholm, Sweden, 2015.

21. Business Sweden. *Bioeconomy in Sweden, Sector Overview. Business Opportunities in a Bioeconomy Growth Market*; Business Sweden: Stockholm, Sweden, 2016.

22. Nordic Council of Ministers. *Nordic Bioeconomy, 25 Cases for Sustainable Change*; Nordic Council of Ministers: Copenhagen, Denmark, 2017.

23. Pfau, S.F.; Hagens, J.E.; Dankbaar, B.; Smits, A.J.M. Visions of Sustainability in Bioeconomy Research. *Sustainability* **2014**, *6*, 1222–1249. [CrossRef]

24. Lewandowski, I. Securing a sustainable biomass supply in a growing bioeconomy. *Glob. Food Secur.* **2015**, *6*, 34–42. [CrossRef]

25. Scarlat, N.; Dallemand, J.-F.; Monforti-Ferrario, F.; Nita, V. The role of biomass and bioenergy in a future bioeconomy: Policies and facts. *Environ. Dev.* **2015**, *15*, 3–34. [CrossRef]

26. Schmidt, O.; Padel, S.; Levidow, L. The bio-economy concept and knowledge base in a public goods and farmer perspective. *Bio-Based Appl. Econ.* **2012**, *1*, 47–63.

27. Staffas, L.; Gustavsson, M.; McCormick, K. Strategies and Policies for the Bioeconomy and Bio-Based Economy: An Analysis of Official National Approaches. *Sustainability* **2013**, *5*, 2751–2769. [CrossRef]

28. Pülzl, H.; Kleinschmit, D.; Arts, B. Bioeconomy—An emerging meta-discourse affecting forest discourses? *Scand. J. For. Res.* **2014**, *29*, 386–393. [CrossRef]

29. Pavone, V.; Goven, J. (Eds.) *Bioeconomies*; Springer International Publishing: Cham, Switzerland, 2017.

30. Bosch, R.; van de Pol, M.; Philp, J. Define biomass sustainability: The future of the bioeconomy requires global agreement on metrics and the creation of a dispute resolution centre. *Nature* **2015**, *523*, 526–528. [CrossRef] [PubMed]

31. Viaggi, D. Towards an economics of the bioeconomy: Four years later. *Bio-Based Appl. Econ.* **2016**, *5*, 101–112.

32. Addeo, F. *Bio-Based Economy for Europe: State of Play and Future Potential—Part 1. Report on the European Commission's Public On-Line Consultation*; European Commission: Brussels, Belgium, 2011.

33. Nebe, S. *Bio-Based Economy in Europe: State of Play and Future Potential—Part 2 Summary of Position Papers Received in Response to the European Commission's Public On-Line Consultation*; European Commission: Brussels, Belgium, 2011.

34. Golembiewski, B.; Sick, N.; Bröring, S. The emerging research landscape on bioeconomy: What has been done so far and what is essential from a technology and innovation management perspective? *Innov. Food Sci. Emerg. Technol.* **2015**, *29*, 308–317. [CrossRef]

35. Bennich, T.; Belyazid, S.; Kopainsky, B.; Diemer, A. The Bio-Based Economy: Dynamics Governing Transition Pathways in the Swedish Forestry Sector. *Sustainability* **2018**, *10*, 976. [CrossRef]

36. Sedlacko, M.; Martinuzzi, A.; Røpke, I.; Videira, N.; Antunes, P. Participatory systems mapping for sustainable consumption: Discussion of a method promoting systemic insights. *Ecol. Econ.* **2014**, *106*, 33–43. [CrossRef]

37. Sterman, J.D. *Business Dynamics: Systems Thinking and Modeling for a Complex World*; Irwin/McGraw-Hill: Boston, MA, USA, 2009.

38. Stave, K.A.; Kopainsky, B. A system dynamics approach for examining mechanisms and pathways of food supply vulnerability. *J. Environ. Stud. Sci.* **2015**, *5*, 321–336. [CrossRef]

39. Lane, D.C. The emergence and use of diagramming in system dynamics: A critical account. *Syst. Res. Behav. Sci.* **2008**, *25*, 3–23. [CrossRef]

40. Bogner, A.; Littig, B.; Menz, W. Introduction: Expert Interviews—An Introduction to a New Methodological Debate. In *Interviewing Experts*; Palgrave Macmillan: London, UK, 2009; pp. 1–13.

41. Meadows, D.H. Whole earth models and systems. *Co-Evol. Q.* **1982**, *Summer*, 98–108.

42. Meadows, D.H. *Thinking in Systems: A Primer*; Chelsea Green Publishing: Hartford, VT, USA, 2008.

43. Senge, P. *The Fifth Discipline: The Art and Practice of the Learning Organization (Revised and Updated)*; Random House, Inc.: New York, NY, USA, 2006.

44. Lelea, M.A.; Roba, G.M.; Christinck, A.; Kaufmann, B. *Methodologies for Stakeholder Analysis—For Application in Transdisciplinary Research Projects Focusing on Actors in Food Supply Chains*; German Institute for Tropical and Subtropical Agriculture (DITSL): Witzenhausen, Germany, 2014.

45. Barriball, K.L.; While, A. Collecting data using a semi-structured interview: A discussion paper. *J. Adv. Nurs.* **1994**, *19*, 328–335. [CrossRef] [PubMed]

46. Harrell, M.C.; Bradley, M.A. *Data Collection Methods: Semi-Structured Interviews and Focus Groups*; RAND Corporation: Santa Monica, CA, USA, 2009.

47. Kim, H.; Andersen, D.F. Building confidence in causal maps generated from purposive text data: Mapping transcripts of the Federal Reserve. *Syst. Dyn. Rev.* **2012**, *28*, 311–328. [CrossRef]

sustainability

MDPI

Article

The Bio-Based Economy: Dynamics Governing Transition Pathways in the Swedish Forestry Sector

Therese Bennich [1,*], Salim Belyazid [1], Birgit Kopainsky [2] and Arnaud Diemer [3]

[1]   Department of Physical Geography, Stockholm University, SE-106 91 Stockholm, Sweden;
     salim.belyazid@natgeo.su.se
[2]   System Dynamics Group, Department of Geography, University of Bergen, Postboks 7802, NO-5020 Bergen,
     Norway; birgit.kopainsky@uib.no
[3]   Center for Studies and Research on Internal Development (CERDI), University of Clermont-Auvergne,
     FR-320, 63009 Clermont-Ferrand, France; arnaud.diemer@uca.fr
*   Correspondence: therese.bennich@natgeo.su.se; Tel.: +46-(0)72-554-9592

Received: 5 March 2018; Accepted: 25 March 2018; Published: 27 March 2018

Abstract: A transition to a bio-based economy would entail change in coupled social–ecological systems. These systems are characterised by complexity, giving rise to potential unintended consequences and trade-offs caused by actions aiming to facilitate a transition process. Yet, many of the analyses to date have been focusing on single and predominantly technological aspects of the bio-based economy. The main contribution of our work is to the development of an integrated understanding of potential future transition pathways, with the present paper focusing specifically on terrestrial biological resources derived from the forestry sector in Sweden. Desired change processes identified include a transition to diversified forest management, a structural change in the forestry industry to enable high-value added production, and increased political support for the bio-based economy concept. Hindrances identified include the ability to demonstrate added values for end consumers of novel biomass applications, and uncertainty linked to a perceived high level of polarisation in the forestry debate. The results outline how these different processes are interrelated, allowing for the identification of high order leverage points and interventions to facilitate a transition to a bio-based economy.

Keywords: bio-based economy; bio-economy; systems analysis; causal loop diagrams; forestry

---

1. Introduction

The development of a bio-based economy may be understood as a transition from a society mainly dependent on fossil-based resources, to an economy built on the use of bio-based resources [1]. The growing interest in the bio-based economy, also referred to as the knowledge-based bio-economy or the bio-economy, can be explained by factors such as its proposed ability to contribute to a resource efficient, competitive, and low-carbon economy, as well as to economic activity and new employment opportunities in rural areas. The bio-based economy has also been conceptualized, similarly to other emerging concepts such as the circular economy or sharing economy, as contributing to the development of a green economy [2].

The premises of a bio-based economy lie partly in its cross-sectoral reach, and a transition would involve multiple actors and interactions across scales and societal domains. The transition process might entail changes in the way bio-based resources are managed and utilized to provide goods and services, enabled by factors such as emerging technologies, new institutions, and shifts in attitudes and values. One way to approach the concept is to employ a coupled social–ecological systems frame [3]. The social in this context refers to all human dimensions relevant to the bio-based economy, including politics, culture, and cooperation. The ecological refers to the biosphere in which all human activity is embedded, and in the case of the bio-based economy more specifically to the dynamics governing the growth and regeneration of bio-based natural resources. Change in social–ecological systems involves complex, simultaneous processes, characterized by multiple and interacting feedbacks, non-linear dynamics, and cause and effect relationships distant in time and space [4]. This complexity gives rise to elements of uncertainty, but also to potential trade-offs and unintended consequences following interventions to facilitate change. In order to fathom the possibility of success of a transition to a bio-based economy, it is crucial to develop an integrated understanding of the seemingly independent processes governing the necessary change in its different components. However, up until now, many of the analyses have been addressing aspects of a transition process in isolation. Moreover, the debate and visions of a bio-based economy have to a large extent been informed by an engineering and technological perspective, potentially overlooking broader ecological, economic, and societal implications [5–8]. This paper seeks to contribute to the development of an integrated structural understanding of the social and ecological processes governing a transition to a bio-based economy. The following guiding research questions served as a starting point for the study:

- What are the social–ecological dynamics currently enabling or hindering a transition to a bio-based economy?
- What actions are proposed to facilitate a transition to a bio-based economy, and what pathways do they form?

In order to answer these overarching research questions, we combine systems analysis and expert interviews in the development of conceptual system maps. We depart from the proposition that such structural analysis can provide a space for learning, a basis for discussion, and a qualitative understanding of the interconnectedness of pathways towards a bio-based economy, allowing for the identification of high order leverage points for intervention. There are several tools that may help visualise how components of systems are interrelated, such as high-level system frameworks, maps of actor networks, subsystem diagrams, stock and flow maps, and policy structure diagrams [9,10]. In this study, the specific diagramming tool of Causal Loop Diagrams (CLDs) is used, with the empirical basis in the expert interviews. CLDs are made up by variables connected by arrows, representing hypotheses about causal relationships. In respect to the purpose of this study, the strength of CLDs lies partly in their ability to close sequences of causes and hypothesized effects, thereby moving from linear to feedback thinking. Aside from identifying higher order leverage points, such analysis may support the identification of unintended consequences and sources of policy resistance, the latter referring to the tendency of interventions to fail due to the response of the system triggered by the intervention itself [10,11].

The study focuses on the case of terrestrial biological resources derived from the forestry and agricultural sectors in Sweden, a country that is promoting a transition on a national scale, and where both biophysical and socio-economic preconditions are perceived as favourable [1,12]. The present paper outlines transition pathways in the forestry sector, highlighting how dynamics directly related to primary production and processing are coupled with a political dimension of the bio-based economy concept. For reasons of space, the results related to dynamics in the agricultural sector are not covered in this paper, but presented in Bennich et al., (in review). The remaining part of the paper is structured as follows: First, the research design and methodological framework are outlined. Thereafter the results in the form of CLDs are presented. These are subsequently used to explore potential transition pathways for the future. The paper ends with a concluding discussion and areas for future research.

2. Research Design and Methodological Framework

2.1. Tools for Systems Analysis: System Dynamics Modelling

The research design is based on the system dynamics modelling process [10]. System dynamics is a method that addresses the structural and endogenous cause of time-dependent system behaviour [13]. Through qualitative causal maps and formal simulation models, dynamic hypotheses about the functioning of complex systems are developed and tested. These models may then be used for policy design and testing, to support management of the systems under study. System dynamics deals explicitly with characteristics of complex systems, such as non-linear behaviour, delays, and feedback processes. The system dynamics modelling process typically consist of the following steps [10]:

(1) Problem identification: What is the aim of the modelling process? What is the behaviour of interest? What are the key variables and concepts giving rise to the behaviour of interest?
(2) Formulation of dynamic hypothesis: What are current theories about the behaviour of interest? What are endogenous consequences of the feedback structures in the system?
(3) Formulation of simulation model to test the hypothesis (e.g., specification of model parameters and initial conditions).
(4) Model testing: Does the model replicate reference data? Is the behavioural output robust? What are the results of the sensitivity analysis?
(5) Policy design and testing: How can proposed policy options be represented in the model and what are their potential impacts? How do different policy options interact?

For the aim of this study, system dynamics was used as a stepwise method for the description of the system and for performing qualitative analysis of proposed interventions and pathways for a successful social–ecological transition to a bio-based economy. In this respect, the study did not extend to a numerical assessment of potential transition pathways, thereby excluding steps three and four and instead focusing on steps one, two, and five of the standard modelling process. The dynamic hypothesis was formulated in the form of conceptual maps, using the tool of CLDs (for further introduction to connotation and use, see Appendix A). CLDs represent causal hypotheses of system structure, and were initially used to communicate formal simulation models, or as an intermediary step between system conceptualisation and the development of a quantitative simulation model [14]. CLDs may however also be used as an analytical tool by their own means. Aside from making hypothesised system structures explicit, potential uses include structuring of problem spaces, identification of research questions, identification of areas of risk and uncertainty, and the development of shared learning and collaboration [15]. Applications of qualitative system dynamics and CLDs are found in a diverse number of areas, where examples specifically related to the sustainability field include ecological economics [16,17], environmental management and governance [18,19], community development [20], and stakeholder participation in environmental decision making [21].

2.2. Expert Interviews

Expert interviews served as the empirical basis for the CLDs. Expert interviews have long been used in research, considered an efficient way of gathering data in an exploratory phase of a research project, a means to obtain knowledge in situations where it might otherwise be difficult to gain access (e.g., where subjects are considered sensitive or taboo), or as a way to identify and reach out to a larger circle of interviewees in a specific context [22,23]. However, the purpose, form, and method for data analysis related to using expert interviews vary. Additionally, multiple perspectives on critical issues such as what constitutes an expert, how expert knowledge is distinguished from other types of knowledge, and on the different types of expert knowledge that exist, are still present in the methodological debate [24].

The main purpose of using expert interviews in this research was to gain exploratory insight into the drivers and hindrances of a transition to a bio-based economy, broadening the debate around potential development pathways for the future. Fourteen experts were selected for participation in the study (for further information about the sample, see Appendix B). The selection was based on an initial literature review and snowball sampling, informed by the approach for actor identification outlined by Lelea et al., (2014) [25]. The interviewees were chosen based on their expertise in terms of experience and knowledge from the forestry and agricultural sectors in Sweden, as well as on their ability to represent larger actor groups of relevance to the bio-based economy. To elicit multiple perspectives on the subject, interviewees were also chosen based on their ability to represent diverse standpoints. In the snowball sample, specific questions were directed towards identifying actors and perspectives that may previously have been overlooked in the broader debate.

Semi-structured interviews were selected as means for data collection, as they are well suited to explore perceptions and views on complex issues, in a sample group with varied professional, educational, and personal backgrounds [26]. The interviews were based on a number of open-ended questions (Appendix C), aiming to elicit central variables, causal relationships, and feedback processes that the interviewees perceived as key to the bio-based economy. The questions were also aiming to support reasoning about desired as well as unintended changes following proposed interventions to facilitate a transition. The interviews were carried out in 2017 and early 2018, and took place either in Stockholm or via Skype. Each interview lasted between 60 and 120 min, and was digitally recorded and transcribed. Issues of validity and reliability in the data collection phase were addressed drawing on the process outlined by Barriball and While (1994) [26]. Specific measures undertaken include the formulation and internal testing of an interview schedule before the actual interviews took place, interviewer training, and continuous and reflexive analysis to identify and avoid any potential ambiguities, leading questions, or inappropriate use of probes throughout the interview process.

2.3. Data Analysis and Development of Dynamic Hypotheses

The process of data analysis and development of the CLDs was based on the method presented by Kim and Andersen (2012) [27]. First, for each interview transcript open coding was used to identify key variables and proposed interventions to facilitate a transition to a bio-based economy. Second, the variables were analysed for thematic content, to define relevant system boundaries. In the next step, the interview transcripts were revisited with the aim of identifying causal links between variables, expressed as causal arguments by the interviewees. Moreover, desired or unintended consequences of proposed interventions were identified, exploring their proposed impacts. The data segments were subsequently translated into CLDs, one for each interviewee.

Next, the CLDs were integrated by maintaining differences while aggregating similarities, in a process utilising elements of axial coding. In this step, as emphasised by Kim and Andersen (2012), there is a leap in conceptualisation [27]. The integrated CLDs combine structures collected from multiple interviews, and specific variable names are replaced with more generalizable terms, in order to compare and combine the diagrams. As an additional step, to ensure that the CLDs were able to represent the mental models of the interviewees, a workbook was sent out to the initial interview sample for confirmation. It contained the integrated CLDs, written descriptions, and a set of questions. The questions were probing additional feedback thinking, clarification, and missing or redundant system structure. After receiving and analysing the response to the workbook, the CLDs were revised accordingly. The final step of model development consisted of simplifying the CLDs, to increase clarity and coherence. The process was documented to allow for traceability between the data segments and model structure. Two system analysts worked in parallel on the development of the CLDs, and the results were compared and discussed to reduce researcher biases.

2.4. The Leverage Points Framework

In order to understand the potential impact of the suggested policies and proposals to facilitate a transition to a bio-based economy, we built on the leverage points framework initially developed by Meadows in 1997 [28]. Leverage points are places in a system where a small initial shift may create large-scale change, and the framework provides a basis for analysing the nature and effectiveness of interventions. Interventions targeting physical parts of a system fall at the lower end of efficiency. Such interventions may entail changing parameter values (e.g., tax rates, minimum wages, and fractions of land set aside for conservation), the size of buffers in the system as compared to the rate at which these buffers change (such as lowering or increasing an inventory), the physical arrangement of the system (e.g., construction of infrastructure), or the length of delays in the feedback processes in the system as compared to the rate of change in the state of the system. Interventions falling at the higher end of efficiency include efforts to change the relative strength of balancing or reinforcing feedbacks (where balancing feedbacks work as control mechanisms in the system and reinforcing feedbacks amplifies system change), and the creation of new information feedbacks (as a lack of information is suggested to be a root cause of systems malfunctioning). Even higher on the scale of effectiveness are actions to change the ability to self-organise, and the overarching rules and goals of the system (acknowledging the role of power, i.e., where the ability to create and change the rules of a system resides). Lastly, the leverage points framework emphasises the role of paradigms as sources of systems, and that the ability to critically reflect on paradigms may prove one of the highest leverage points for system change [28].

3. Results

While parallel visions and understandings of the bio-based economy exist, the results provide an integrated understanding of the change processes perceived as key by the interviewed experts. The results, in the form of CLDs, are presented under the following themes: (1) The dynamics directly linked to primary production and processing in the forestry sector, and (2) The political dimension of the bio-based economy concept. The thematic separation is made for reasons of clarity, but interlinkages are indicated and further elaborated in Section 3.4. Variable names are designated by quotation marks in the text.

3.1. Dynamics Governing a Transition in Primary Production and Processing

During the interviews, key desired change processes identified were a "Shift to diversified forestry" in primary production, and a "Shift to high value-added production" in processing (variables in capital letters, Figure 1). The two processes are described in the following subsections.

3.1.1. Drivers of a Transition in Primary Production

A "Shift to diversified forestry" was identified as essential to a transition to a bio-based economy. It may be understood as a shift away from the currently dominant industrial forestry management model based on mono-cultures, volume maximization, and clear-cuts. This shift would imply change in multiple areas, such as management practices, the number of productive species, and the characteristics of the forest biomass. The larger the "Shift to diversified forestry", the higher the "Nature values", encompassing productive, ecological, recreational, and cultural forest values. Four reinforcing feedbacks govern the potential "Shift to diversified forestry". The higher the "Nature values", the more "Public engagement", which in turn gives rise to higher "Demand for non-market values", reflecting the use value of the forest. Different underlying factors explain this proposed relationship, such as proximity to urban areas (demand increases when proximity increases), and accessibility (demand increases with accessibility). As the "Demand for non-market values" increases, so does the "Forest owner self-confidence and motivation" to take on an active management role. The impact of changes in demand for non-market values on the forest owner's self-confidence and motivation depends on factors such as the sense of community in the region (the higher the sense of community, the larger the considerations of public interests in forest management). With active forest management, the overall diversity of management practices increases, thereby supporting the "Shift to diversified forestry" (reinforcing feedback, R1, Figure 1).

The larger the "Shift to diversified forestry" and the higher the "Nature values", the higher the forest owners' "Perceived ability to provide non-market values". Similarly, with more diversified forestry, the "Ability to meet demand for high quality biomass" increases. Successfully providing these diverse services has an important positive impact on the self-motivation and confidence of the forest owner. The higher the fulfillment of management goals, the higher the self-motivation and confidence, leading to a larger "Shift to diversified forestry" (reinforcing feedbacks, R2-3, Figure 1). This effect is further enhanced by diversified forestry contributing to higher "Soil quality", which in the long-term supports the attainment of productivity related management goals (reinforcing feedback, R4, Figure 1).

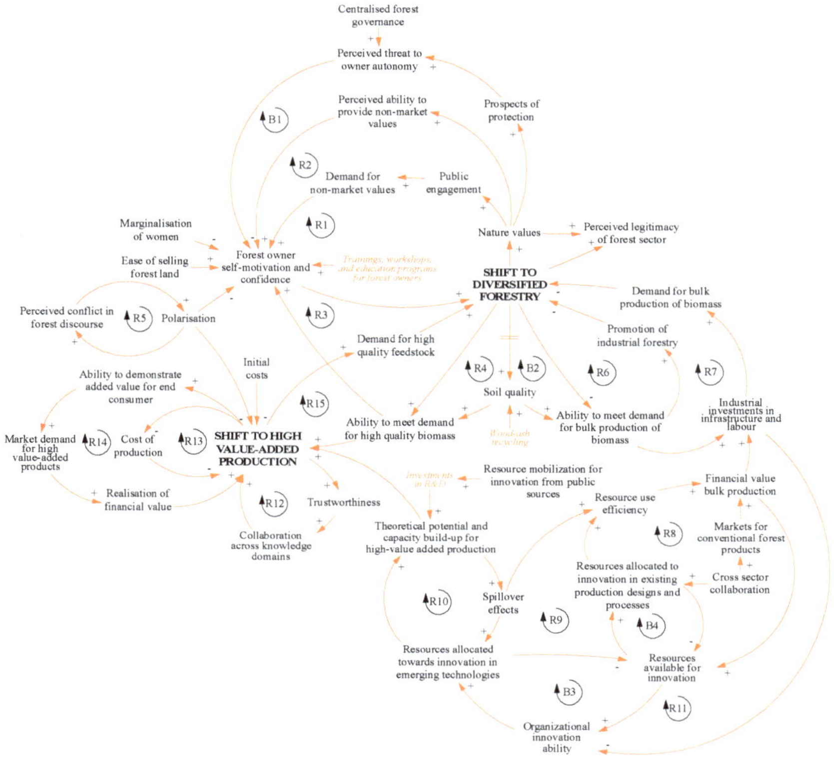

Figure 1. Feedback processes governing key elements of a transition process in the forestry sector. Capitalized variables represent desired change processes, and variables in italic denote proposed interventions. Causal Loop Diagrams (CLDs) make use of arrows to indicate causal relationships between system variables. These relationships can be either positive (represented by a plus sign) or negative (represented by a negative sign). A positive relationship implies that, if variable X is connected to variable Y, they move in the same direction (an increase in X will lead to an increase in Y, and a decrease in X will lead to a decrease in Y). A negative relationship suggests that the variables move in opposite directions (an increase in X will lead to a decrease in Y, and a decrease in X will lead to an increase in Y). Feedbacks may be either reinforcing or balancing (denoted by a R and B, respectively). For a further introduction to CLDs, see Appendix A.

3.1.2. Hindrances of a Transition in Primary Production

One balancing and one reinforcing feedback counteract the shift to diversified forestry. First, when "Nature values" increase, so do the "Prospects of protection", that is, the prospects of conservation of land areas with high nature values. These prospects increase the "Perceived threats to owner autonomy", causing the "Forest owner self-motivation and confidence" to decrease, thereby hampering the shift to more diversified forestry management models (balancing feedback, B1, Figure 1). Second, the discourse in the forestry sector is described as polarized. The higher the "Perceived polarization in forest discourse", the higher the "Polarization", and vice versa (reinforcing feedback, R5, Figure 1). Polarization erodes "Forest owner self-motivation and confidence", through creating uncertainty regarding the relative sustainability and financial viability of different management options.

In addition, there are lock-in effects created by feedbacks linked to the current structure of the forestry industry, which have an impact on the likelihood of a shift to diversified forestry. Currently, the "Ability to meet demand for bulk production of biomass" is high, promoting the industrial use of forests, and decreasing the "Shift to diversified forestry". The promotion of industrial forestry can be explained by factors such as high confidence in the current model as long as it remains profitable, and actor mobilization creating a larger ability to protect industrial interests. The less forest land allocated to diversified modes of production, the larger the ability to meet demand for bulk production of biomass (reinforcing feedback, R6, Figure 1). Second, as the "Ability to meet demand for bulk production of biomass" increases, so does "Industrial investments in infrastructure and labor". The larger these investments are, that is, the higher the production capacity of the forestry industry, the higher the demand for "Bulk production of biomass", and the lower the "Shift to diversified forestry" (reinforcing feedback, R7, Figure 1). However, taking into account a longer time horizon, low diversity in the forestry industry can lower the "Soil quality". This can be explained partly by the industrialized mode of production having negative environmental impacts, but also by an increasing demand for biomass leading to pressure to utilize harvest residues to a larger extent. Lower soil quality, in turns, erodes the "Ability to meet demand for bulk production of biomass", thus weakening the ability to maintain a highly industrialized mode of forest management over time (balancing feedback, B2, Figure 1).

3.1.3. Drivers of a Transition Linked to Innovation in the Processing Stage

Two change processes have been identified as important with regards to the innovation potential of the forestry sector. First, the process of the conventional industrial production becoming more advanced and resource efficient, and second the potential shift towards high value-added production built on economies of scope rather than economies of scale.

The process of making the current industrial structure more advanced depends on the "Resources allocated to innovation in existing production designs and processes". The more resources allocated towards these ends, the larger the "Resource use efficiency", in turns leading to higher "Financial value of bulk production". The "Financial value of bulk production", that is, the profit streams from conventional production, determines the "Resources available for innovation", further enabling resource allocation towards innovation in existing product designs and processes (reinforcing feedback, R8, Figure 1). An additional factor identified as important in this respect is the "Cross sector collaboration". The more collaboration across sectors, the higher the innovation potential in conventional forestry production, and the larger the "Markets for conventional forest products", thereby contributing to both the "Resources allocated to innovation in existing production designs and processes" and the "Financial value of bulk production".

There is a choice between allocating resources towards existing production processes, or towards completely novel modes of production. The latter may be described as a new form of forestry, where biomass is utilized based on qualitative characteristics, rather than on bulk. More "Resources allocated towards innovation in emerging technologies" makes the "Theoretical potential and capacity build-up for high value-added production" increase. As long as this entails "Spillover effects", increasing "Resource use efficiency" in conventional production processes (i.e., contributing to and being compatible with the existing industrial structure), the "Resources allocated towards innovation in emerging technologies" will increase (reinforcing feedbacks, R9-10, Figure 1). The choice of allocating resources towards emerging technologies depends on the "Organizational innovation ability", referring to the ability of an organization to innovate when phasing pressures. The higher the "Organizational innovation ability", the larger the tendency to allocate resources towards innovation in new areas. Counteracting such developments is a reinforcing feedback working through the "Financial value of bulk production" and "Industrial investments in infrastructure and labor". The larger the industrial investments in existing production processes, the larger the lock-in effect, and the lower the "Organizational innovation ability" (reinforcing feedback, R11, Figure 1). The balancing feedbacks B3 and B4 (Figure 1) represent the fact that resources are limited, and that the more that is allocated either to innovation in emerging technologies or to strengthening current production processes, the less remains to spend elsewhere.

The "Shift to high value-added production" depends on the "Theoretical potential and capacity build-up for high value-added production", as well as the "Ability to meet demand for high quality biomass" in primary production. Four feedbacks reinforce a potential transition process. As the "Shift to high value-added production" starts to unfold, the "Trustworthiness" of the actor increases, facilitating new forms of collaboration across the value chain and related knowledge domains. Collaboration strengthens innovation capabilities, and creates knowledge about new markets. It also allows for learning effects through interaction, and for validation of production processes. Thus, the "Collaboration across knowledge domains" supports the shift to high value-added production (reinforcing feedback, R12, Figure 1). Another reinforcing feedback is the cost reduction loop. As the "Shift to high value-added production" occurs, the cost of production decreases, for example through learning effects. With a lower "Cost of production", a "Shift to high value-added production" is further supported (reinforcing feedback, R13, Figure 1). Market development has been identified as a main hindrance in the transition process towards high value-added production in the Swedish forestry sector. However, it has also been noted that it is the "Shift to high value-added production" itself that holds the potential to create new markets for these products. This development hinges on the "Ability to demonstrate added value for end consumer". Unless these new applications of biomass are able to showcase better performance quality wise, markets are not likely to materialize. If the "Market demand for high value-added products" increases, so will the "Realization of financial value", further supporting a "Shift towards high value-added production" (reinforcing feedback, R14, Figure 1). High initial costs, as well as "Polarization" in the forestry debate, are hindering factors in a transition process. Lastly, the larger the "Shift to high-value added production", the greater the "Demand for high quality feedstock", driving the "Shift to diversified forestry". The larger the shift to diversified forestry, the higher the "Ability to meet demand for high quality biomass", further enabling the "Shift to high value-added production" (reinforcing feedback, R15, Figure 1).

3.2. The Political Dimension of the Bio-Based Economy Concept

In addition to the dynamics linked directly to primary production and processing, a political dimension to the transition to a bio-based economy was highlighted during the interviews.

3.2.1. Key Feedbacks Governing Political Support for the Bio-Based Economy

In order for a transition process to be facilitated, "Political support for the bio-based economy" is central. Eight reinforcing feedbacks have been identified as important in this regard. The higher the political support for the bio-based economy, the higher the "Resource mobilization for innovation". Both directly, in terms of allocation of public funding to, for instance, research programs, and indirectly by the means of reducing political uncertainty (currently perceived as high), thereby attracting resources from private funding sources. Resource mobilization for innovation enables the "Development of flagship products with high symbolic value", which in turns increases the "Awareness among decision makers", creating more political support for the bio-based economy (reinforcing feedback, R1, Figure 2). Flagship products with high symbolic value could also create "Public support", ultimately increasing legitimacy and political support (reinforcing feedback, R2, Figure 2). Political support may also facilitate initiatives of other kinds, such as collaborative programs and platforms for dialogue, engaging actors from different sectors of the bio-based economy. Such programs would ensure a "Continuity of actor dialogue over time", enabling the development of a "Shared definition and understanding of the bio-based economy". With a shared understanding and consensus on the bio-based economy concept and its objectives, even stronger political support can be facilitated (reinforcing feedback, R3, Figure 2). Similarly, clarity on the meaning of the bio-based economy would increase "Public support", thereby reinforcing the legitimacy of the bio-based economy and thus further make the political support increase (reinforcing feedback, R4, Figure 2). Additionally, political support could enable "Investments in green jobs and traineeship programs", suggested as a means to create "Public support". The underlying idea is to highlight the ability of the bio-based economy to contribute to addressing societal challenges such as unemployment, integration, and inequality. Yet again, when "Public support" increases, so does the political support (reinforcing feedback, R5, Figure 2). Moreover, political support enables "Investments in measurement and follow-up" of developments towards a bio-based economy, increasing the "Perceived contribution of the bio-based economy to the total economy". As the perceived importance grows, so does the political support (reinforcing feedback, R6, Figure 2). With investments in the measurement and follow-up, the ability to develop "Indicators for communication purposes" also increases, contributing to a transition process through strengthening public support for the bio-based economy (reinforcing feedback, R7, Figure 2). These indicators would for instance be designed to broaden the public understanding of the bio-based economy, so that the concept is perceived to encompass more than bio-energy, and to communicate specific environmental gains (*e.g.*, reductions in greenhouse gas emissions following a transition). This development is further strengthened by "Investments in measurements and follow-up" from actors other than the government, enabled by the "Shared definition and understanding of a bio-based economy" (reinforcing feedback, R8, Figure 2). The relative strength of the feedbacks in Figure 2 are potentially affected by factors not part of the feedbacks themselves. One such driver is the "Novelty of bio-based economy concept", creating a political window of opportunity which increases the "Political support for the bio-based economy". Another example is the variable "Service-based share of the bio-based economy", encompassing activities related to health, recreation, and tourism, having a positive impact on the "Perceived contribution of bio-based economy to total economy". It was suggested that the potential to further develop the service-based share of the bio-based economy is relatively high, but currently overlooked in the debate.

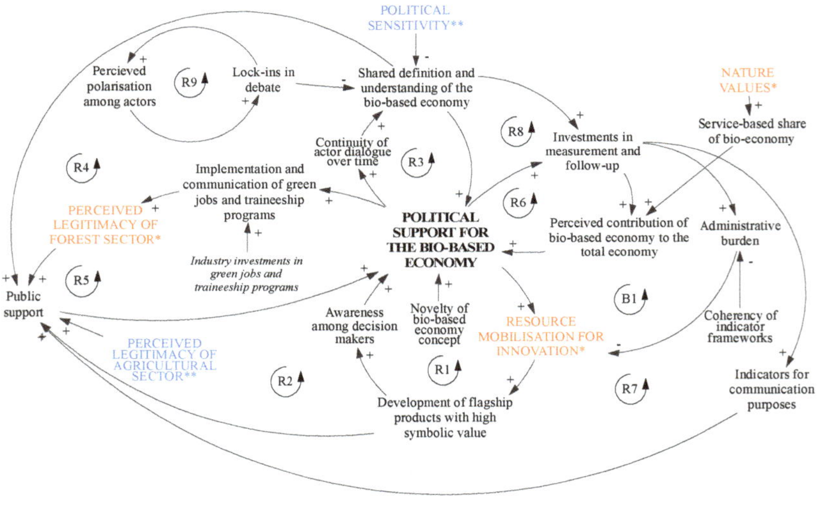

Figure 2. The political dimension of the bio-based economy concept, coupled with developments in primary production and processing. Desired change includes an increase in the political support for the bio-based economy. Variables in italics represent proposed interventions, and capitalized variables in orange/blue the connections between sectors. * Developments linked to forestry sector (see Figure 1). ** Developments linked to dynamics in the agricultural sector, see Bennich et al., (in review).

3.2.2. Dynamics Eroding Political Support for the Bio-Based Economy

Two feedbacks counteract the build-up of political support for the bio-based economy. The attainment of a shared understanding of the bio-based economy is hindered by a "Perceived polarization among actors", creating "Lock-ins in the debate". The more lock-ins, the higher the perceived polarization (reinforcing feedback, R9, Figure 2). It was emphasized that these lock-ins make the debate evolve around details rather than the greater picture or systemic vision of the bio-based economy, that it makes it difficult find common ground to continue the dialogue, and that polarization currently constitutes a significant hindrance for the overall transition process. Another potential hindrance that may become increasingly pressing is linked to administration. Measurements and follow-up are considered important, but do also increase the "Administrative burden" on actors of the bio-based economy, thereby reducing "Resource mobilization for innovation" (balancing feedback, B1, Figure 2). The underlying assumption is that the greater the administrative burden, the more resources need to be allocated towards meeting reporting requirements, thereby reducing the willingness and ability to innovate. The higher the "Coherency of indicator frameworks" for the bio-based economy, the lower the "Administrative burden".

3.3. Proposed Leverage Points and Interventions

Proposed interventions (variables in italic, Figures 1 and 2) target four different leverage points: The "Forest owner self-motivation and confidence", the "Theoretical potential and capacity build-up for high value-added production", the "Soil quality", and the "Implementation and communication of green jobs and traineeship programs". The specific intervention proposed to increase forest owner self-motivation and confidence is to implement "Trainings, workshops, and education programs for forest owners", reversing a perceived decline in such activities. This would increase the ability to shift to diversified forestry, directly strengthening the reinforcing feedbacks R1, R2, R3, and R4 (Figure 1). With independent decision making, the overall ability to cope with the uncertainty created by the polarized debate is also expected to increase, weakening the effect of the reinforcing feedback R5 (Figure 1). Taken together, these developments could support the "Shift to diversified forestry". The intervention suggested to increase the "Theoretical potential and capacity build-up for high value-added production" is to ensure "Investments in R&D", thereby creating greater "Resource efficiency" in the current industrial structure, as well as higher potential for a "Shift to high-added value production". The proposal to increase levels of "Wood-ash recycling" is emphasizing the need to increase nutrient circularity, in order to improve "Soil quality".

The proposal to facilitate "Industry investments in green jobs and traineeship programs" is targeting the fourth leverage point, "Implementation and communication of green jobs and traineeship programs". By this means, developments within the forestry sector could serve as a way to address societal challenges such as inequality, unemployment, and segregation. This would strengthen the perceived legitimacy of the forestry sector, and so contribute to the build-up of both "Public support" and "Political support for the bio-based economy" (Figure 2).

3.4. Interconnectedness

During the interviews, it was highlighted that developments in primary production are tightly coupled with a political dimension of the bio-based economy concept. For example, political developments have an impact on the access to financial capital, as the "Political support for the bio-based economy" enables "Resource mobilization for innovation", thereby supporting "Investments in R&D" in the forestry sector. Political support, in turns, is partly dependent on perceptions among the public. The "Perceived legitimacy of the forest sector" makes the "Public support" for the bio-based economy increase. The perceived legitimacy of the forestry sector is built on the ability of the forestry sector to provide nature values (including production as well as use-values of the forest), and additionally on the level of alignment between societal values opposing industrial use of the forest and actual forest management practices. Moreover, "Nature values" constitute the basis for the "Service-based share of the bio-based economy". The more "Nature values", the higher the ability of the "Service-based share of the bio-based economy" to expand. The larger the "Service-based share of the bio-based economy", the higher the "Perceived importance of bio-based share of all economy", ultimately having a positive impact on the "Political support for the bio-based economy".

In addition to the interconnectedness between developments in the forestry sector and the political dimension of the bio-based economy, linkages between the forestry and agricultural sector were identified. A specific example is the proposal to introduce environmental taxes as a means to support the adoption of more environmentally friendly practices in the agricultural sector (Bennich, et al., in review). While this proposal provides an opportunity to support the attainment of the objectives of a bio-based economy in the agricultural sector, it is suggested to simultaneously increase the "Prospects of loss of jobs in fossil-based sectors", making the "Political sensitivity" of the bio-based economy concept grow, thereby reducing the ability to create a "Shared understanding and definition of the bio-based economy", and so lowering the "Political support for the bio-based economy". Another factor highlighted during the interviews is that the perceived legitimacy of the agricultural sector, just as the "Perceived legitimacy of the forestry sector", has an impact on "Public support". While the legitimacy in the forestry sector to a large extent is determined by the ability to provide nature values, the legitimacy in the agricultural sector is suggested to be based on the provision of food for human consumption (Bennich, et al., in review). The lower the adherence to the food first principle, the lower the perceived legitimacy of the agricultural sector, and the less the "Public support" for the bio-based economy. Hence, the interconnectedness of different dimensions of the bio-based economy could be seen as an opportunity, where developments within a single sector can support the broader transition process. It may however also become a hindrance, as developments perceived as undesirable in one sector might have negative spillover effects in another, thereby impeding the transition.

4. Transition Pathways towards a Bio-Based Economy

4.1. Summary of Proposed Interventions

Transition pathways towards a bio-based economy can be explored by asking "what-if" questions, departing from the CLDs and interventions proposed during the interviews. In the forestry sector, proposed interventions rank relatively high on the scale of effectiveness in accordance with the leverage points framework, as they directly target the relative strength of balancing and reinforcing feedbacks, and the goals of the system. Worth noting is that the feedbacks identified as drivers of a transition may also work in the opposite direction, moving the system away from the desired state or outcome, and that many of the proposed interventions are aiming to ensure shifts in loop dominance to avoid this. Table 1. provides a summary of suggested interventions and their desired impact, as well as examples of controversies, uncertainties and questions remaining to be explored. The summary is limited to suggestions and factors specifically brought up during the interviews.

Additional leverage points identified during the interviews include the variable "Centralisation of forest governance" (where actions to decentralise forest governance is expected to lower the perceived threat to owner autonomy, and thereby increase self-motivation among forest owners), the "Ease of selling forest land" (where efforts to keep the markets well-functioning are assumed to lead to a larger proportion of engaged forest owners), and the "Marginalisation of women" (where actions to ensure the inclusion of women would have a positive impact on overall levels of active decision making in the forestry sector). One of the proposed interventions identified and currently carried out is the "Investments in R&D". However, it was emphasised that efforts should be redirected, from interventions targeting the theoretical potential and capacity build-up through research investments, to interventions targeting the commercialisation of novel biomass applications. No specific proposals were identified, but such a shift could supposedly entail addressing any of the variables surrounding the "Shift to high value-added production", such as the "Market demand for high value-added production", the "Ability to demonstrate added value for end consumer", or the "Initial costs" (Figure 1). Another proposed leverage point, linked to the political dimension of the transition to a bio-based economy, is the "Shared definition and understanding of the bio-based economy concept" (Figure 2).

Table 1. Summary of proposed interventions in the forestry sector.

Proposed Intervention	Desired Change	Potential Unintended Consequences	Additional Leverage Points	Uncertainties and Examples of Related Questions
"What if training programs, workshops, and courses for forest owners are implemented?"	Through increasing forest owner self-motivation and confidence, the shift to diversified forestry would be supported.	As long as the linkage between nature values, prospects of protection, and the perceived threat to the autonomy of forest owners remain, the resulting balancing feedback might hinder the continued shift to diversified farming.	"Centralization of forest governance" (where decentralization is assumed to support a shift to diversified forestry) "Ease of selling forest land" (where better functioning markets are hypothesized to increase the fraction of active forest owners) "Marginalization of women" (where efforts to ensure the inclusion of women in the forest sector would make the fraction of active forest owners increase)	How to overcome the policy resistance created by the perceived conflict between centralized and decentralized forms of forest governance?
"What if industrial investments in green jobs traineeship programs were enabled?"	Investing in and communicating these activities would make the perceived legitimacy of the forestry sector increase, ultimately supporting a broader shift to a bio-based economy through a build-up of public and political support.		The "Shared definition and understanding of the bio-based economy concept" (the more developed the shared definition and understanding, the higher the public and political support for the bio-based economy)	What other measures can be taken to recognize societal values and ensure that forest management practices are aligned with these? Are negative attitudes towards mono-cultures predominant also for species other than spruce and pine (e.g., willow)?
"What if higher levels of wood-ash recycling were achieved?"	Nutrient circularity would be enhanced, preserving soil quality and thereby supporting production values in both diversified and industrial forestry.	Potential negative environmental impacts, such as an increasing presence of pollutants in the forest environment.		What factors govern the decision to increase wood-ash recycling? How to avoid potential negative environmental impacts?
"What if investments in R&D are supported?"	Theoretical potential and capacity of the forest industry are built up, in order to facilitate a shift to high-value added production.	A lock-in effect might be created and reinforced, through innovation strengthening the current industrial structure, thereby lowering the ability for structural change.	"Ability to demonstrate added value for end consumer" (the greater the ability to demonstrate added values, the higher the shift to high-value added production) "Market demand for high value-added products" (the larger the market demand, the higher the shift to high-value added production) Cost of production (the higher the cost of production, the lower the shift to high-value added production) Initial costs (the higher the upfront cost, the lower the ability to facilitate a shift to high-value added production)	How to leverage the theoretical potential of the bio-based economy through bringing novel applications to the market? What should forest biomass ideally be used for?

4.2. Synergies and Trade-Offs

As suggested by these results, there are many leverage points and resulting ways in which the forestry sector can contribute to a transition to a bio-based economy, and large potential for developing efficient and coherent bundles of proposals to facilitate change. Yet, the multiple change processes identified as desirable during the interviews result in different priorities in terms of action. Are these change processes compatible, in the sense that achieving one of the objectives of the bio-based economy supports, or at least does not hinder, the attainment of other goals? In the forestry sector, there seems to be a trade-off between diversified and industrial modes of production. Diversified forestry promotes both productive and use-values of the forest in the long-term, while industrial use promotes productive values in the short-term. However, diversified forestry hinges on the emergence of markets valuing qualitative aspects of the biomass, enabled by a shift to high value-added production in the processing stage. This shift is partly dependent on the current industrial structure, as it generates and directs capital to the build-up of theoretical potential and capacity for high-value added production. Thus, the currently dominant industrial use of forests both hinders and enables the shift to diversified forestry and high value-added production. A shift in loop dominance may facilitate an overall change in the system, but such developments might be dependent on either timing or, as suggested during the interviews, actors other than the forestry industry taking the lead in the process.

5. Conclusions and Future Research

The transition to a bio-based economy has been described as a solution to multiple societal challenges, be they social, economic, or environmental. However, change in coupled social–ecological systems entails great complexity and uncertainty, and multiple and sometimes contradictory views on the objectives and priorities of a transition still exist. Yet, until now many of the analyses have mainly been focused on isolated and, to a large extent, technological transition pathways. This paper attempts to contribute to a more systemic and integrated understanding of transition pathways for a bio-based economy, with a specific focus on the forestry sector in Sweden. Our approach combines qualitative system dynamics and expert interviews in the development of conceptual system maps, depicting the interplay of key change processes suggested to underpin a transition to a bio-based economy.

The findings make explicit the prevailing diversity in aims and priorities of the bio-based economy held by different actors in Sweden. Objectives, as expressed by the interviewees, include a shift to diversified forest management, a structural change in the forestry industry to focus on high-value added production, and the creation of stronger political support for the bio-based economy. While recognizing that objectives and priorities differ, the contribution of our study is an integrated causal theory of change towards meeting these aims. Enabling dynamics identified during the interviews include the build-up of forest owners' self-motivation and confidence to take on an active management role, and the emergence of markets for high-quality feedstock. Hindrances include a perceived uncertainty about the relative sustainability of forest management practices, and a low ability to demonstrate added values of novel biomass applications.

The results also highlight a number of leverage points and proposed interventions. Some of these interventions may have synergetic effects, as in the example of efforts directed towards achieving forest owners' active participation in management processes, which are expected to contribute to both production and environmental management goals. There are also processes that create change that could potentially inhibit the attainment of other objectives. One example might be investments in R&D, leading to a larger potential for high-value added production but also making the current structure of the forestry industry more advanced, thereby creating lock-in effects. The identification of potential lock-ins supports the possibility to redirect efforts to other points of intervention. Additionally, the results point to interventions that could result in unintended consequences and policy resistance, such as centralized decisions to promote conservational efforts without simultaneously intervening to ensure self-motivation and active decision making among forest owners. Finally, and perhaps unexpectedly, many of the proposed leverage points and interventions are addressing values, beliefs, and attitudes, for instance related to perceptions about risk, uncertainty, and conflict, as well as to expectations about future market developments, awareness about characteristics of novel biomass applications among consumers, and the build-up of trust between actors in the bio-based economy.

The examination of transition pathways serving as a basis for this study is exploratory, and there is room to further test and discuss the propositions made. Avenues for future research may include re-examining the causalities identified, to deepen the understanding of the conditions under which they hold. The proposed system structure is also limited to the accounts made by the selected experts, and further research could entail identifying additional system structures within the boundaries of this study that for different reasons may not have been identified during the interviews. For example, it was recognised that developments in the agricultural and forestry sectors are interrelated through the political dimension of the bio-based economy concept. One way to further explore transition pathways could include identifying additional interlinkages, in primary production as well as other sectors, and how these relationships are perceived among different actors.

In terms of proposals suggested to facilitate change, next steps could entail an analysis of the feasibility of options, identifying where power and responsibilities lie, as well as the causal structure linked to the implementation phase. Moreover, the CLDs presented in this paper may be used as a basis to discuss additional proposals and their potential impacts. One example of an area that might currently be overlooked is the perceived conflict in the forest discourse and the resulting high level of polarisation. Constituting a barrier to both the shift to diversified forestry and to structural change enabling high-value added production, interventions lowering the perceived level of conflict may be crucial to facilitate a transition process. A second example might be the political dimension of the bio-based economy, where there is large potential to explore additional interventions.

Another possibility for future research lies in further capturing interactions across scales. Each variable and feedback structure presented in this paper may be disaggregated and analyzed in sub-systems, adjusted to the level of detail relevant to specific decision-making contexts of actors in the bio-based economy. An additional aspect to consider is how the dynamics identified at the regional and national scale relate to broader geographical dynamics, linking to fundamental questions of distribution of resources, fairness, market powers, and overall levels of consumption. By accounting for interactions not only between domains but also across scales, different pathways and their potential implications may be discussed in an explicit way.

Lastly, while the results highlight key feedbacks and stress that the order of intervention matter in a transition process, they do not allow for any inference about the relative strength of feedbacks, potential shifts in feedback dominance, or how the speed of change in different parts of the system affects goal attainment. We therefore foresee room for further work in the direction of quantification and simulation, allowing for rigorous inference and learning about system behaviour over time. Deliberate reflection on the use of combinations of qualitative and quantitative modelling in analysing the bio-based economy could also contribute to the discussion on how modelling tools may be used to better understand and manage sustainability transitions in a broader context.

Acknowledgments: We are grateful to the participants in this study, for contributing their time and knowledge throughout the research process. This work received funding from the European Union's Horizon 2020 research and innovation programme under the Marie Skłodowska-Curie grant agreement no. 675153 (ITN JD AdaptEconII).

Author Contributions: All authors contributed to the research design, the methodological framework, and to the writing of the paper. T.B. carried out the main data collection, S.B. and T.B. jointly worked on data analysis and development of the conceptual models.

Conflicts of Interest: The authors declare no conflict of interest. The founding sponsors had no role in the design of the study; in the collection, analyses, or interpretation of data; in the writing of the manuscript, and in the decision to publish the results.

Appendix A. An introduction to Causal Loop Diagrams (CLDs)

CLDs are graphical representations of system structure, and can be used to conceptualize a system, for communication purposes, or as an analytical tool. In the context of this research, they are used as a medium for analyzing transition pathways to a bio-based economy in Sweden (where the present paper outlines results linked to the forestry sector, while dynamics linked to the agricultural sector are presented in Bennich, et al., in review). CLDs display key variables and their interactions through causal relationships and feedback loops. Each link included in the diagrams represent a hypothesis about the causal structure of the system under study. In our work, the CLDs are based on causal relationships elicited from expert interviews. Table A1 provides an overview of the CLD connotation.

Table A1. An introduction to the use of CLDs.

Graphical Representation:	Denotes:	Interpretation:
Variable X — Variable Y (+)	The link represents a causal relationship between variable X and variable Y. The (+) suggests that the relationship is positive.	If X goes up (down), then Y will go up (down). If there is a change in X, then Y will change in the same direction. The relationship is one-directional, a change in Y has no effect on X.
Variable X — Variable Y (-)	The link represents a causal relationship between variable Y and variable X. The (-) suggests that the relationship is negative.	If Y goes up (down), then X will go down (up). If there is a change in Y, then X will change in the opposite direction. The relationship is one-directional, a change in X has no effect on Y.
Variable X — R — Variable Y (+)(+)	The figure displays a reinforcing feedback loop.	A feedback loop is reinforcing if there are no (-) signs, or if the number of (-) signs is even. This type of feedback reinforces an initial change in the system, and is a source of growth, erosion, and collapse.
Variable X — B — Variable Y (+)(-)	The figure displays a balancing feedback loop.	A feedback loop is balancing if the number of (-) signs is odd. If a variable in a balancing loop changes, the feedback effect opposes and may reverse the initial change. Balancing feedback loops are self-correcting.
Variable Y — Variable X (+)	The mark on the arrow indicates a delay in the system.	This mark denotes that the causal effect of a change in variable Y on variable X is significantly delayed in time. It is not the usual convention to make delays explicit in CLDs, unless they are significant in relation to other causalities in the CLD.

Appendix B. Overview of the Area Expertise, Current Position, and Educational Background of Interviewees

Interviewee No.	Specific Area of Expertise	Current Position	Educational Background
1.	Climate change, air pollution, and agriculture	Advisor at non-governmental organisation	Environmental Engineering
2.	Energy, climate change, and the bio-based economy	Expert and policy advisor, working at interest and business organisation representing the green industries in Sweden	Economics
3.	Business development, policy design, and the bio-based economy	Senior adviser at the Ministry of Enterprise and Innovation	Biology
4.	Gender equality in the forest sector, forest management models among private forest owners	Committee member, Forest owner association	Economics
5.	Sustainable food systems, policy making in the agricultural sector	Research coordinator and consultant for municipalities	Agronomy

Interviewee No.	Specific Area of Expertise	Current Position	Educational Background
6.	Business development in the green industries, sustainable agricultural production systems	Head of corporate social responsibility at agricultural cooperative	Environmental Science
7.	The bio-based economy and policy development at the EU level	International coordinator, The Swedish Forest Agency	Environmental communication
8.	Sustainable agriculture, environmental communication	Associate professor, Örebro University, Sweden	Sustainability Science
9.	The bio-based economy, innovation in the forest industry	Consultant for governments and industry in forestry related issues	Forest sector and policy analysis
10.	Sustainable forestry	Senior lecturer, Lund University, Sweden	Environmental Science
11.	Innovation policy, sustainability transitions, bio-refinery development	Associate senior lecturer, Lund University, Sweden	Economic Geography
12.	Decision making among private forest owners, policy development for the bio-based economy	Senior lecturer in environmental management, Stockholm University, Sweden	Ecology
13.	Policy development and collaboration for the bio-based economy	Coordinator for the circular and bio-based economy innovation and partnership platform at the Swedish Government	Agronomy
14.	Bio-energy, policy frameworks for the bio-based economy	Desk officer, Government Offices of Sweden	Chemistry

Appendix C. Interview Guide for Semi-Structured Interviews

a) *Welcoming and gathering of participant information*

1. Introduction to the AdaptEcon research project, researcher background, research process.
2. Tell me about your background and your current position?

b) *Interview questions*

3. Are you familiar with the bio-based economy concept from before? How would you define a bio-based economy/how do you understand the concept?
4. In what ways do you/your organisation/employer work with a bio-based economy?
5. Can you describe a desirable development that would follow from a transition to a bio-based economy? What is the desired change that a transition would bring (short term/long term)?
6. What would be desirable effects on the sectorial (agriculture/forestry) level, and on a national level?
7. What indicators could be used to trace/measure this development?
8. Can you give examples of actions or proposals to implement in order to facilitate a transition process?
9. What are the main challenges to overcome in order to facilitate a transition process?
10. Can you come to think of any unintended consequences following a transition process?
11. Can you give examples of uncertainties or areas of risk linked to a transition process?
12. What measures could reduce this uncertainty/risk?
13. What actors should take lead in the transition process?
14. Can you give examples of actors or perspectives relevant to the bio-based economy, but currently being overlooked in the general debate?

c) *Closing of interview*

    15.   Other questions/comments?

    16.   Thanking of participant and snowball sampling.

References

1. Formas. *Swedish Research and Innovation Strategy for a Bio-Based Economy*; Swedish Research Council for Environment, Agricultural Sciences and Spatial Planning; Formas: Stockholm, Sweden, 2012.
2. Skånberg, K.; Olsson, O.; Hallding, K. *Den Svenska Bioekonomin: Definitioner, Nulägesanalys och Möjliga Framtider*; SEI: Stockholm, Sweden, 2016.
3. Berkes, F.; Folke, C.; Colding, J. *Linking Social and Ecological Systems: Management Practices and Social Mechanisms for Building Resilience*; Cambridge University Press: Cambridge, UK, 2000.
4. Folke, C.; Biggs, R.; Norström, A.; Reyers, B.; Rockström, J. Social–ecological resilience and biosphere-based sustainability science. *Ecol. Soc.* **2016**, *21*, 41. [CrossRef]
5. Bugge, M.; Hansen, T.; Klitkou, A. What Is the Bioeconomy? A Review of the Literature. *Sustainability* **2016**, *8*, 691. [CrossRef]
6. Staffas, L.; Gustavsson, M.; McCormick, K. Strategies and Policies for the Bioeconomy and Bio-Based Economy: An Analysis of Official National Approaches. *Sustainability* **2013**, *5*, 2751–2769. [CrossRef]
7. Golembiewski, B.; Sick, N.; Bröring, S. The emerging research landscape on bioeconomy: What has been done so far and what is essential from a technology and innovation management perspective? *Innov. Food Sci. Emerg. Technol.* **2015**, *29*, 308–317. [CrossRef]
8. Pülzl, H.; Kleinschmit, D.; Arts, B. Bioeconomy—An emerging meta-discourse affecting forest discourses? *Scand. J. For. Res.* **2014**, *29*, 386–393. [CrossRef]
9. Holtz, G.; Alkemade, F.; de Haan, F.; Köhler, J.; Trutnevyte, E.; Luthe, T.; Halbe, J.; Papachristos, G.; Chappin, E.; Kwakkel, J.; et al. Prospects of modelling societal transitions: Position paper of an emerging community. *Environ. Innov. Soc. Transit.* **2015**, *17*, 41–58. [CrossRef]
10. Sterman, J.D. *Business Dynamics: Systems Thinking and Modeling for a Complex World*, Nachdr; Irwin/McGraw-Hill: Boston, MA, USA, 2009.
11. Meadows, D.H. Whole Earth Models and Systems. Available online: http://donellameadows.org/wp-content/userfiles/Whole-Earth-Models-and-Systems.pdf (accessed on 23 March 2018).
12. Regeringskansliet. Innovation partnership programmes—Mobilising new ways to meet societal challenges. *Regeringskansliet*, 19 July 2016. Available online: http://www.government.se/articles/2016/07/innovation-partnership-programmes--mobilising-new-ways-to-meet-societal-challenges/ (accessed on 5 April 2017).
13. Richardson, G.P. Reflections on the foundations of system dynamics: Foundations of System Dynamics. *Syst. Dyn. Rev.* **2011**, *27*, 219–243. [CrossRef]
14. Lane, D.C. The emergence and use of diagramming in system dynamics: A critical account. *Syst. Res. Behav. Sci.* **2008**, *25*, 3–23. [CrossRef]
15. Stave, K.A.; Kopainsky, B. A system dynamics approach for examining mechanisms and pathways of food supply vulnerability. *J. Environ. Stud. Sci.* **2015**, *5*, 321–336. [CrossRef]
16. Van den Belt, M.; Kenyan, J.R.; Krueger, E.; Maynard, A.; Roy, M.G.; Raphael, I. Public sector administration of ecological economics systems using mediated modeling. *Ann. N. Y. Acad. Sci.* **2010**, *1185*, 196–210. [CrossRef] [PubMed]
17. Videira, N.; Schneider, F.; Sekulova, F.; Kallis, G. Improving understanding on degrowth pathways: An exploratory study using collaborative causal models. *Futures* **2014**, *55*, 58–77. [CrossRef]
18. Dawson, L.; Elbakidze, M.; Angelstam, P.; Gordon, J. Governance and management dynamics of landscape restoration at multiple scales: Learning from successful environmental managers in Sweden. *J. Environ. Manag.* **2017**, *197*, 24–40. [CrossRef] [PubMed]
19. Elbakidze, M.; Dawson, L.; Andersson, K.; Axelsson, R.; Angelstam, P.; Stjernquist, I.; Teitelbaum, S.; Schlyter, P.; Thellbro, C. Is spatial planning a collaborative learning process? A case study from a rural–urban gradient in Sweden. *Land Use Policy* **2015**, *48*, 270–285. [CrossRef]
20. Hovmand, P.S. *Community Based System Dynamics*; Springer: New York, NY, USA, 2014.

21. Stave, K. Participatory System Dynamics Modeling for Sustainable Environmental Management: Observations from Four Cases. *Sustainability* **2010**, *2*, 2762–2784. [CrossRef]

22. Ford, D.N.; Sterman, J. *Expert Knowledge Elicitation to Improve Mental and Formal Models*; Working Paper; Sloan School of Management, Massachusetts Institute of Technology: Cambridge, MA, USA, 1997.

23. Bogner, A.; Littig, B.; Menz, W. Introduction: Expert Interviews—An Introduction to a New Methodological Debate. In *Interviewing Experts*; Palgrave Macmillan: London, UK, 2009; pp. 1–13.

24. Bogner, A.; Littig, B.; Menz, W. (Eds.) *Interviewing Experts*; Palgrave Macmillan UK: London, UK, 2009.

25. Lelea, M.A. Universität Kassel, and Wissenschaftliche Betriebseinheit Tropenzentrum. In *Methodologies for Stakeholder Analysis for Application in Transdisciplinary Research Projects Focusing on Actors in Food Supply Chains: Reload Reducing Losses Adding Value*; DITSL: Witzenhausen, Germany, 2014.

26. Barriball, K.L.; While, A. Collecting data using a semi-structured interview: A discussion paper. *J. Adv. Nurs.* **1994**, *19*, 328–335. [CrossRef] [PubMed]

27. Kim, H.; Andersen, D.F. Building confidence in causal maps generated from purposive text data: Mapping transcripts of the Federal Reserve: H. Kim and D. F. Andersen: Building Confidence in Causal Maps. *Syst. Dyn. Rev.* **2012**, *28*, 311–328. [CrossRef]

28. Meadows, D.H. *Thinking in Systems: A Primer*; Chelsea Green Publishing: Hartford, VT, USA, 2008.

sustainability

MDPI

Article

Impact of Bio-Based Plastics on Current Recycling of Plastics

Luc Alaerts *, Michael Augustinus and Karel Van Acker

Department of Materials Engineering, KU Leuven, Kasteelpark Arenberg 44, 3001 Leuven, Belgium;
michael.augustinus@student.kuleuven.be (M.A.); karel.vanacker@kuleuven.be (K.V.A.)
* Correspondence: luc.alaerts@kuleuven.be

Received: 11 April 2018; Accepted: 4 May 2018; Published: 9 May 2018

Abstract: Bio-based plastics are increasingly appearing in a range of consumption products, and after use they often end up in technical recycling chains. Bio-based plastics are different from fossil-based ones and could disturb the current recycling of plastics and hence inhibit the closure of plastic cycles, which is undesirable given the current focus on a transition towards a circular economy. In this paper, this risk has been assessed via three elaborated case studies using data and information retrieved through an extended literature search. No overall risks were revealed for bio-based plastics as a group; rather, every bio-based plastic is to be considered as a potential separate source of contamination in current recycling practices. For PLA (polylactic acid), a severe incompatibility with PET (polyethylene terephthalate) recycling is known; hence, future risks are assessed by measuring amounts of PLA ending up in PET waste streams. For PHA (polyhydroxy alkanoate) there is no risk currently, but it will be crucial to monitor future application development. For PEF (polyethylene furanoate), a particular approach for contamination-related issues has been included in the upcoming market introduction. With respect to developing policy, it is important that any introduction of novel plastics is well guided from a system perspective and with a particular eye on incompatibilities with current and upcoming practices in the recycling of plastics.

Keywords: bio-based plastics; recycling; circular economy; policy measures; market uptake; PLA (polylactic acid); PHA (polyhydroxy alkanoate); PEF (polyethylene terephthalate)

---

1. Introduction

Bio-based plastics appear already in a broad array of consumption goods. Production of bio-based plastics currently comprises ca. 1% of total plastics production and this share is expected to rise [1]. The Nova Institute has estimated this growth of overall production of bio-based plastics will increase by ca. 50% in 2021 (Figure 1) [2]. Hence the share of bio-based plastics would then increase towards ca. 1.5%, the exact figure depending on the growth of fossil-based plastics. The development and growth of bio-based plastics fit into the search for alternatives to crude oil as a feedstock of organic compounds. Crude oil is a finite feedstock, and today most of the products made from it end up as carbon dioxide in the atmosphere, contributing to global warming. While the production of bio-based compounds is not at all sustainable by definition, the primary raw material source has the potential to be renewable if sufficient care is taken in the development of harvesting and production processes.

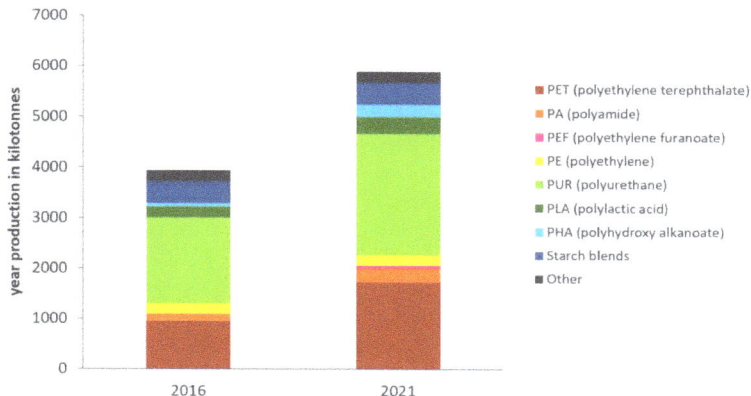

Figure 1. Global production capacities of bio-based plastics in 2016 and estimations for 2021 (data from [2]).

There are many types of bio-based plastics, and further subdivisions can be made; for instance, they differ based on their degree of biodegradability, or on their molecular similarity with existing fossil-based plastics. For example, plastics like bio-based PET (polyethylene terephthalate) or bio-based PE (polyethylene) are essentially identical to their fossil-based counterparts PET or PE, and are called 'drop-in' bio-based plastics for this reason. The only difference is in the production processes of the building blocks of these plastics, as the primary raw materials are different. There are also bio-based plastics with building blocks of a particular basis that are much more easily derived from plant-based feedstocks and from which no fossil-based counterparts have been developed, for reasons of molecular chemistry. Examples of such plastics are PLA (polylactic acid), PHA (polyhydroxy alkanoate) and PEF (polyethylene furanoate), which really amount to being new materials with new properties. As such, they offer the opportunity to compete with fossil-based plastics based on performance and not simply on price alone. An example is the application of PEF for the packaging of carbonated beverages [3]. It is for this reason that further growth of their market share is generally expected: Figure 1 shows the highest relative increases for bio-based PET and PHA, and in 2021 PEF is expected to be a newcomer to the market.

When bio-based plastics will be increasingly used for common applications like bottles, trays, packaging etc. they will also end up in waste streams and as such enter the established recycling processes for fossil-based plastics. As explained in the previous paragraph, a number of bio-based plastics have to be considered as new materials. Hence there may be risks that in some cases, from a certain minimum occurrence, they might prove to be incompatible with these processes, leading to a decreased quality of the recycled plastic stream in which the bio-based plastics have ended up [4]. If this is the case, this would hamper the closure of material cycles in plastics recycling, which is particularly relevant given the current policy focus on the circular economy and on the recycling of plastics, as reflected in the recent launch of a strategy for plastics in the circular economy by the European Commission [5].

In this paper, a review of the risks associated with the increased occurrence of bio-based plastics made from novel building blocks in existing recycling processes will be provided. The results of this analysis will, on the one hand, allow detection of knowledge gaps in this area. On the other hand, this analysis will serve policy developments in the field of the circular economy and, in particular, plastics recycling. In this way, the paper provides an outlook as to if and how policy makers should be prepared for the increased occurrence of bio-based plastics.

2. Materials and Methods

The premise for setting up the analysis is that the preferred end-of-life scenario for bio-based plastic is to collect it as plastic waste and send it for recycling. Hence, for those bio-based plastics that claim to be biodegradable, the possibilities for collection together with organic waste for composting or digestion purposes are not included. From the perspective of keeping material cycling in the economy, mixing these two types of plastics is less desirable. Also, it is logical to organise plastic recycling per product category to the extent that this is feasible.

In order to undertake the risk analysis, we started by outlining the currently applied recycling processes. In the first place, we have considered the recycling of PET and HDPE (high-density polyethylene) bottles. These two plastics display, for the moment, the best outlook in terms of the production of high-quality recyclates: selective collection is in place in many countries, efficient mechanical recycling processes have been developed, and there are examples of the high-grade application of recyclates [6–8]. Next, we expand on the recycling of mixed household packaging waste, given the increased overall focus on plastics recycling as explained above.

In a next step, the impact of small amounts of bio-based plastics made from new building blocks was assessed. As we felt that this group of plastics is too heterogeneous in nature, we have chosen a case-by-case approach by considering subsequently the impact of three examples of such plastics, PLA, PHA and PEF, in order to develop a more general perspective on the risks. As preparation for these exercises, we first considered the impact of polyvinyl chloride (PVC) in PET recycling. PVC is known as an unwanted contaminant even in very low concentrations [8], and gathering the available information on the case of PVC in PET recycling was considered instructive as a preparation for assessing the impact of the selected bio-based plastics. Given the very small amounts of these plastics appearing on the market, setting up separate collection is not viable and hence they will act as contaminants whose impact on recycling processes and products has to be analysed [9]. The analysis starts by considering the physical, chemical and other, more practical, properties (for instance related to the particular application or appearance) of such contamination. Then, these are compared with the properties of PET and HDPE bottles or household packaging waste constituting the main flows. Eventually, this results in the identification of the possible pathways that particular bio-based plastics can follow and the possible impact that may arise from their presence in certain amounts.

The data at the base of this analysis was retrieved from research papers, policy documents, publications from sector organizations and websites. A deliberate and clear choice was made at the beginning of the research not to limit the consulted sources to peer-reviewed academic literature only. We realised that a significant part of this analysis pertains to unit operations interacting in a system, which is an area under focus by many other actors besides academic researchers. In fact, the reality that much of the work (including more conceptual aspects) in circular economy research is driven by non-academic actors has been reported before [10]. In that way, one of the aims of this paper is also to stimulate and define areas for further knowledge building.

3. Results

3.1. Current Recycling Processes

3.1.1. Polyethylene Terephthalate (PET) and High-Density Polyethylene (HDPE) Bottles

In this section, we focus on the currently operational series of unit operations to recover PET and HDPE from bottles for recycling. For many years PET and HDPE bottles have been routinely separated and collected in many European countries. A high-level representation of the recycling system is shown in Figure 2.

The overview in Figure 2 starts from bottles as they have been isolated from household waste—depending on how the selective collection of household waste has been organised, they may be collected separately or combined with other waste streams like metal cans and/or tetra packs. Further separation of PET and HDPE bottles takes place in sorting centres, either manually or using optical recognition with near-infrared (NIR) spectroscopy, estimated to be in use at 60% of the sorting centres in 2006 [6]. In such equipment, the positive detection of polymer types takes place, and subsequent separation from the flow is done using air pulses. In a next step, a series of unit operations converts the bottles into small, cleaned fragments. These unit operations comprise for instance of washing to remove (residues of) labels, glue, paper etc., removal of caps, fragmenting, and further purification steps like, for instance, metal separation and flotation. The latter technique takes advantage of the density differences of HDPE and PET compared to water: HDPE flakes float while PET flakes sink to the bottom (see Table 1). This allows PET flakes to be obtained without HDPE and vice versa; also, other impurities can be removed if the density difference is suitable. The separate plastic fragment streams are then fed to the respective mechanical recycling processes. Here, the flakes are dried, molten and eventually pellets of rPET (recycled PET) and rHDPE (recycled HDPE) are extruded. During mechanical recycling, the chemical polymer structure itself is essentially maintained, although some degradation will always occur to a certain extent. For this reason, a direct full recycling into the same level of application is not possible. Therefore, it is not possible to obtain bottles solely from rPET with the same quality as bottles from virgin PET, and hence in practice downcycling is taking place, for instance in the production of bottles with a certain fraction of recycled content, the production of bottles for lower grade applications, e.g., soap bottles from beverage bottles, or the production of applications other than bottles, e.g., rPET textile and fillings or rHDPE storage boxes, tubings and cable trays [7].

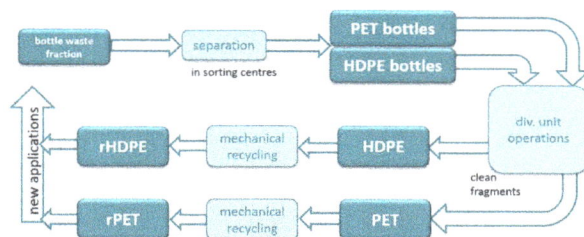

Figure 2. High-level overview of the unit operations from bottle collection to recyclates of polyethylene terephthalate (PET) and high-density polyethylene (HDPE).

Table 1. Densities and melting points of a number of plastics.

Plastic Type	ρ (kg/m^3)	T_m ($^\circ$C)
Polyethylene terephthalate (PET)	1350–1390	255
High-density polyethylene (HDPE)	930–970	125
Polyvinyl chloride (PVC)	1100–1450	210
Polylactic acid (PLA)	1200–1450	155–165
Poly-3-hydroxybutyrate (PHB)	1300	180
Polyethylene furanoate (PEF)	1400–1550	225

The ranges are based on the results of a general search on the internet, completed by data retrieved in the references consulted for this paper.

3.1.2. Mixed Plastic Waste

This kind of waste comprises all plastic household packaging like foils, cans, jars, margarine containers, yoghurt pots, flower pots etc. A focus on separate collection for this fraction has appeared recently, either together with or without bottles (in the former case, such separation can take place in sorting centres). This fraction is intrinsically much more heterogeneous in nature, due to the following [11,12]:

- Objects made from different polymer types can have very similar appearances.
- A same polymer type can be used in very different ways given the application, e.g., with different additives, crystallization degrees, intrinsic viscosities etc. Compared to bottles, there is much less uniformity per plastic type among the encountered applications.
- The occurrence of all kinds of add-ons like labels, inks, impurities, combinations etc.
- Different behavior in sorting unit operations depending on the final application; consider, for instance, the different aerodynamics of foils vs. yoghurt containers vs. clamshells that become apparent in wind shifting.
- Different objects may clump in the case of bale packing that is too dense.

Such mixed plastic streams are evidently much more challenging with respect to recycling. At present, there is only marginal production of recyclates of any particular plastic type from streams in operation today. In fact, there are quite a few applications of recyclates in objects occurring in these kinds of streams (e.g., PET-thermoforms containing up to 50% rPET from bottles) [13]. Some examples exist of composite products derived from mixed plastics, like garden tables or traffic infrastructure [14].

3.1.3. Impact of Polyvinyl Chloride (PVC) on Recycled PET (rPET)

The negative impact of PVC (polyvinyl chloride) in PET bottle recycling has been known for many years in the field. As both PVC and PET have a density higher than water, PVC impurities that are not taken out in the first unit operations automatically end up in the PET fraction obtained after flotation and are sent for mechanical recycling. Even with PVC contaminations as low as 0.005% (coming down to e.g., one bottle of PVC in 20,000 PET bottles) the obtained rPET is not applicable for most applications any more. The melt temperature of PVC is much lower than that of PET (see Table 1). This means that at the temperatures applied in the mechanical recycling of PET in a molten state, PVC contamination is at temperatures much higher than their melting point for a prolonged time. In such conditions it is typical that degradation starts to occur, leading to chain scissions and/or the release of functional groups. In the case of PVC, hydrochloric acid is released, chemically breaking the polymer chains. The decomposed PVC assumes a yellow to brown discoloration and the occurrence of black spots in the obtained rPET. It is clear that the resulting rPET is unacceptable and has inferior material properties [8,15,16].

Due to the above, a number of modifications to the chain of unit operations has been implemented already to protect the quality of rPET from the negative influence of PVC. At first, the occurrence of PVC in bottle applications decreased considerably over the years, e.g., PVC is currently not used for food and beverage packaging. The only possible occurrence may still be in all kinds of add-ons to bottles, like sleeves, labels etc. In fact, the PET recycling sector actively discourages any introduction of PVC-containing add-ons to PET bottles [8]. PVC can be detected by NIR, so sorting installations equipped with this technology are able to reject PVC objects, PVC bottles or PET and HDPE bottles with sleeves, labels or other add-ons made from PVC. The European Pet Bottle Platform (EPBP) website mentions a separation accuracy level of 85–96%.

In summary, the issue of PVC in PET recycling is known and has been taken up by the field. The occurrence of PVC in the current bottle fraction is already very low. Together with the application of NIR separation technology, this should ensure that today PVC does not hamper the mechanical recycling of PET and HDPE bottles. Let us consider now how the analogy with newly introduced bio-based polymers works.

3.2. Impact of Polylactic Acid (PLA) on Plastics Recycling

PLA or polylactic acid claims to be a fully bio-based plastic and also to be biodegradable. The building block of PLA is lactic acid, obtained by fermentation of plant-derived sugars. PLA is applied in transparent bottles, containers, packaging and foils. Given the high permeability of PLA to water, it is an interesting material for the packaging of food products like lettuce or bread. Its application in bottles is rather modest for the same reason [6]. With respect to bottle applications, starting from Figure 2 and considering properties and existing knowledge of PLA we can come to a rough analysis of the behaviour of a PLA bottle in the current recycling chain. At first, visual discrimination based on appearance between a bottle made from PET and from PLA is not possible: both materials are transparent and very similar. So, both at the level of consumers and of manual sorting, separation is not possible unless an extra element is introduced e.g., via labelling [6]. Next, at the level of sorting centres, if NIR technology is available for the positive detection of PET bottles, then at least the majority of PLA bottles would become separated. The EPBP estimates an efficiency of 86–95% for this. Other sources report numbers up to 99.6% (see Table 2). Also, PLA is denser than water so in the flotation tank any PLA fragments will eventually follow the PET stream towards mechanical recycling [17]. Hence, the impact of PLA on rPET quality will especially have to be assessed.

Table 2. Separation efficiencies of PLA bottles from PET bottles by near-infrared (NIR) spectroscopy.

Separation Efficiency (%)	Reference
86–95	[8]
96	[18]
97	[19]
90–98	[20]
99.6	[21]

3.2.1. Occurrence of PLA in Plastic Waste

Starting from available numbers on the current and expected occurrence of PLA, we have calculated the level of PLA estimated to occur in the bottle and the mixed plastics fractions from household waste. This approach allows sufficient accuracy to provide an order-of-magnitude estimation of PLA occurrence. The raw numbers at the basis of these calculations are available in Appendix A, Table A1. The next paragraph describes the most important steps and numbers in the calculations.

We started from general numbers describing the European market. In first instance we have assessed the total amounts of PET, as PLA contaminations are especially relevant with respect to this fraction (see above). Next we subdivided bottles and mixed plastics. For PET, we started from reported production data for the year 2015 and estimated production data for the year 2021, and assumed that 71% is applied in bottles and 10% in objects ending up in the mixed plastics fraction, based on a number of sources. For PLA, we started from reported production data for the year 2016 and estimated production data for the year 2021. We found a proportion of 60% of PLA applied for bottles and other packaging [22]; for bottles only, we did not find any proportion and estimated a range of 5–35% of total PLA production used for bottles. Taken these proportions together (and assuming as such that all plastics are retrieved via the existing collection systems), estimations of PLA concentrations in waste fractions have been obtained as shown in Table 3.

Table 3. Amounts of PET and PLA applied in bottles and in other packaging in Europe, and ratio of PLA vs. PET (data sources: see Appendix A, Table A1).

Application	PET (Mton)		PLA (kton)		PLA/PET Ratio (%)	
	2015	2021	2016	2021	2015/2016	2021
Bottles	3.3	3.9	2.9–20	4.2–30	0.09–0.6	0.1–0.8
Other packaging	0.47	0.55	14–32	21–46	3–7	4–8

The overall numbers in the table reflect the fact that production of both PET and PLA is expected to increase towards 2021, with the increase of PLA relatively higher than that of PET. This is reflected in the expected (slight) increase of the concentrations of PLA in PET. The ranges appearing in the table are the result of a chosen range for the total PLA production directed to bottle production. These ranges are to be considered as a reflection of the order of magnitude in which to consider current and future potential contaminations by PLA.

3.2.2. Impact of PLA in Current Recycling of Bottles

The different behavior of PLA at higher temperatures has already been reported to cause a number of issues in the drying and processing steps of the mechanical recycling of PET [11,15,17,23,24]:

- The difference in melting temperatures between PLA and PET is ca. 100 °C (see Table 1). This means that at the temperatures applied in mechanical PET recycling, implying PET is in a molten state, PLA contamination is at temperatures much above its melting point for a prolonged time. In such conditions degradations start to occur, in a similar way as explained above for PVC. PLA degradation leads to a yellowing of the resulting product. Also, in the pretreatment of the feed to the mechanical recycling of PET, issues can occur during drying of the flakes, as the lower softening point of PLA causes the PLA fragments to become sticky, resulting in an agglomeration of flakes and fouling of the drying installation [17].
- PLA has a different glass-transition temperature, resulting in opaqueness or haziness when processing rPET into pellets, as PLA and PET will undergo phase separation as they are not miscible in the solid state [15,25,26].
- Also, upon further processing of contaminated rPET pellets, the pellets with PLA contamination could stick to surrounding PET pellets, resulting in the formation of clusters, hampering further operations [15].

The first of the above issues starts to occur from contamination of 2% and higher. The issues with respect to transparency and discoloration are evidently present already from contamination of 0.1% or even lower (see Table 4). There is some discussion in literature about where the safe threshold for having no quality impact at all is situated; a number of claims have been published, both by entities marketing PLA and by bodies defending the PET recycling industries. These claims need to be checked further with respect to the exact experimental conditions (lab scale vs. full-size process; to which extent is virgin PET added) and the exact application (e.g., PLA contamination up to 5 wt % does not affect a bottle-to-fibre application [20]). An overview of the available data on the impact of PLA contamination in PET is given in Table 4. However it is clear that PLA contamination should be kept under stringent control if rPET quality and the ability to process is to be protected, especially with respect to bottle applications.

Table 4. Reported threshold concentrations for PLA in PET and impact.

Concentration		Impact	Test Details	Ref.
%	ppm			
0.05	453	"No visual deviations in terms of colour and transparency"	Foil of 1.3 mm thickness obtained by extrusion tests with 70:30 ratio of rPET:PET	[23]
0.05	453	"Not any significant difference in colour and haze"	Plaque let-down study with plaques between 0.063 and 3 mm	[18]
<0.1	<1000	"Makes any rPET resin unsuitable"	"For most applications"—no further details obtained after request	[24]
0.1	1000	significant opacification of recycled PET	Plaque test	[17,20]
0.1	1000	PET recyclate unusable for many end-products	n.a.	[27]
0.3	3000	"Lowers the onset of crystallization and retards recondensation"	Injection molding	[25]
>0.3	>3000	PLA causes yellowing of PET	Plaque test	[17]
2	20,000	Besides lower quality resin, also agglomeration and sticking to dryer walls	n.a.	[17]
5	50,000	Besides lower quality resin, also agglomeration and sticking to dryer walls	n.a.	[17]

In our analysis of the market penetration of PLA, we estimated contamination of collected PET bottles by PLA to be 0.8% at the highest in 2021 (including the assumption that all PLA bottles are collected via the existing systems). With NIR separation technology, assuming a sorting efficiency of 90% (a conservative estimation, see Table 2), this assumed contamination level would come down to 0.08%, which is still quite close to the 0.1% threshold to be really sure for having smooth rPET quality. Then, it is clear that further removal of PLA may be very well required in the recycling chain in order to assure high-quality rPET. The following options can be considered if additional purification of the PET bottle stream is necessary:

- Adjustment of NIR sorting equipment for decreasing the number of sorting mistakes leading to positive selection of PLA bottles or PET bottles with PLA add-ons. This may, however, also lead to a larger stream of rejected bottles, decreasing the rPET yield and the eventual financial return of the recycling process [6].
- Installation of additional NIR sorting equipment into the fragment stream going to flotation: this would lead to a further decrease of 90% of the amount of PLA contamination and would also allow removal of all kinds of add-ons, as these are in that stage of the process also occurring as fragments.
- Perhaps the low softening temperature of PLA (see above) would allow running a separation process based on a hot conveyor belt or a rotating drum, as has been described already for the removal of PVC from PET [7,16,28].
- Other options are explicit labeling of PLA bottles, either for the purpose of communication at the consumer level (sorting message), or for instance using chemical markers, allowing smooth detection in automatic sorting [29].

Any of these options would lead to an increased cost for the overall recycling process, either because of the installation of new hardware or because of yield losses. In fact, this situation of PLA contamination in PET is very reminiscent to that of PVC in PET, as described above, with the PET recycling sector being concerned that even small market penetrations would seriously hinder the existing recycling infrastructure and actively discourage the use of PLA add-ons of any kind for PET bottles [8,24]. The difference with PVC is that PLA is becoming more popular for use, hence the trend is towards increasing amounts of PLA ending up in waste streams.

3.2.3. Impact of PLA in Current Recycling of Mixed Plastic Waste

For the moment, the applications already developed for mixed plastic waste (see above) have not yet revealed issues to the best of our knowledge—given the expected increase of PLA applications, the question is whether from a certain moment, for instance, the lower melting and degradation temperature of PLA could hamper the existing processes, or that the biodegradability could impact the quality of the products over time, for instance, when applied outdoors.

As explained above, there are currently no operational processes that extract PET from mixed plastic waste for production of rPET. If in the future such attempts would be undertaken (PET is one of the common polymers found in mixed plastics waste, occurring in different forms), it is clear that the same issues with respect to rPET quality would arise. Our estimations have revealed that contamination of PET by PLA could be as high as 8% by 2021, showing that a very large supplementary effort would be needed to get below the safe limit of 0.1%.

3.3. Impact of Polyhydroxy Alkanoate (PHA) on Plastics Recycling

PHA (polyhydroxy alkanoate) is a collective name for a group of polymers made of chemically similar building blocks. PHB (poly-3-hydroxybutyrate) is the most widespread member of the PHA family. PHA plastics are produced in a biological process induced by micro-organisms. The polymers obtained are biodegradable. While the different PHA types differ in properties depending on their chemical composition, a particular characteristic of PHA is its biocompatibility, making them suitable for medical applications. PHA also has good barrier properties, of interest for food product packaging. For these reasons, applications of PHA are found, for instance, in single-use packaging films, bags, containers, paper coatings, agricultural foils, biodegradable carriers for long-term dosage of compounds like drugs or fertilizers, and medical applications like surgical pins, sutures, wound dressings, bone and blood-vessel replacements etc. [30].

3.3.1. Occurrence of PHA in Plastic Waste

Data on the occurrence of PHA were difficult to retrieve. Some overall numbers have been presented already in Figure 1: PHA shows an estimated fourfold increase between 2016 and 2021; in the latter year world production is estimated to be 0.25 Mton and European production 0.065 Mton. With respect to applications, the consulted data sources stress the wide range of applications for PHA, with most of the applications in soft packaging. However, we did not retrieve numbers allowing estimation of the amounts ending up in the bottle or mixed plastic fractions. Hence, it is assumed that current bottle applications are marginal, and that the end-of-life scenarios for many PHA applications are determined by the application itself, e.g., agricultural foils will biodegrade in the field and medical waste in general does not end up in the selectively collected household waste streams destined for recycling [30].

3.3.2. Impact of PHA on Current Plastic Recycling

The few data that were retrieved on properties of PHB allow us to conclude that if it were to end up in the feed of rPET production (e.g., via the bottle fraction), similar issues as encountered for PLA with respect to the mechanical recycling of PET may occur. ue to its higher density compared to water, PHB may eventually end up in the feed to rPET production. Moreover, its melting temperature is 180 °C, much lower than that of PET. The current application of PHA for bottles is estimated to be negligible. Although a steep increase of applications is expected, the assumption is that the main developments are in the direction of more complex applications and not bottles [30].

With respect to mixed plastics, most probably some PHA applications are already ending up there. Given the state-of-the-art of knowledge with respect to the occurrence of the much more common PLA in this stream (see above), for the moment no further conclusions can be drawn on its impacts.

3.4. Impact of Polyethylene Furanoate (PEF) on Plastic Recycling

PEF (polyethylene furanoate) is produced by the building blocks ethylene glycol and furane dicarbonic acid. It is obtained from fructose sugar, which can be obtained from many plant feedstocks. The most particular feature of PEF is its enhanced barrier properties, for instance, for carbon dioxide and oxygen, making the material very interesting for soft drink bottles and food packaging [3]. Currently, PEF is not available on the market, but its entrance is very near given the announced startup of Synvina in Antwerp, being a joint venture of BASF and Avantium.

With respect to the future PEF bottles, starting from Figure 2 and considering the properties and existing knowledge of PEF, we can come to a rough analysis of their behaviour in the current recycling chain of bottles. Visual discrimination based on appearance between a bottle made from PET and from PEF is not possible: both materials are transparent and very similar. So both at the level of consumers and of manual sorting, separation is not possible unless an extra effort were to be introduced, e.g., via labelling. Next, at the level of sorting centres, if NIR technology is available for positive detection of PET bottles, it is probable that the majority of the PEF bottles would be separated [31]. Also, PEF is denser than water (see Table 1) so after flotation PEF fragments will follow the PET stream towards mechanical recycling. Hence, quality needs to be assessed, especially the impact of PEF on rPET.

3.4.1. Occurrence of PEF in Plastic Waste

The market entrance of PEF is imminent in coming years. For the year 2021, a yearly production in Europe of ca. 18 kton is expected by the Nova Institute [2]. It is difficult to estimate further production growth beyond 2021. Besides the overall growth in plastics produced for packaging and the expected increase in the share of bio-based plastics, there is also a scenario that PEF might replace PET at an accelerated rate given the expected superior barrier properties of PEF [3], allowing beverage bottles to be made with much less material and/or increasing the shelf life of the products.

As PEF contamination is especially relevant with respect to the PET bottle fraction (see above), we have tried to estimate the relative amounts of PEF compared to the amount of PET used for bottles (see Table 5). In the calculations for PET, we have used the same assumptions as above for assessing the market penetration of PLA; for PEF, we have used the estimated production from the Nova Institute and further assumed that all PEF goes into bottles. It appears from these calculations that the occurrence of PEF in PET could be as high as 0.45% by 2021.

Table 5. Amounts of PET and PLA applied in bottles and in other packaging in Europe, and ratio of PLA vs. PET (data sources: see Appendix A, Table A1).

Application	PET (Mton)		PEF (kton)		PEF/PET Ratio (%)	
	2015	2021	2016	2021	2016	2021
Bottles	3.3	3.9	0	0.067	0	0.45

3.4.2. Expected Impact of PEF on Current Recycling of Bottles

The impact of PEF on the mechanical recycling of PET has been investigated: one source states that contamination of up to at least 2% has been demonstrated not to lead to any negative impact on rPET quality, e.g., negative impact on haze, colour and other properties [31]. Another source states that no negative effect on haze occurs up to 5% [32]. The absence of haze could suggest that PET and PEF are miscible in the solid state at least up to these amounts [24]. Compared with the estimations above in Table 5, it appears that at least until 2021 the amount of PEF available on the market will be far below these levels.

In this context it is relevant to learn that Synvina has proactively approached the EPBP to assess the impact of PEF on the existing recycling chain of PET bottles, and has obtained a conditional 'approval' for market entrance from this entity. Based on dedicated studies, the EPBP concluded that until a market penetration of 2% is reached there will be no issues with PEF ending up in PET streams. For larger market penetrations, either additional tests or the development of a separate collection and recycling system for PEF bottles will need to be in place in order to further anticipate any negative impact on rPET quality [31].

3.4.3. Expected Impact of PEF in Current Recycling of Mixed Plastic Waste

Given the limited relative occurrence of PEF on the market in the near future, the absence of negative impacts on PET recycling processes and the main application of PEF expected to be in first instance in bottles, the amounts ending up in mixed plastic waste will most probably be sufficiently small to exclude any negative effects in coming years, both in processing mixed plastics as one single fraction as such and in any attempt to obtain pure plastic waste streams like PET from this fraction.

4. Discussion and Conclusions

The approach followed in the three cases above has shown that the introduction of bio-based plastics on the current operational range of recycling processes should be considered as the introduction of a number of types of novel plastics. In the first instance, every new introduction starts as a contamination, and with respect to further evolutions the following questions are most relevant:

- Which incompatibilities may occur? From which amounts do they become noticeable?
- How strong are the current recycling processes that have been developed?
- By which amounts is the development of dedicated collection and recycling rewarding?

One aspect of assessing the impact of a contamination comes down to the probability of a novel plastic ending up in the final products of current plastics recycling. This is a consequence of the properties of the contamination compared to the main flow of plastics, and the organisation and technologies of sorting and recycling that deal with this main flow in a certain manner. Starting with the estimated, measured or known market penetration, it is possible to assess the pathway of a contamination and to obtain probable concentrations in the recycling processes and at the recyclate level, and comparing these with the lowest levels on which negative impacts have been observed or demonstrated—this is the second aspect of the impact of a contamination. This can be measured with tests; here the definition of what is being tested is crucial, as an incompatibility may be less noticeable in one application compared to another one. For instance the impact for an rPET bottle destined for beverages is different compared to the impact for an opaque rPET bottle destined to contain soap.

The latter two of the above questions pertain to the current recycling of PET: processes for high-grade rPET destined for bottle production have been established and sector organizations have been founded in particular to protect these activities, e.g., by promoting the design for recycling. This has resulted in published compatibility lists, e.g., discouraging particular materials for sleeves and add-ons, with the ability to produce high-grade rPET as the reference. Hence any negative impacts of contaminations have a higher chance of being discovered right in this chain, so it is not a coincidence that PET recycling features so prominently in the analyses above. To a certain extent it is beneficial that compatibility with the established recycling processes is being strived for, because it may lead to smoother introductions of novel plastics and the detection of possible issues well in advance. For instance, the market entrance of PEF has been anticipated in this way. On the other hand there is the risk that the mere power exerted by the incumbent actors in recycling inhibits the introduction of novel plastics for too long a time, even if such introductions would be beneficial for other reasons.

In this paper, a review has been carried out for three concrete bio-based plastics. For PLA, the evidence is clear that its presence, even in small amounts, is detrimental to the quality of rPET: contamination in the feed to mechanical recycling should be maintained well below 0.1% in order to protect rPET quality. For the bottle fraction, our estimations for now and coming years till 2021 have shown that state-of-the-art separation (equipped with NIR technology) might lead to a contamination in the feed to mechanical recycling of PET currently and in the near future not too far below this 0.1% threshold. In order to further elaborate this analysis, a next step could be to obtain better reference values by checking the numbers of the estimation, e.g., by sample measurements of bottle waste streams (also considering sleeves and add-ons). These actions could then be repeated to keep track of evolutions in PLA concentrations over time. Also, the current occurrence of any quality issues related to PLA in PET in the field could be checked, e.g., by interviewing companies. The outcome of these actions should allow an assessment of when the set of currently applied unit operations for separation and recycling would not be able to lead to rPET of sufficient quality in the longer term. If this were to be done, then further options to consider are investments in extra unit operations and/or developing adequate labeling of PLA bottles. The value of the latter option will also be dependent on any further developments in the direction of bringing value to a separate PLA stream for recycling PLA. It appears that there are several possibilities for additional technologies for PLA separation, like an extra separation of flakes using NIR technology, or technologies that are already known for PVC removal.

With respect to PLA contamination of mixed plastics, our estimations have shown that contamination may be in the range of several percent. Although any issues were not revealed in the current study, perhaps due to the application development in this area that is only emerging, with further sorting and separation technology and application developments the estimated higher concentration of PLA in mixed plastics might very well give rise to issues (e.g., due to its lower heat stability compared to many fossil-based plastics). Hence here also sample measurements and interviews with companies using mixed plastics as an input to their production could be helpful to establish reference values and to monitor evolution. In this respect, any plans to implement, extend or modify post-consumer plastics collection should be thoroughly evaluated with respect to the creation of possible contamination pathways of PLA into rPET.

For PHA, besides the different kinds of polymers considered, there is also a whole range of applications and only few data available, making analysis of the current situation difficult. Similar issues as encountered with PLA might occur, given the rather low heat stability of PHB. For the moment, no issues are known or expected as the main current applications of PHA do not seem to lead to end-of-life scenarios hampering the existing high-end mechanical recycling processes. There are no indications that this would change in the short term, but any trends in application of PHA are to be monitored given the expected steep increase in PHA production in coming years.

With respect to PEF, no issues are known with respect to the impact on mechanical recycling of common plastics. For the impact on rPET this has been effectively tested for contamination up to 2%. As PEF still has to be launched on the market, this allows us to conclude that in coming years no issues are to be expected. What could happen in the longer term is not clear; much is also dependent on to which extent a separate collection and recycling system for PEF would be operational and successful. Anyhow, this has gained the attention of both the (future) producer of PEF and the sector organization of PET recycling, hence it is very probable that any risk will be anticipated well. In fact the producer's approach to assess in advance compatibility with the existing recycling landscape is to be encouraged as it is a clear demonstration of the necessary system thinking in this field.

Summarizing the three case studies, for PLA the facts are known so future risks can be assessed by measuring amounts. For PHA, it will be crucial to monitor future application development, and for PEF, a particular approach for contamination-related issues has been an element of project management. Hence the study did not reveal bottlenecks or negative impacts generally valid for all bio-based plastics. One of the next questions could then be what would be the next bio-based candidate material to appear in post-consumer plastics waste.

Overall, the challenge with respect to bio-based plastics is a matter of guiding well both their introduction together with developments in the recycling landscape, with a particular eye on their incompatibilities e.g., with process conditions or combinations with other plastics. The story in this paper is, therefore, fully written within the context of the current state of the art of applied recycling technologies and, in general, how post-consumer plastic collection and sorting have been organized. Hence, the findings are to be seen fully in the light of the current situation. With any future changes of the recycling landscape, the analysis has to be repeated. For instance, if there were to be initiatives in extended producer responsibility, maybe the waste streams obtained would become much more pure and some issues may simply disappear, if for example PLA bottles would then not end up with PET bottles any more.

With respect to developing policy advice, a number of suggestions for preparing next steps can be made. First, it is important that introduction of novel plastics is guided well, with a clear focus on the whole system; see as an example the way the introduction of PEF is being anticipated. Next, it is important that all plastic types occurring as contamination in current waste streams are considered in the context of any changes in the recycling landscape. On the other hand, with the supply of plastic types constantly changing and more abrupt changes to be expected in the (near) future, it has to be considered that from a certain moment the recycling landscape itself would need a reorganisation; such an operation requires a realignment of many actors and is, hence, complex, but it would avoid desirable developments in the production of plastics being blocked for an unnecessarily long time.

Finally, the current analysis did not aim to draw any conclusions about the mere desirability of bio-based plastics and/or future increases in these plastics. Such developments should, anyhow, not be steered too much by the concrete implications for the recycling landscape; as long as these plastics can be recycled well by themselves, the recycling landscape should be able to accommodate them over time, and here policy has the option to support or even guide this process by carefully managing the new entries temporarily as contaminants.

Author Contributions: L.A. conceived the approach for the study, arranged interactions with policy officers of the Flemish administration, and wrote the paper; M.A. performed the literature research; K.V.A. reviewed the paper and guided the overall process from phrasing the research questions to elaborating the policy advice.

Acknowledgments: The authors are very grateful for full financial support received from the Flemish administration via the Steunpunt Circulaire Economie (Policy Research Centre Circular Economy). This publication contains the opinions of the authors, not that of the Flemish administration. The Flemish administration will not carry any liability with respect to the use that can be made of the produced data or conclusions. The authors are also grateful to policy officers of the Waste Agency Flanders (OVAM) for extended discussions and comments.

Conflicts of Interest: The authors declare no conflicts of interest.

Appendix A

Table A1. Data used to estimate penetration of PLA, PHA and PEF in post-consumer plastics.

Data	Value	Unit	Reference
World plastics production 2015	322	Mton	[33]
Yearly growth of world plastics production	8	Mton/y	[33]
World plastics production 2021	370	Mton	based on previous
World PET production 2015	27.8	Mton	[33]
World PET production 2021	32.2	Mton	based on previous
Share of PET in world plastics production	9	%	calculated 2015 data
Share of PET production in Europe	17	%	[33]
Share of PET used for bottles	71	%	[33]
Share of PET used for non-bottle packaging	10	%	[11]
World production of bioplastics 2016	4.16	Mton	[2]
World PLA production 2016	5.1	%	[2]
World production of bioplastics 2021	6.11	Mton	[2]
World PLA production 2021	5.3	%	[2]
European bioplastics production 2016	27.1	%	[2]
European bioplastics production 2021	26	%	[2]
Share of PLA used for bottles and other packaging	60	%	[22]
Share of PLA used for bottles	5–35	%	own estimation
World PHA production 2016	1.6	%	[2]
World PHA production 2021	4.1	%	[2]
World PEF production 2016	0	%	[2]
World PEF production 2021	1.1	%	[2]

References

1. Van den Oever, M.; Molenveld, K.; van der Zee, M.; Bos, H. *Bio-Based and Biodegradable Plastics—Facts and Figures*; Report no. 1722; Wageningen Food & Biobased Research: Wageningen, The Netherlands, 2017; Available online: http://edepot.wur.nl/408350 (accessed on 6 April 2018).

2. Nova-Institute. European Bioplastics. 2016. Available online: http://www.european-bioplastics.org/market/ (accessed on 31 August 2017).

3. Lauwers, B. Avantium Werkt aan de Drankverpakking van de toekomst—Dit is Bijzonder Disruptief. *Trends*, 9 February 2017.

4. Giljum, S.; Bruckner, M.; Gözet, B.; de Schutter, L. *Land under Pressure: Global Impacts of the EU Bioeconomy*; Friends of the Earth Europe: Brussels, Belgium, 2016; Available online: https://www.foeeurope.org/sites/default/files/resource_use/2016/land-under-pressure-report-global-impacts-eu-bioeconomy.pdf (accessed on 6 April 2018).

5. European Commision. A European Strategy for Plastics in a Circular Economy. Available online: http://ec.europa.eu/environment/circular-economy/pdf/plastics-strategy.pdf (accessed on 6 April 2018).

6. Bioplastics. OVAM, Projectgroep Bioplastics. 2015. Available online: http://www.ovam.be/sites/default/files/atoms/files/Rapport%20Bioplastics.pdf (accessed on 6 April 2018).

7. ASG Environmental Science Research & Development Institute. Available online: http://www.petbottlewashingline.com (accessed on 6 April 2018).

8. EPBP Website. Design Guidelines. Available online: http://www.epbp.org/design-guidelines (accessed on 6 April 2018).

9. Greene, K.L.; Tonjes, D.J. Degradable plastics and their potential for affecting solid waste systems. *WIT Trans. Ecol. Environ.* **2014**, *180*. [CrossRef]

10. Kirchherr, J.; Reike, D.; Hekkert, M. Conceptualizing the circular economy: An analysis of 114 definitions. *Resour. Conserv. Recycl.* **2017**, *127*, 221–232. [CrossRef]

11. NAPCOR. Report on Postconsumer PET Container Recycling Activity. 2017. Available online: https://www.plasticsrecycling.org/images/pdf/resources/reports/NAPCOR-APR_2016RateReport_INAL.pdf (accessed on 6 April 2018).

12. Schedler, M.; Eagles, K. *Moving Forward on PET Thermoform Recycling*; Resource Recycling Inc.: Portland, OR, USA, 2011; pp. 26–28.

13. Petcore. The Recycling of Thermoforms: An Important Challenge for the PET Value Chain. 2016. Available online: https://petcore-europe.prezly.com/the-recycling-of-thermoforms-an-important-challenge-for-the-pet-value-chain (accessed on 6 April 2018).

14. Het ECO-OH! Recycling Proces. ECO-OH! 2017. Available online: https://www.eco-oh.com/nl/eco-oh-groep (accessed on 6 April 2018).

15. Dvorak, R.; Kosior, E.; Fletcher, J. WRAP Final Report—Improving Food Grade rPET Quality for Use in UK Packaging. 2013. Available online: http://www.wrap.org.uk/sites/files/wrap/rPET%20Quality%20Report.pdf (accessed on 6 April 2018).

16. Hurd, D. *Best Practices and Industry Standards in PET Plastic Recycling*; Washington State Department of Community, Trade and Economic Development: Washington, DC, USA, 2000. Available online: http://www.seas.columbia.edu/earth/RRC/documents/best_practise_pet_recycling.pdf (accessed on 6 April 2018).

17. Eco Entreprises Québec. Fact Sheet Impact of Packaging on Curbside Recycling Collection and Recycling System: PLA Bottle. 2012. Available online: http://www.eeq.ca/wp-content/uploads/PLA-bottles.pdf (accessed on 6 April 2018).

18. NatureWorks. Using NIR Sorting to Recycle PLA Bottles. 2009. Available online: https://www.natureworksllc.com/~/media/The_Ingeo_Journey/EndofLife_Options/mech_recycling/20090708_NatureWorks_UsingNIRSortingtoRecyclePLABottles_pdf.pdf (accessed on 6 April 2018).

19. European Bioplastics: Fact Sheet Mechanical Recycling. 2015. Available online: http://docs.european-bioplastics.org/publications/bp/EUBP_BP_Mechanical_recycling.pdf (accessed on 6 April 2018).

20. COTREP. Fiche Technique 35. 2007. Available online: http://www.cotrep.fr/fileadmin/contribution/mediatheque/avis-generaux/francais/corps-de-l-emballage-et-additifs/Cotrep_AG35_bouteille_PLA.pdf (accessed on 6 April 2018).

21. Gertman, R.; Shireman, B.; Pfahl, D.; Disa, A. *The Bioplastics Sorting Project—Final Grant Report of Technical Findings*; California's Department of Resources Recycling and Recovery: St. Sacramento, CA, USA, 2013. Available online: http://www.calrecycle.ca.gov/publications/Documents/1464/20131464.pdf (accessed on 6 April 2018).

22. Detzel, A.; Kauertz, B.; Derreza-Greeven, C. *Study of the Environmental Impacts of Packagings Made of Biodegradable Plastics*; Report no. 001643/E; Umwelt Bundesambt: Dessau-Roßlau, Germany, 2013; Available online: http://www.uba.de/uba-info-medien/4446.html (accessed on 6 April 2018).

23. STW Foodgate. Recyclage van PET- en PLA-flessen na NIR-Sortering. 2009. Available online: http://www.pack4food.be/files/documents/nieuwsbrieven/Recyclage%20van%20PET-%20en%20PLA-flessen%20na%20NIR-sortering.pdf (accessed on 6 April 2018).

24. Petcore. Petcore Position Statement. Polylactic Acid (PLA). Petcore Evaluation of Polylactic acid (PLA). 2008. Available online: http://www.emballagedigest.fr/blog.php?2008/02/22/6818-petcore-evaluation-of-polylactic-acid-pla (accessed on 6 April 2018).

25. Thoden van Velzen, E.U.; Brouwer, M.T.; Molenveld, K. *Technical Quality of rPET*; Report no. 1661; Food & Biobased Research: Wageningen, The Netherlands, 2016; Available online: http://library.wur.nl/WebQuery/wurpubs/fulltext/392306 (accessed on 6 April 2018).

26. McLauchlin, A.R.; Ghita, O.R. Studies on the thermal and mechanical behavior of PLA-PET blends. *J. Appl. Polym. Sci.* **2016**, *133*. [CrossRef]

27. Hollstein, F.; Wohllebe, M.; Sixto, A.; Manjon, D. Identification of bio-plastics by NIR-SWIR-Hyperspectral-Imaging. In Proceedings of the 2nd Optical Characterization of Materials, Karlsruhe, Germany, 18–19 March 2015; Beyerer, J., León, F.P., Längle, T., Eds.; KIT Scientific Publishing: Karlsruhe, Germany, 2015.

28. Bedekovic, G.; Salopek, B.; Sobota, I. Electrostatic separation of PET/PVC mixture. *Tech. Gaz.* **2011**, *18*, 261–266.

29. Kosior, E.; Mitchell, J.; Davies, K.; Kay, M.; Ahmad, R.; Billiet, E.; Silver, J. Plastic packaging recycling using intelligent separation technologies for materials (PRISM). In Proceedings of the ANTEC Conference, Anaheim, CA, USA, 8–10 May 2017; pp. 500–506.

30. Bugnicourt, E.; Cinelli, P.; Lazzeri, A.; Alvarez, V. Polyhydroxyalkanoate (PHA): Review of synthesis, characteristics, processing and potential applications in packaging. *eXPRESS Polym. Lett.* **2014**, *8*, 791–808. [CrossRef]

31. European Pet Bottle Platform. European PET Bottle Platform Technical Opinion—Synvina—Poly(ethylene 2,5-furandicarboxylate) Resin. 2017. Available online: http://www.epbp.org/download/319/interim-approval-synvinas-polyethylene-25-furandicarboxylate-or-pef (accessed on 6 April 2018).

32. De Jong, E.; Dam, M.A.; Sipos, L.; Gruter, G.-J.M. Furandicarboxylic Acid (FDCA), A Versatile Building Block for a Very Interesting Class of Polyesters. In *Biobased Monomers, Polymers, and Materials*; ACS Symposium Series; Smith, P.B., Gross, R.A., Eds.; American Chemical Society: Washington, DC, USA, 2012. [CrossRef]

33. Website Plastics Insight. Available online: https://www.plasticsinsight.com/global-pet-resin-production-capacity/ (accessed on 6 April 2018).

sustainability

MDPI

Article

Exploring the Dedicated Knowledge Base of a Transformation towards a Sustainable Bioeconomy

Sophie Urmetzer [1],*, Michael P. Schlaile [1,2], Kristina B. Bogner [1], Matthias Mueller [1] and Andreas Pyka [1]

[1]   Department of Innovation Economics (520i), University of Hohenheim, Wollgrasweg 23,
      70599 Stuttgart, Germany; schlaile@uni-hohenheim.de (M.P.S.); kristina.bogner@uni-hohenheim.de (K.B.B.);
      m_mueller@uni-hohenheim.de (M.M.); a.pyka@uni-hohenheim.de (A.P.)
[2]   Center for Applied Cultural Evolution, 1776 Millrace Drive, Eugene, OR 97403, USA
*   Correspondence: sophie.urmetzer@uni-hohenheim.de; Tel.: +49-711-459-24482

Received: 16 April 2018; Accepted: 18 May 2018; Published: 23 May 2018

Abstract: The transformation towards a knowledge-based bioeconomy has the potential to serve as a contribution to a more sustainable future. Yet, until now, bioeconomy policies have been only insufficiently linked to concepts of sustainability transformations. This article aims to create such link by combining insights from innovation systems (IS) research and transformative sustainability science. For a knowledge-based bioeconomy to successfully contribute to sustainability transformations, the IS' focus must be broadened beyond techno-economic knowledge. We propose to also include systems knowledge, normative knowledge, and transformative knowledge in research and policy frameworks for a sustainable knowledge-based bioeconomy (SKBBE). An exploration of the characteristics of this extended, "dedicated" knowledge will eventually aid policymakers in formulating more informed transformation strategies.

Keywords: sustainable knowledge-based bioeconomy; innovation systems; sustainability transformations; dedicated innovation systems; economic knowledge; systems knowledge; normative knowledge; transformative knowledge; bioeconomy policy

1. Introduction

In the light of so-called *wicked problems* (e.g., [1,2]) underlying the global challenges that deeply affect *social*, *environmental*, and *economic* systems, fundamental transformations are required in all of these sustainability dimensions. Therefore, solution attempts need to be based on a systemic consideration of the dynamics, complementarities, and interrelatedness of the affected systems [3].

A relatively new and currently quite popular approach to sustainability transformations addressing at least some of these problems is the establishment of a bio-based economy: the bioeconomy concept relies on novel and future methods of intelligent and efficient utilization of biological resources, processes, and principles with the ultimate aim of substituting fossil resources (e.g., [4–11]). It is therefore frequently referred to as *knowledge-based bioeconomy* [11–13]. Whereas the idea of a bioeconomy is promoted both by academia and in policy circles, it remains unclear what exactly it is comprised of, how to spur the transformation towards a knowledge-based bioeconomy, and how it will affect sustainable development [14,15]. While the development and adoption of novel technologies that help to substitute fossil resources by re-growing biological ones certainly is a condition *sine qua non*, a purely technological substitution process will hardly be the means to confront the global challenges [3,16–20]. It must be kept in mind that a transformation towards a sustainable bioeconomy is only one important contribution to the overall transformation towards sustainability. We explicitly acknowledge that unsustainable forms of bio-based economies are conceivable and even—if left unattended—quite likely [21]. All the more, we see the necessity of finding ways to intervene in the already initiated transformation processes to afford their sustainability.

For successful interventions in the transformation towards a more sustainable bioeconomy, a systemic comprehension of the underlying dynamics is necessary. The *innovation system* (IS) perspective developed in the 1980s as a research concept and policy model [22–26] offers a suitable framework for such systemic comprehension. In the conventional understanding, according to Gregersen and Johnson, an IS "can be thought of as a system which creates and distributes knowledge, utilizes this knowledge by introducing it into the economy in the form of innovations, diffuses it and transforms it into something valuable, for example, international competitiveness and economic growth" ([27], p. 482). While welcoming the importance attributed to knowledge by Gregersen and Johnson and other IS researchers (e.g., [28–32]), particularly in the context of a *knowledge*-based bioeconomy, in this article, we aim to re-evaluate the role and characteristics of knowledge generated and exploited through IS. We argue that knowledge is not just utilized by and introduced in *economic* systems, but it also shapes (and is shaped by) *societal* and *ecological* systems more generally. Consequently, especially against the backdrop of the required transformation towards a *sustainable knowledge-based bioeconomy* (SKBBE), that which is considered as "something valuable" goes beyond an economic meaning (see also [33], on a related note). For this reason, it is obvious that the knowledge base for an SKBBE cannot be a purely techno-economic one. We rather see a need for exploring additional types of knowledge and their characteristics necessary for fostering the search for truly transformative innovation [16].

From the sustainability literature, we know that at least three types of knowledge are relevant for tackling (wicked) problems related to transformations towards sustainability: *Systems* knowledge, *normative* knowledge, and *transformative* knowledge [34–38]. Undoubtedly, these knowledge types need to be centrally considered and fostered for a transformation towards an SKBBE.

In the course of this paper, we aim to clarify the meaning and the characteristics of knowledge necessary for sustainability-oriented interventions in the transformation towards a bioeconomy. To reach this aim, we will explore the following research questions:

- Based on a combination of IS research with the sustainability science perspectives, what are the characteristics of knowledge that are instrumental for a transformation towards an SKBBE?
- What are the policy-relevant implications of this extended perspective on the characteristics of knowledge?

The article is structured as follows: Section 2 sets the scene by reviewing how knowledge has been conceptualized in economics. Aside from discussing in which way the understanding of the characteristics of economic knowledge has influenced innovation policy, we introduce the three types of knowledge (systems, normative, and transformative) relevant for governing sustainability transformations. Section 3 specifies the general meaning of these three types of knowledge, highlights their relevance and instrumental value for transformations towards an SKBBE, and relates them to the most prevalent characteristics of knowledge. Subsequently, Section 4 presents the policy-relevant implications that can be derived from our previous discussions. The concluding Section 5 summarizes our article and proposes some avenues for further research.

2. Knowledge and Innovation Policy

The understanding of knowledge and its characteristics varies between different disciplines. Following the Oxford Dictionaries, knowledge can be defined as "[f]acts, information, and skills acquired through experience or education" or simply as "theoretical or practical understanding of a subject" [39]. The Cambridge Dictionary defines knowledge as the "understanding of or information about a subject that you get by experience or study, either known by one person or by people generally" [40]. A more detailed definition by Zagzebski ([41], p. 92) states that "[k]nowledge is a highly valued state in which a person is in cognitive contact with reality. It is, therefore, a relation.

On one side of the relation is a conscious subject, and on the other side is a portion of reality to which the knower is directly or indirectly related". Despite this multitude of understandings of knowledge, most researchers and policymakers probably agree with the statement that knowledge "is a crucial economic resource" ([28], p. 27). Therefore, the exact understanding and definition of knowledge and its characteristics strongly affect how researchers and policymakers tackle the question of how to best deal with and make use of this resource. Policymakers intervene in IS to improve the three key processes of knowledge creation, knowledge diffusion, and knowledge use (its transformation into something valuable). Policy recommendations derived from an incomplete understanding and representation of knowledge, however, will not be able to improve the processes of knowledge flow in IS and can even counteract the attempt to turn knowledge into something genuinely valuable.

2.1. Towards a More Comprehensive Conceptualization of Knowledge

A good example that highlights the importance of how we define knowledge is the understanding and treatment of knowledge in mainstream neoclassical economics. Neoclassical economists describe knowledge as an intangible good with public good features (non-excludable, non-rivalrous in consumption). Due to the (alleged) non-excludable nature of knowledge, new knowledge flows freely from one actor to another (spillover) such that other actors can benefit from new knowledge without investing in its creation (free-riding) [42]. In this situation, the knowledge-creating actors cannot fully benefit from the value they created, that is, the actors cannot appropriate the returns that resulted from their research activity (appropriability problem) [43]. There is no need for learning since knowledge instantly diffuses from one actor to another and the transfer of knowledge is costless. As Solow is often accredited with pointing out, knowledge falls "like manna from heaven" (see, e.g., [44,45] with reference to [46,47]), and it can instantly be acquired and used by all actors [48].

In contrast to mainstream neoclassical economics, (evolutionary or neo-Schumpeterian) innovation economists and management scholars consider other features of knowledge, thus, providing a much more appropriate analysis of knowledge creation and innovation processes. Innovation economists argue that knowledge can rather be seen as a *latent public good* [48] that exhibits many non-public good characteristics relevant for innovation processes in IS. Since these more realistic knowledge characteristics strongly influence knowledge flows, their consideration improves the understanding of the three key processes of knowledge creation, knowledge diffusion, and knowledge use (transforming knowledge into something valuable) [27]. In what follows, we present the latent public good characteristics of knowledge and structure them according to their relevance for these key processes in IS. Note that for the agents creating, diffusing, and using knowledge, we will use the term *knowledge carrier* in a similar sense as Dopfer and Potts ([49], p. 28), who wrote that "the micro unit in economic analysis is a knowledge carrier . . . acquiring and applying knowledge".

Characteristics of knowledge that are most relevant in the knowledge creation process are the *cumulative* nature of knowledge (e.g., [50,51]), *path dependency* of knowledge (e.g., [52,53]), and knowledge *relatedness* (e.g., [54,55]). As the creation of new knowledge or innovation results from the (re-)combination of previously unconnected knowledge [56,57], knowledge has a cumulative character and can only be understood and created if actors already have a knowledge stock they can relate the new knowledge to [54,58]. The more complex and industry-specific knowledge gets, the higher the importance of prior knowledge and knowledge relatedness (see also the discussions in [55,59]).

Characteristics of knowledge that are especially important for the knowledge diffusion process are *tacitness*, *stickiness*, and *dispersion*. Knowledge is not equal to information [60,61]. In fact, as Morone [62] also explains, information can be regarded as that part of knowledge that can be easily partitioned and transmitted to someone else; information requires knowledge to become useful. Other parts of knowledge are *tacit* [63], that is, very difficult to be codified and to be transported [64]. Tacit knowledge is excludable and, therefore, not a public good [65]. So, even if the knowledge carrier is willing to share, tacitness makes it impossible sometimes to transfer this knowledge [66]. In addition, knowledge and its transfer can be *sticky* [67,68], which means that the transfer of this knowledge requires significantly more effort than the transfer of other knowledge. According to Szulanski [67], both knowledge and the process of knowledge exchange can be sticky. The reasons may be the kind and amount of knowledge itself but also attributes of the knowledge carriers. Finally, the *dispersion* of knowledge also influences the possibility of diffusing knowledge. Galunic and Rodan [64] explain dispersed knowledge by using the example of a jigsaw puzzle. The authors state that knowledge is distributed if all actors receive a photocopy of the picture of the jigsaw puzzle. In contrast, knowledge is dispersed if every actor receives one piece of the jigsaw puzzle, meaning that everybody only holds pieces of the knowledge but not the 'whole' picture. Dispersed knowledge (or systems-embedded knowledge) is difficult to be transferred from one to the other actor (as detecting dispersed knowledge can be problematic, too [64]), thus hindering knowledge diffusion.

Characteristics of knowledge (and knowledge carriers) that influence the possibility to use the knowledge within an IS, that is, to transform it into something valuable, are the *context specificity* and *local* characteristics of knowledge. Even if knowledge is freely available in an IS, the public good features of knowledge are not necessarily decisive, and it might be of little or no use to the receiver. We have to keep in mind that knowledge itself has no value; it only becomes valuable to someone if the knowledge can be used, for example, to solve certain problems [69]. Assuming that knowledge has different values for different actors, more knowledge is not always better. Actors need the right knowledge in the right context at the right time and have to be able to combine this knowledge in the right way to utilize the knowledge. The "resource" knowledge might only be relevant and of use in the narrow context for and in which it was developed [64]. Moreover, to understand and use new knowledge, agents need *absorptive capacities* [70,71]. These capacities vary with the disparity of the actors exchanging knowledge: the larger the *cognitive distance* between them, the more difficult it is to exchange and internalize knowledge. Hence, the cognitive distance can be critical for learning and transforming knowledge into something valuable [72,73].

Note that while we have described the creation, diffusion, and use of knowledge in IS as rather distinct processes, this does not imply any linear character or temporal sequence of these processes. Quite the contrary, knowledge creation, diffusion, and use and the respective characteristics of knowledge may overlap and intertwine in a myriad of ways. For example, due to the experimental nature of innovation in general and the fundamental uncertainty involved, there are path dependencies, lock-ins (for example, in terms of stickiness), and feedback that lead to evolutionary cycles of variation/recombination, selection, and transmission or retention of knowledge. Moreover, the vast literature on knowledge *mobilization*, knowledge *translation*, and knowledge *transfer* (e.g., [74–77]) suggests that there can be various obstacles between the creation, diffusion, and use of knowledge, and that so-called *knowledge mediators* or *knowledge brokers* may be required to actively guide these interrelated processes (see also [78], on a related discussion). Consequently, we caution against reading the "trichotomy" of creation, diffusion, and use as connoting that knowledge will be put to good use by the carriers in the end so long as the conditions, such as social network structures, for diffusion are right. In fact, the notion of "optimal" network structures for diffusion may be misguided against the backdrop of the (in-)compatibility of knowledge, cognitive distance, and the dynamics underlying the formation of social networks [58].

2.2. How Knowledge Concepts Have Inspired Innovation Policy Making

Depending on the underlying concept of knowledge, different schools of thought influenced innovation policies in diverse ways (see also [60,79,80]). Following the mainstream neoclassical definition, the (alleged) public good characteristics of knowledge may result in market failure and the appropriability problem. As a consequence, policies have mainly focused on the mitigation of potential externalities and the elimination of inefficient market structures. This was done, for example, by incentive creation (via subsidies or intellectual property rights), the reduction of market entry barriers, and the production of knowledge by the public sector [81]. As Smith also states, "policies of block funding for universities, R&D subsidies, tax credits for R&D etc. [were] the main instruments of post-war science and technology policy in the OECD area" ([82], p. 8).

Policies changed (at least to a certain extent) when the understanding of knowledge changed. Considering knowledge as a *latent* public good, the main rationale for policy intervention is not market failure, but rather systemic problems [81,83]. Consequently, it can be argued that the mainstream neoclassical perspective neglects the importance (and difficulty) of facilitating knowledge creation, knowledge diffusion, and knowledge use in IS (see also [79,80], on a related note). Western innovation policies are often based on the IS approach and inspired by the more comprehensive understanding of knowledge and its implications for innovation. They generally aim at solving inefficiencies in the system (for example, infrastructural, transition, lock-in/path dependency, institutional, network, and capabilities failures as summarized by [83]). These inefficiencies are tackled, for example, by supporting the creation and development of different institutions in the IS as well as fostering networking and knowledge exchange among the system's actors [81]. Since "knowledge is created, distributed, and used in social systems as a result of complex sets of interactions and relations rather than by isolated individuals" ([84], p. 2), *network science* [85] especially has provided methodological support for policy interventions in innovation networks [86–88].

It is safe to state that innovation policies have changed towards a more realistic evaluation of innovation processes over the last decades [89], although in practice, they often still fail to adequately support processes of knowledge creation, diffusion, and use. Even though many policymakers nowadays appreciate the advanced understanding of knowledge and innovation, what Smith wrote more than two decades ago is arguably still valid to some extent, namely that "linear notions remain powerfully present in policy thinking, even in the new innovatory context" ([82], p. 8). Such a non-systemic way of thinking is also reflected by the strongly disciplinary modus operandi which is most obviously demonstrated by the remarkable difficulties still present in concerted actions at the level of political departments.

2.3. Knowledge Concepts in Transformative Sustainability Science

Policy adherence to the specific knowledge characteristics identified by economists has proven invaluable for supporting IS to produce innovations. However, to what end? So far, innovation has frequently been implicitly regarded as desirable per se [3,90,91] and, by default, creating something valuable. However, if IS research shall be aimed at contributing to developing solution strategies to global sustainability challenges, a mere increase in innovative performance by improving the flow of economically relevant knowledge will not suffice [3]. In times of globally effective wicked problems challenging our current production and consumption patterns, it is evident that research into knowledge creation and innovation cannot be a task for economists or any other isolated discipline alone (see also [92], on a related discussion). Additional types of knowledge particularly relevant for addressing wicked problems have been proposed by sustainability science in general and *transformational* sustainability research in particular [36]. Solution options for the puzzle of reconciling economic development with sustainability goals have been found to require three kinds of knowledge: First, *systems knowledge*, which relates to the understanding of the dynamics and processes of ecological and social systems (including IS); second, *normative knowledge*, which determines the desired (target) states of a system; and third, *transformative knowledge*, which builds on systems and normative knowledge to inform the development of strategies for changing systems towards the desired state [34–38]. Although there are alternative terms for these three types of knowledge (such as *explanatory* knowledge, *orientation* knowledge, and *action-guiding* knowledge, as used in [93]), for the sake of terminological consistency with most recent publications, we adopt the terms systems knowledge, normative knowledge, and transformative knowledge.

The fundamental significance of these three kinds of knowledge (systems, normative, and transformative) for sustainability transformations has been put forward by a variety of research strands from theoretical [34,35] to applied planning perspectives [94,95]. Explorations into the specific characteristics in terms of how such knowledge is created, diffused, and used within IS, however, are missing so far. For the particular case of a dedicated transformation towards an SKBBE, we seek to provide some clarification as a basis for an improved governance towards desired ends.

3. Dedicated Knowledge for an SKBBE Transformation

A dedicated transformation towards an SKBBE can be framed with the help of the newly introduced concept of *dedicated innovation system* (DIS) [3,16,96], which goes beyond the predominant focus on technological innovation and economic growth. DIS are dedicated to *transformative innovation* [97,98], which calls for experimentation and (co-)creation of solution strategies to overcome systemic inertia and the resistance of incumbents. In the following, we specify in what ways the IS knowledge needs to be complemented to turn into *dedicated knowledge* instrumental for a transformation towards an SKBBE. Such dedicated knowledge will thus have to comprise economically relevant knowledge as regarded in IS as well as systems knowledge, normative knowledge, and transformative knowledge. Since little is known regarding the meaning and the nature of the latter three knowledge types, we need to detail them and illuminate their central characteristics. This will help to fathom the processes of knowledge creation, diffusion, and use, which will be the basis for deriving policy-relevant implications in the subsequent Section 4.

3.1. Systems Knowledge

Once the complexity and interdependence of transformation processes on multiple scales is acknowledged, systemic boundaries become quite irrelevant. In the context of an SKBBE, systems knowledge must comprise more than the conventional understanding of IS in terms of actor configurations, institutions, and interrelations. As already stressed by Grunwald ([99], p. 154), "sufficient insight into natural and societal systems, as well as knowledge of the interactions between society and the natural environment, are necessary prerequisites for successful action in the direction of sustainable development". Although the IS literature has contributed much to systems knowledge about several levels of economic systems, including technological, sectoral, regional, national, and global IS, the interplay between IS, the Earth system (e.g., [100,101]), and other relevant (sub-)systems (e.g., [102–106]) must also be regarded as a vital part of systems knowledge in the context of sustainability and the bioeconomy. On that note, various authors have emphasized the importance of understanding systemic thresholds and tipping points (e.g., [107–110]) and network structures (e.g., [111,112]), which can thus be considered important elements of systems knowledge. In this regard, it may also be important to stress that systems knowledge is (and must be) subject to constant revision and change, because, as Boulding ([113], p. 9) already emphasized, "we are not simply acquiring knowledge about a static system which stays put, but acquiring knowledge about a whole dynamic process in which the acquisition of the knowledge itself is a part of the process".

To give a prominent example which suggests a lack of systems knowledge in bioeconomy policies, we may use the case of biofuels and their adverse effects on land-use and food supply in some of the least developed countries [114,115]. In this case, the wicked problem addressed was climate change due to excessive CO_2 emissions, and the solution attempt was the introduction of bio-based fuel for carbon-reduced mobility. However, after the first boom of biofuel promotion, emissions savings were at best underwhelming or negative since the initial models calculating greenhouse gas savings had insufficiently considered the effects of the biofuel policies on markets and production: whereas the carbon intensity of biofuel crop cultivation was taken into account, the overall expansion of the agricultural area and the conversion of former grasslands and forests into agricultural land was not [114,115]. These *indirect land-use change* (ILUC) effects are estimated to render the positive effects of biofuel usage more than void, which represents a vivid example for how (a lack of) comprehensive systems knowledge can influence the (un)sustainability of bioeconomy transformations.

In accordance with much of the IS literature's focus on knowledge and the common intellectual history of IS and evolutionary economics (e.g., [23]), it becomes clear that an economic system, in general, and a (knowledge-based) bioeconomy, in particular, may also be regarded as "a coordinated system of distributed knowledge" ([69], p. 413). Potts posits that "[k]nowledge is the solution to problems. A solution will consist of a *rule*, which is a generative system of connected components" ([69], p. 418f.). The importance of rules is particularly emphasized by the so-called *rule-based approach* (RBA) to evolutionary economics developed by Dopfer and colleagues (e.g., [49,116–121]). According to the RBA, a "rule is defined as the idea that organizes actions or resources into operations. It is the element of knowledge in the knowledge-based economy and the locus of evolution in economic evolution" ([49], p. 6). As Blind and Pyka also elucidate, "a rule represents knowledge that enables its carrier to perform economic operations, i.e., production, consumption and transactions. The distinction between generic rules and operations based on these rules is essential for the RBA" ([122], p. 1086). According to the RBA, these generic rules may be further distinguished into subject and object rules: subject rules are the cognitive and behavioral rules of an economic agent, whereas object rules are social and technical rules that represent the organizing principles for social and technological systems [49,118]. The latter include, for example, *Nelson-Winter organizational routines* [123] and *Ostrom social rules* (e.g., [124–126]). From this brief summary of the RBA, it already becomes clear that an understanding of the bioeconomic systems' rules and their interrelations is an instrumental element of systems knowledge. Or, as Meadows puts it, "[p]ower over the rules is real power" ([127], p. 158).

Since it can be argued that the creation, diffusion, and use of systems knowledge is the classical task of the sciences [93,99], most of the characteristics of latent public goods (as outlined above) can be expected to also hold for systems knowledge in terms of its relatedness, cumulative properties, and codifiability. Special features to be considered when dealing with systems knowledge in the context of a transformation towards an SKBBE will be twofold: First, systems knowledge may be quite sticky, that is, it may require much effort to be transferred. This is owed to the fact that departing from linear cause-and-effect thinking and starting to think in systems still requires quite some intellectual effort on the side of the knowledge carrier (see also [128], on a related note). Second, systems knowledge can be expected to be strongly dispersed among different disciplines and knowledge bases of great cognitive distances, such as—with recourse to the example of ILUC—economics, agricultural sciences, complexity science, and other (social and natural) sciences.

3.2. Normative Knowledge

According to Abson and colleagues ([35], p. 32), "[n]ormative knowledge encompasses both knowledge on desired system states (normative goals or target knowledge. . .) and knowledge related to the rationalization of value judgements associated with evaluating alternative potential states of the world (as informed by systems knowledge. . .)". In the context of an SKBBE, it becomes clear that normative knowledge must refer not only to directionality, responsibility, and legitimacy issues in IS (as discussed in [3]) but also to the targets of the interconnected physical, biological, social, political, and other systems (e.g., [102]). Thereby, for the transformation of knowledge into "something valuable" within IS (cf. [27]), the dedication of IS to an SKBBE also implies that the goals of "international competitiveness and economic growth" (cf. [27]) must be adjusted and re-aligned with what is considered something valuable in conjunction with the other interconnected (sub-)systems (for example, social and ecological ones) (see also [129,130] on the related discussion about orientation failure in IS).

Yet, one of the major issues with prior systemic approaches to sustainability transformations, in general, seems to be that they tend to oversimplify the complexity of normative knowledge and value systems by presuming a consensus about the scale and importance of sustainability-related goals and visions [131]. As, for instance, Miller and colleagues [132] claim, "[i]nquiries into values are largely absent from the mainstream sustainability science agenda" ([132], p. 241). However, sustainability is a genuinely normative phenomenon [93] and knowledge related to norms, values, and desired goals that indicate the necessity for and direction of change is essential for the successful systemic change towards a *sustainable* bioeconomy (and not just any bioeconomy for the sake of endowing the biotechnology sector). Norms, values, and narratives of sustainability are regularly contested and contingent on diverse and often conflicting and (co-)evolving worldviews [3,131–138].

Similar ambiguity can be observed in the context of the bioeconomy (e.g., [15,139]). When taking the complexity of normative knowledge seriously, it may even be impossible to define globally effective rules, norms, or values (in terms of a universal paradigm for an SKBBE) [21]. Arguably, it may be more important to empower actors within IS to "apply, negotiate and reconcile norms and principles based on the judgements of multiple stakeholders" ([140], p. 12). The creation of normative knowledge for an SKBBE can thus be expected to depend on different initial conditions such as the cultural context, whereas the diffusion of a globally effective canon of practices for an SKBBE is highly unlikely (see also [141]). Normative knowledge for an SKBBE is, therefore, intrinsically local in character despite the fact that sustainable development is a global endeavor.

Moreover, the creation of normative knowledge is shaped by *cultural evolutionary processes* (e.g., [138,142–149]). This means, for example, that both subject rules that shape the sustainability goals of the individual carriers (for example, what they consider good or bad) and object rules that determine what is legitimate and important within a social system or IS are subject to path dependence, competition, and feedback at the level of the underlying ideas (e.g., [58,131,150,151]). The diffusion of normative knowledge about the desired states of a system is therefore always contingent on its context specificity and dependent on cultural evolution. In Boyd and Richerson's words, "people acquire beliefs, attitudes, and values both by teaching and by observing the behavior of others. Culture is not behavior; culture is information ... that, together with individuals' genes and their environments, determines their behavior" ([145], p. 74). While many object rules are codifiable as laws and formal institutions, most subject rules can be assumed to remain tacit so that normative knowledge consists of a combination of tacit and codified knowledge. Of course, "people are not simply rule bound robots who carry out the dictates of their culture" ([145], p. 72), but rules can often work subconsciously to evolve institutions (e.g., [152]) and shared paradigms that span the "bounded performative space" of an IS (see, e.g., [3], on a related note).

Consequently, when referring to normative knowledge and the constituting values and belief systems, we are not only dealing with the competition and evolution of knowledge at the level of rules and ideas driven by (co-)evolutionary processes across the societal sub-systems of individuals, the market, the state, civil society, and nature [106]. To a great extent, the cognitive distances of competing carriers *within* sub-systems and their conflicting strategies can also pose serious impediments to normative knowledge creation, diffusion, and use. This complex interrelation may, thus, be understood from a multilevel perspective with feedback between worldviews, visions, paradigms, the Earth system, regimes, and niches [153].

3.3. Transformative Knowledge

Transformative knowledge can, in the context of this article, be understood as knowledge about how to accelerate and influence the ongoing transformation towards an SKBBE. As, for instance, Abson and colleagues [35] explain, this type of knowledge is necessary for the development of tangible strategies to transform systems (based on systems knowledge) towards the goals derived from normative knowledge. Theoretical and practical understanding must be attained to afford transitions from the current to the desired states of the respective system(s), which will require a mix of codified and tacit elements. Creating transformative knowledge will encompass the acquisition of skills and knowledge about how to effect systemic changes, or, as Almudi and Fatas-Villafranca put it, how to deliberately shape the evolutionary processes in other sub-systems (a mechanism referred to as *promotion* [106]). Although wicked problems that necessitate these changes are most often global in nature, their solution strategies will have to be adapted to the local conditions [97]. While global concepts and goals for a bioeconomy may be relatively easy to agree upon, the concrete measures and resource allocation will be negotiated and disputed at the regional and local scales [154]. This renders transformative knowledge in IS exceptionally local.

In line with the necessity for a change of goals and values, scholars of the educational sciences argue that effective transformative knowledge will also require a revision of inherited *individual* value frames and assumptions on the side of the knowledge carriers themselves [155]. This process of fundamentally challenging personal worldviews inherent in the absorption of truly transformative knowledge makes this type of knowledge extremely sticky and inhibited by lock-ins and path dependence. For a transformation from a fossil to a bio-based economy, the collective habituation to a seemingly endless and cheap supply of fossil resources and the ostensibly infinite capacity of ecosystems to absorb emissions and waste must be overcome. In line with findings from cultural evolution and the RBA, sustainability education research has also pointed to the importance of acknowledging that human action is driven not only by cognitive knowledge but also unconsciously by "deeper" levels of knowing such as norms, assumptions, values, or beliefs [156]. Consequently, only when being effective on these different levels of consciousness can transformative knowledge unfold its full potential to enable its carriers to induce behavioral change in themselves, a community, or the society. Put differently, the agents of sub-systems will only influence the replication and selection processes according to sustainable values in other sub-systems (via promotion) if they expect advantages in individual and social well-being [106].

Besides systems and normative knowledge, transformative knowledge thus requires the skills to affect deeper levels of knowing and meaning, thereby influencing more immediate and conscious levels of (cognitive and behavioral) rules, ideas, theories, and action [157,158]. Against this backdrop, it may come as no surprise that the prime minister of the German state of Baden-Wuerttemberg, member of the green party, has so far failed to push state policies towards a mobility transformation away from individual transport on the basis of combustion technology. In an interview, he made it quite clear that although he has his chauffeur drive him in a hybrid car on official trips, in his private life he "does what he considers right" by driving "a proper car"—namely a Diesel [159].

From what we have elaborated regarding the characteristics of transformative knowledge, we must conclude that its creation requires a learning process on multiple levels. It must be kept in mind that it can only be absorbed if the systemic understanding of the problem and a vision regarding the desired state are present, that is, if a certain level of capacity to absorb transformative knowledge is given. Furthermore, Grunwald [93] argues that the creation of transformative knowledge must be reflexive. In a similar vein, Lindner and colleagues stress the need for reflexivity in IS, and they propose various quality criteria for reflexive IS [130]. In terms of its diffusion and use, transformative knowledge is thought to become effective only if it is specific to the context and if its carriers have internalized the necessity for transformation by challenging their personal assumptions and values. Consequently, since values and norms have evolved via cultural evolution, transformative knowledge also needs to include knowledge about how to influence the cultural evolutionary processes (e.g., [133,160–163]). To take up Brewer's *culture design* approach, "change processes can only be guided if their evolutionary underpinnings are adequately understood. This is the role for approaches and insights from cultural evolution" ([160], p. 69).

4. Policy-Relevant Implications

4.1. Knowledge-Related Gaps in Current Bioeconomy Policies

The transformation towards an SKBBE must obviously be guided by strategies derived from using transformative knowledge which is, by definition, based on the other relevant types comprising dedicated knowledge. We suspect that the knowledge which guided political decision-makers in developing and implementing current bioeconomy policies so far has, in some respect, not been truly transformative. Important processes of creating, diffusing, and using systems and, especially normative knowledge, have not sufficiently been facilitated. We propose how more detailed insights into the characteristics of dedicated knowledge can be used to inform policymakers in improving their transformative capacities. Based on the example of two common issues of critique in the current bioeconomy policy approaches, we will substantiate our knowledge-based argument. Bioeconomy policies have been identified (i) to be biased towards economic goals and, therefore, take an unequal account of all three dimensions of sustainability [21,164–168]; and, to some extent related to it; (ii) to only superficially integrate all relevant stakeholders into policy making [21,165,169–173].

Bioeconomy policies brought forward by the European Union (EU) and several nations have been criticized for a rather narrow techno-economic emphasis. While using the term sustainable as an attribute to a range of goals and principles frequently, the EU bioeconomy framework, for example, still overemphasizes the economic dimension. This is reflected by the main priority areas of various political bioeconomy agendas which remain quite technocratic: keywords include biotechnology, eco-efficiency, competitiveness, innovation, economic output, and industry in general [14,164]. The EU's proposed policy action along the three large areas (i) the investment in research, innovation and skills; (ii) the reinforcement of policy interaction and stakeholder engagement; and (iii) the enhancement of markets and competitiveness in bioeconomy sectors ([174], p. 22), reveals a strong focus on fostering *economically relevant* and technological knowledge creation. In a recent review [175] of its 2012 Bioeconomy Strategy [174] the European Commission (EC) did indeed observe some room for improvement with regard to more comprehensive bioeconomy policies by acknowledging that "the achievement of the interlinked bioeconomy objectives requires an integrated (i.e., cross-sectoral and cross-policy) approach within the EC and beyond. This is needed in order to adequately address the issue of multiple trade-offs but also of synergies and interconnected objectives related to bioeconomy policy (e.g., sustainability and protection of natural capital, mitigating climate change, food security)" ([175], p. 25).

An overemphasis on economic aspects of the bioeconomy in implementation strategies is likely to be rooted in an insufficient stock of systems knowledge. If the bioeconomy is meant to "radically change [Europe's] approach to production, consumption, processing, storage, recycling and disposal of biological resources" ([174], p. 8) and to "assure over the long term the prosperity of modern societies" ([4], p. 2), the social and the ecological dimension have to play equal roles. Furthermore, the systemic interplay between all three dimensions of sustainability must be understood and must find its way into policy making via systems knowledge. While the creation of systems knowledge within the individual disciplines does not seem to be the issue (considering, for example, advances in Earth system sciences, agriculture, and political sciences), its interdisciplinary diffusion and use seem to lag behind (see also [160], on a related note). The prevalent characteristics of this knowledge relevant for its diffusion have been found to be stickiness and dispersal (see Section 3.1 above). To reduce the stickiness of systems knowledge and, thus, improve its diffusion and transfer, long-term policies need to challenge the fundamental principles still dominating in education across disciplines and across school levels: linear cause-and-effect thinking must be abandoned in favor of systemic ways of thinking. To overcome the wide dispersal of bioeconomically relevant knowledge across academic disciplines and industrial sectors, policies must encourage inter- and transdisciplinary research even more and coordinate knowledge diffusion across mental borders. This, in turn, calls for strategies that facilitate connecting researchers across disciplines and with practitioners as well as translating systems knowledge for the target audience (e.g., [74]). Only then can systems knowledge ultimately be used for informing the creation processes of transformative knowledge.

This brings us to the second issue of bioeconomy policies mentioned above: the failure of bioeconomy strategies to involve all stakeholders in a sincere and open dialogue on goals and paths towards (a sustainable) bioeconomy [169,170,173]. Their involvement in the early stages of a bioeconomy transformation is not only necessary for receiving sufficient acceptance of new technologies and the approval of new products [168,170]. These aspects—which, again, mainly affect the short-term *economic* success of the bioeconomy—are addressed well across various bioeconomy strategies. However, "[a]s there are so many issues, trade-offs and decisions to be made on the design and development of the bioeconomy, a commitment to participatory governance that engages the general public and key stakeholders in an *open* and *informed dialogue* appears vital" ([168], p. 2603; italics added). From the perspective of dedicated knowledge, there is a reason why failing to integrate the knowledge, values, and worldviews of the people affected will seriously impede the desired transformation: the processes of creation, diffusion, and use of normative knowledge and transformative knowledge are contingent on the input of a broad range of stakeholders—basically, of everyone who will eventually be affected by the transformation. The use of normative knowledge (that is, the agreement upon common goals), as well as the use of transformative knowledge (that is, the definition of transformation strategies), have both been identified to be intrinsically local and context-specific (see Sections 3.2 and 3.3). A policy taking account of these characteristics will adopt mechanisms to enable citizens to take part in societal dialogue which must comprise three tasks: offering suitable participatory formats, educating people to become responsible citizens, and training transdisciplinary capabilities to overcome cognitive distances between different mindsets as well as to reconcile global goals with local requirements. In this respect, there has been a remarkable development at the European level: while the German government is still relying on the advice of a Bioeconomy Council representing only the industry and academia for developing the bioeconomy policy [154], the recently reconstituted delegates of the European Bioeconomy Panel represent a variety of societal groups: "business and primary producers, policymakers, researchers, and civil society organisations" ([175], p. 13). Unsurprisingly, their latest publication, the bioeconomy stakeholders' *manifesto*, gives some recommendations that clearly reflect the broad basis of stakeholders involved, especially concerning education, skills, and training [176].

For a structured overview of the elements of dedicated knowledge and their consideration by current bioeconomy policy approaches, see Table 1.

4.2. Promising (But Fragmented) Building Blocks for Improved SKBBE Policies

Although participatory approaches neither automatically decrease the cognitive distances between stakeholders nor guarantee that the solution strategies agreed upon are based on the most appropriate (systems and normative) knowledge [95], an SKBBE cannot be achieved in a top-down manner. Consequently, the involvement of stakeholders confronts policymakers with the roles of coordinating agents and knowledge brokers [74,75,77,177]. Once a truly systemic perspective is taken up, the traditional roles of different actors (for example, the state, non-governmental organizations, private companies, consumers) become blurred (see also [178–180]), which has already been recognized in the context of environmental governance and prompted Western democracies to adopt more participatory policy approaches [181]. A variety of governance approaches exist, ranging from *adaptive governance* (e.g., [182–184]) and *reflexive governance* (e.g., [130,185]) to *Earth system governance* (e.g., [18,101,186]) and various other concepts (e.g., [107,187–190]). Without digressing too much into debates about the differences and similarities of systemic governance approaches, we can already contend that the societal roots of many of the sustainability-related wicked problems clearly imply that social actors are not only part of the problem but must also be part of the solution. Against this background, transdisciplinary research and participatory approaches such as co-design and co-production of knowledge have recently gained momentum with good reason (e.g., [37,191–198]) and are also promising in the context of the transformation towards an SKBBE. Yet, the question remains why only very few, if any, bioeconomy policies have taken participatory approaches and stakeholder engagement seriously (see, e.g., [170,199], on a related discussion).

Table 1. The elements of dedicated knowledge in the context of SKBBE policies.

Central Knowledge Types as Elements of Dedicated Knowledge	General Meaning	Sustainability and Bioeconomy-Related Instrumental Value	Most Prevalent Characteristics Regarding Creation, Diffusion and Use	Consideration by Current Bioeconomy Policy Approaches
Economically relevant knowledge	Knowledge necessary to create economic value.	Knowledge necessary to create economic value in line with the resources, processes, and principles of biological systems.	Latent public good, depending on the technology in question.	Adequately considered.
Systems knowledge	Descriptive, interdisciplinary understanding of relevant systems.	Understanding of the dynamics and interactions between biological, economic, and social systems.	Sticky and strongly dispersed between disciplines.	Insufficiently considered.
Normative knowledge	Knowledge about desired system states to formulate systemic goals.	(Knowledge of) Collectively developed goals for sustainable bioeconomies.	Intrinsically local, path-dependent, and context-specific; but sustainability as a global endeavor.	Partially considered.
Transformative knowledge	Know-how for challenging worldviews and developing tangible strategies to facilitate the transformation from current system to target system.	Knowledge about strategies to govern the transformation towards an SKBBE.	Local and context-specific, strongly sticky, and path-dependent.	Partially considered.

To better acknowledge the characteristics of dedicated knowledge, we can propose a combination of four hitherto rather fragmented but arguably central frameworks that may be built on to improve bioeconomy policy agendas in terms of creating, diffusing, and using dedicated knowledge (note that the proposed list is non-exhaustive but may serve as a starting point for developing more adequate knowledge-based bioeconomy policies):

- Consider the roles of policymakers and policy making from a co-evolutionary perspective (see also [138]), where the "state" is conceived as one of several sub-systems (for example, next to the individuals, civil society, the market, and nature) shaping contemporary capitalist societies [106]. Through the special co-evolutionary mechanism of *promotion*, political entities are able to deliberately influence the propagation (or retention) of certain knowledge, skills, ideas, values, or habits within other sub-systems and, thereby, trigger change in the whole system [106].
- Take up insights from *culture design* (e.g., [133,160–163,200]) and findings on transmission and learning biases in cultural evolution (e.g., [201–203]) that may help to explain and eventually overcome the stickiness and locality of both systems and normative knowledge and thereby increase the absorptive capacities of DIS actors for dedicated knowledge.
- Use suggestions from the literature on *adaptive governance* such as the combination of indigenous knowledge with scientific knowledge (to overcome path dependencies), continuous adaptation of transformative knowledge to new systems knowledge (to avoid lock-ins), embracing uncertainty (accepting that the behavior of systems can never be completely understood and anticipated), and the facilitation of self-organization (e.g., [183,184]) by empowering citizens to participate in the responsible co-creation, diffusion, and use of dedicated knowledge.
- Apply *reflexive governance* instruments as guideposts for DIS, including principles of transdisciplinary knowledge production, experimentation, and anticipation (creating systems knowledge), participatory goal formulation (creating and diffusing normative knowledge), and interactive strategy development (using transformative knowledge) ([130,204]) for the bioeconomy transformation.

In summary, we postulate that for more sustainable bioeconomy policies, we need more adequate knowledge policies.

5. Conclusions

Bioeconomy policies have not effectively been linked to findings and approved methods of sustainability sciences. The transformation towards a bioeconomy, thus, runs into the danger of becoming an unsustainable and purely techno-economic endeavor. Effective public policies that take due account of the knowledge dynamics underlying transformation processes are required. In the context of sustainability, it is not enough to just improve the capacity of an IS for creating, diffusing, and using economically relevant knowledge. Instead, the IS must become more goal-oriented and dedicated to tackling wicked problems [3,205]. Accordingly, for affording such systemic dedication to the transformation towards an SKBBE, it is central to consider dedicated knowledge (that is, a combination of the understanding of economically relevant knowledge with systems knowledge, normative knowledge, and transformative knowledge).

Drawing upon our insights into such dedicated knowledge, we can better understand why current policies have not been able to steer the bioeconomy transformation onto a sustainable path. We admit that recent policy revision processes (e.g., [173,175,176,206–208])—especially in terms of viewing the transition to a bioeconomy as a societal transformation, a focus on participatory approaches, and a better coordination of policies and sectors—are headed in the right direction. However, we suggest that an even stronger focus on the characteristics of *dedicated* knowledge and its creation, diffusion, and use in DIS is necessary for the knowledge-based bioeconomy to become truly sustainable. These characteristics include stickiness, locality, context specificity, dispersal, and path dependence. Taking dedicated knowledge more seriously entails that the currently most influential players in bioeconomy governance (that is, the industry and academia) need to display a serious willingness to learn and acknowledge the value of opening up the agenda-setting discourse and allow true participation of all actors within the respective DIS. Although in this article, we focus on the role of knowledge, we are fully aware of the fact that in the context of an SKBBE, other points of systemic intervention exist and must also receive appropriate attention in future research and policy endeavors [127,209].

While many avenues for future inter- and transdisciplinary research exist, the next steps may include

- enhancing systems knowledge by analyzing which actors and network dynamics are universally important for a successful transformation towards an SKBBE and which are contingent on the respective variety of a bioeconomy,
- an inquiry into knowledge mobilization and, especially the role(s) of knowledge brokers for the creation, diffusion, and use of dedicated knowledge (for example, installing regional bioeconomy hubs),
- researching the implications of extending the theory of knowledge to other relevant disciplines,
- assessing the necessary content of academic and vocational bioeconomy curricula for creating bioeconomy literacy beyond techno-economic systems knowledge,
- applying and refining the RBA to study which subject rules and which object rules are most important for supporting sustainability transformations,
- and many more.

Author Contributions: S.U., M.P.S. and K.B.B. contributed equally as lead authors. K.B.B. mainly authored Section 2 with contributions from M.M. S.U. and M.P.S. mainly authored Sections 3 and 4. M.M. and A.P. were involved in the overall supervision and provided input to all sections of the draft. All authors read carefully and approved the final version of the manuscript.

Funding: This research received no specific grant but A.P. and M.M. gratefully acknowledge financial support from the Dieter Schwarz Stiftung.

Acknowledgments: The authors would like to thank the participants of the 10th European Meeting on Applied Evolutionary Economics (EMAEE), 31 May–3 June 2017, in Strasbourg for valuable comments on an earlier draft. Moreover, the authors are grateful for Lukas Zuschrott's help with formatting the final article and references.

Conflicts of Interest: The authors declare no conflict of interest.

References

1. Rittel, H.W.J.; Webber, M.M. Dilemmas in a general theory of planning. *Policy Sci.* **1973**, *4*, 155–169. [CrossRef]
2. Pohl, C.; Truffer, B.; Hirsch Hadorn, G. Addressing wicked problems through transdisciplinary research. In *The Oxford Handbook of Interdisciplinarity*; Frodeman, R., Klein, J.T., Pacheco, R.C.S., Eds.; Oxford University Press: Oxford, UK, 2017; pp. 319–331.
3. Schlaile, M.P.; Urmetzer, S.; Blok, V.; Andersen, A.; Timmermans, J.; Mueller, M.; Fagerberg, J.; Pyka, A. Innovation systems for transformations towards sustainability? Taking the normative dimension seriously. *Sustainability* **2017**, *9*, 2253. [CrossRef]
4. BMBF; BMEL. *Bioeconomy in Germany: Opportunities for a Bio-Based and Sustainable Future*; Federal Ministry of Education and Research & Federal Ministry of Food and Agriculture: Berlin/Bonn, Germany, 2015.
5. Dabbert, S.; Lewandowski, I.; Weiss, J.; Pyka, A. (Eds.) *Knowledge-Driven Developments in the Bioeconomy*; Springer: Cham, Switzerland, 2017.
6. Lewandowski, I. (Ed.) *Bioeconomy*; Springer International Publishing: Cham, Switzerland, 2018.
7. Philp, J. The bioeconomy, the challenge of the century for policy makers. *New Biotechnol.* **2018**, *40*, 11–19. [CrossRef] [PubMed]
8. von Braun, J. Bioeconomy: The New Transformation of Agriculture, Food, and Bio-Based Industries, Implications for Emerging Economies. 2017. Available online: http://www.ifpri.org/event/bioeconomy-%E2%80%93-new-transformation-agriculture-food-and-bio-based-industries-%E2%80%93-implications (accessed on 15 April 2018).

9.   The White House. *National Bioeconomy Blueprint*; The White House: Washington, DC, USA, 2012.

10.   El-Chichakli, B.; von Braun, J.; Lang, C.; Barben, D.; Philp, J. Five cornerstones of a global bioeconomy. *Nature* **2016**, *535*, 221–223. [CrossRef] [PubMed]

11.   Virgin, I.; Fielding, M.; Fones Sundell, M.; Hoff, H.; Granit, J. Benefits and challenges of a new knowledge-based bioeconomy. In *Creating Sustainable Bioeconomies: The Bioscience Revolution in Europe and Africa*; Virgin, I., Morris, E.J., Eds.; Routledge: Abingdon, UK, 2017; pp. 11–25.

12.   Pyka, A.; Prettner, K. Economic growth, development, and innovation: The transformation towards a knowledge-based bioeconomy. In *Bioeconomy*; Lewandowski, I., Ed.; Springer International Publishing: Cham, Switzerland, 2018; pp. 331–342.

13.   DECHEMA Gesellschaft für Chemische Technik und Biotechnologie e.V. on behalf of the German Presidency of the Council of the European Union. En Route to the Knowledge-Based Bio-Economy—Cologne Paper. 2007. Available online: http://dechema.de/dechema_media/Cologne_Paper-p-20000945.pdf (accessed on 1 April 2018).

14.   Staffas, L.; Gustavsson, M.; McCormick, K. Strategies and policies for the bioeconomy and bio-based bconomy: An analysis of official national approaches. *Sustainability* **2013**, *5*, 2751–2769. [CrossRef]

15.   Bugge, M.; Hansen, T.; Klitkou, A. What is the bioeconomy? A review of the literature. *Sustainability* **2016**, *8*, 691. [CrossRef]

16.   Pyka, A. Dedicated innovation systems to support the transformation towards sustainability: Creating income opportunities and employment in the knowledge-based digital bioeconomy. *J. Open Innov.* **2017**, *3*, 385. [CrossRef]

17.   Pyka, A.; Buchmann, T. Die Transformation zur wissensbasierten Bioökonomie. In *Technologie, Strategie und Organisation*; Burr, W., Stephan, M., Eds.; Springer: Wiesbaden, Germany, 2017; pp. 333–361.

18.   Patterson, J.; Schulz, K.; Vervoort, J.; van der Hel, S.; Widerberg, O.; Adler, C.; Hurlbert, M.; Anderton, K.; Sethi, M.; Barau, A. Exploring the governance and politics of transformations towards sustainability. *Environ. Innov. Soc. Trans.* **2017**, *24*, 1–16. [CrossRef]

19.   Morone, P. The times they are a-changing: Making the transition toward a sustainable economy. *Biofuels Bioprod. Biorefin.* **2016**, *10*, 369–377. [CrossRef]

20.   Westley, F.; Olsson, P.; Folke, C.; Homer-Dixon, T.; Vredenburg, H.; Loorbach, D.; Thompson, J.; Nilsson, M.; Lambin, E.; Sendzimir, J.; et al. Tipping toward sustainability: Emerging pathways of transformation. *Ambio* **2011**, *40*, 762–780. [CrossRef] [PubMed]

21.   Pfau, S.; Hagens, J.; Dankbaar, B.; Smits, A. Visions of sustainability in bioeconomy research. *Sustainability* **2014**, *6*, 1222–1249. [CrossRef]

22.   Freeman, C. *Technology Policy and Economic Performance: Lessons from Japan*; Pinter: London, UK, 1987.

23.   Freeman, C. *Systems of Innovation: Selected Essays in Evolutionary Economics*; Edward Elgar: Cheltenham, UK, 2008.

24.   Lundvall, B.-Å. (Ed.) *National Systems of Innovation: Towards a Theory of Innovation and Interactive Learning*; Pinter: London, UK, 1992.

25.   Nelson, R.R. (Ed.) *National Innovation Systems: A Comparative Study*; Oxford University Press: New York, NY, USA, 1993.

26.   Edquist, C. (Ed.) *Systems of Innovation*; Routledge: London, UK, 1997.

27.   Gregersen, B.; Johnson, B. Learning economies, innovation systems and European integration. *Reg. Stud.* **1997**, *31*, 479–490. [CrossRef]

28.   Lundvall, B.-Å.; Johnson, B. The learning economy. *J. Ind. Stud.* **1994**, *1*, 23–42. [CrossRef]

29.   Lundvall, B.-Å. The economics of knowledge and learning. In *Product Innovation, Interactive Learning and Economic Performance*; Christensen, J.L., Lundvall, B.-Å., Eds.; Elsevier JAI: Amsterdam, The Netherlands, 2004; pp. 21–42.

30.   Lundvall, B.-Å. Post script: Innovation system research—Where it came from and where it might go. In *National Systems of Innovation: Toward a Theory of Innovation and Interactive Learning*; Lundvall, B.-Å., Ed.; Anthem Press: London, UK, 2010; pp. 317–349.

31.   OECD. *Knowledge Management in the Learning Society: Education and Skills*; OECD: Paris, France, 2000.

32.   Edquist, C. Systems of innovation: Perspectives and challenges. In *The Oxford Handbook of Innovation*; Fagerberg, J., Mowery, D.C., Nelson, R.R., Eds.; Oxford University Press: Oxford, UK, 2005; pp. 181–208.

33.   Martin, B.R. Twenty challenges for innovation studies. *Sci. Public Policy* **2016**, *43*, 432–450. [CrossRef]

34. ProClim. *Research on Sustainability and Global Change—Visions in Science Policy by Swiss Researchers*; ProClim: Berne, Switzerland, 2017.

35. Abson, D.J.; von Wehrden, H.; Baumgärtner, S.; Fischer, J.; Hanspach, J.; Härdtle, W.; Heinrichs, H.; Klein, A.M.; Lang, D.J.; Martens, P.; et al. Ecosystem services as a boundary object for sustainability. *Ecol. Econ.* **2014**, *103*, 29–37. [CrossRef]

36. Wiek, A.; Lang, D.J. Transformational sustainability research methodology. In *Sustainability Science*; Heinrichs, H., Martens, P., Michelsen, G., Wiek, A., Eds.; Springer: Dordrecht, The Netherlands, 2016; pp. 31–41.

37. von Wehrden, H.; Luederitz, C.; Leventon, J.; Russell, S. Methodological challenges in sustainability science: A call for method plurality, procedural rigor and longitudinal research. *Chall. Sustain.* **2017**, *5*. [CrossRef]

38. Knierim, A.; Laschewski, L.; Boyarintseva, O. Inter- and transdisciplinarity in bioeconomy. In *Bioeconomy*; Lewandowski, I., Ed.; Springer International Publishing: Cham, Switzerland, 2018; pp. 39–72.

39. English Oxford Living Dictionaries. Definition of Knowledge in English. Available online: https://en.oxforddictionaries.com/definition/knowledge (accessed on 13 April 2018).

40. Cambridge University Press. Meaning of "Knowledge" in the English Dictionary. 2018. Available online: https://dictionary.cambridge.org/dictionary/english/knowledge (accessed on 16 April 2018).

41. Zagzebski, L. What is knowledge? In *The Blackwell Guide to Epistemology*; Greco, J., Sosa, E., Eds.; Blackwell Publishing Ltd.: Malden, MA, USA, 1999; pp. 92–116.

42. Pyka, A.; Gilbert, N.; Ahrweiler, P. Agent-based modelling of innovation networks: The fairytale of spillover. In *Innovation Networks: New Approaches in Modelling and Analyzing*; Pyka, A., Scharnhorst, A., Eds.; Springer: Heidelberg, Germany, 2009; pp. 101–126.

43. Arrow, K. Economic welfare and the allocation of resources for invention. In *The Rate and Direction of Inventive Activity: Economic and Social Factors*; Nelson, R.R., Ed.; Princeton University Press: Princeton, NJ, USA, 1962; pp. 609–626.

44. Audretsch, D.B.; Leyden, D.P.; Link, A.N. Regional appropriation of university-based knowledge and technology for economic development. *Econ. Dev. Q.* **2013**, *27*, 56–61. [CrossRef]

45. Acs, Z.J.; Audretsch, D.B.; Lehmann, E.E. The knowledge spillover theory of entrepreneurship. *Small Bus. Econ.* **2013**, *41*, 757–774. [CrossRef]

46. Solow, R.M. A Contribution to the theory of economic growth. *Q. J. Econ.* **1956**, *70*, 65. [CrossRef]

47. Solow, R.M. Technical change and the aggregate production function. *Rev. Econ. Stat.* **1957**, *39*, 312. [CrossRef]

48. Nelson, R.R. What is private and what is public about technology? *Sci. Technol. Hum. Values* **1989**, *14*, 229–241. [CrossRef]

49. Dopfer, K.; Potts, J. *The General Theory of Economic Evolution*; Routledge: London, UK, 2008.

50. Boschma, R. Proximity and innovation: A critical assessment. *Reg. Stud.* **2005**, *39*, 61–74. [CrossRef]

51. Foray, D.; Mairesse, J. The knowledge dilemma in the geography of innovation. In *Institutions and Systems in the Geography of Innovation*; Feldman, M.P., Massard, N., Eds.; Springer: New York, NY, USA, 2002; pp. 35–54.

52. Dosi, G. Technological paradigms and technological trajectories. *Res. Policy* **1982**, *11*, 147–162. [CrossRef]

53. Rizzello, S. Knowledge as a path-dependence process. *J. Bioecon.* **2004**, *6*, 255–274. [CrossRef]

54. Morone, P.; Taylor, R. *Knowledge Diffusion and Innovation: Modelling Complex Entrepreneurial Behaviours*; Edward Elgar: Cheltenham, UK, 2010.

55. Vermeulen, B.; Pyka, A. The role of network topology and the spatial distribution and structure of knowledge in regional innovation policy: A calibrated agent-based model study. *Comput. Econ.* **2017**, *11*, 23. [CrossRef]

56. Arthur, W.B. The structure of invention. *Res. Policy* **2007**, *36*, 274–287. [CrossRef]

57. Schumpeter, J.A. *Theorie der wirtschaftlichen Entwicklung*; Duncker & Humblot: Leipzig, Germany, 1911.

58. Schlaile, M.P.; Zeman, J.; Mueller, M. It's a match! Simulating compatibility-based learning in a network of networks. *J. Evol. Econ.* **2018**, in press.

59. Frenken, K.; van Oort, F.; Verburg, T. Related variety, unrelated variety and regional economic growth. *Reg. Stud.* **2007**, *41*, 685–697. [CrossRef]

60. Rooney, D.; Hearn, G.; Mandeville, T.; Joseph, R. *Public Policy in Knowledge-Based Economies: Foundations and Frameworks*; Edward Elgar: Cheltenham, UK, 2003.

61. Adolf, M.; Stehr, N. *Knowledge*; Routledge: London, UK, 2014.

62. Morone, P. Knowledge, innovation and internationalisation: A roadmap. In *Knowledge, Innovation and Internationalization*; Morone, P., Ed.; Taylor and Francis: Hoboken, NJ, USA, 2013; pp. 1–13.

63. Polanyi, M. *The Tacit Dimension*; University of Chicago Press: Chicago, IL, USA, 1966.

64. Galunic, D.C.; Rodan, S. Resource recombinations in the firm: Knowledge structures and the potential for Schumpeterian innovation. *Strat. Mgmt. J.* **1998**. [CrossRef]

65. Antonelli, C. The evolution of the industrial organisation of the production of knowledge. *Camb. J. Econ.* **1999**, *23*, 243–260. [CrossRef]

66. Nonaka, I. A dynamic theory of organizational knowledge creation. *Organ. Sci.* **1994**, *5*, 14–37. [CrossRef]

67. Szulanski, G. *Sticky Knowledge: Barriers to Knowing in the Firm*; SAGE: London, UK, 2003.

68. von Hippel, E. "Sticky information" and the locus of problem solving: Implications for innovation. *Manag. Sci.* **1994**, *40*, 429–439. [CrossRef]

69. Potts, J. Knowledge and markets. *J. Evol. Econ.* **2001**, *11*, 413–431. [CrossRef]

70. Cohen, W.M.; Levinthal, D.A. Innovation and learning: The two faces of R & D. *Econ. J.* **1989**, *99*, 569–596.

71. Cohen, W.M.; Levinthal, D.A. Absorptive capacity: A new perspective on learning and innovation. *Admin. Sci. Q.* **1990**, *35*, 128–152. [CrossRef]

72. Nooteboom, B.; van Haverbeke, W.; Duysters, G.; Gilsing, V.; van den Oord, A. Optimal cognitive distance and absorptive capacity. *Res. Policy* **2007**, *36*, 1016–1034. [CrossRef]

73. Bogner, K.; Mueller, M.; Schlaile, M.P. Knowledge diffusion in formal networks: The roles of degree distribution and cognitive distance. *Int. J. Comput. Econ. Econom.* **2018**, in press.

74. Bennet, A.; Bennet, D. *Knowledge Mobilization in the Social Sciences and Humanities: Moving from Research to Action*; MQI Press: Marlinton, VI, USA, 2007.

75. Jacobson, N.; Butterill, D.; Goering, P. Development of a framework for knowledge translation: Understanding user context. *J. Health Serv. Res. Policy* **2003**, *8*, 94–99. [CrossRef] [PubMed]

76. Szulanski, G. The process of knowledge transfer: A diachronic analysis of stickiness. *Organ. Behav. Hum. Dec. Process.* **2000**, *82*, 9–27. [CrossRef]

77. Mitton, C.; Adair, C.E.; McKenzie, E.; Patten, S.B.; Waye Perry, B. Knowledge transfer and exchange: Review and synthesis of the literature. *Milbank Q.* **2007**, *85*, 729–768. [CrossRef] [PubMed]

78. Adomßent, M. Exploring universities' transformative potential for sustainability-bound learning in changing landscapes of knowledge communication. *J. Clean. Prod.* **2013**, *49*, 11–24. [CrossRef]

79. Nyholm, J.; Normann, L.; Frelle-Petersen, C.; Riis, M.; Torstensen, P. Innovation policy in the knowledge-based economy: Can theory guide policy making? In *The Globalizing Learning Economy*; Archibugi, D., Lundvall, B.-Å., Eds.; Oxford University Press: Oxford, UK, 2001; pp. 239–272.

80. Lundvall, B.-Å. Innovation policy in the globalizing learning economy. In *The Globalizing Learning Economy*; Archibugi, D., Lundvall, B.-Å., Eds.; Oxford University Press: Oxford, UK, 2001; pp. 273–291.

81. Chaminade, C.; Edquist, C. Rationales for public policy intervention in the innovation process: A systems of innovation approach. In *The Theory and Practice of Innovation Policy: An International Research Handbook*; Smits, R.E., Kuhlmann, S., Shapira, P., Eds.; Edward Elgar: Cheltenham, UK, 2010; pp. 95–114.

82. Smith, K. Interactions in Knowledge Systems: Foundations, Policy Implications and Empirical Methods, STEP Report R-10. 1994. Available online: https://brage.bibsys.no/xmlui/bitstream/handle/11250/226741/STEPrapport10-1994.pdf?sequence=1 (accessed on 16 April 2018).

83. Klein Woolthuis, R.; Lankhuizen, M.; Gilsing, V. A system failure framework for innovation policy design. *Technovation* **2005**, *25*, 609–619. [CrossRef]

84. Rooney, D.; Hearn, G.; Ninan, A. Knowledge: Concepts, policy, implementation. In *Handbook on the Knowledge Economy*; Rooney, D., Hearn, G., Ninan, A., Eds.; Edward Elgar: Cheltenham, UK, 2005; pp. 1–16.

85. Barabási, A.-L. *Network Science*; Cambridge University Press: Cambridge, UK, 2016.

86. Ahrweiler, P.; Keane, M.T. Innovation networks. *Mind Soc.* **2013**, *12*, 73–90. [CrossRef]

87. Buchmann, T.; Pyka, A. Innovation networks. In *Handbook on the Economics and Theory of the Firm*; Dietrich, M., Krafft, J., Eds.; Edward Elgar: Cheltenham, UK, 2012; pp. 466–482.

88. Scharnhorst, A.; Pyka, A. (Eds.) *Innovation Networks: New Approaches in Modelling and Analyzing*; Springer: Berlin/Heidelberg, Germany, 2009.

89. Edler, J.; Fagerberg, J. Innovation policy: What, why, and how. *Oxf. Rev. Econ. Policy* **2017**, *33*, 2–23. [CrossRef]

90. Soete, L. Is innovation always good? In *Innovation Studies: Evolution and Future Challenges*; Fagerberg, J., Martin, B.R., Andersen, E.S., Eds.; Oxford Univ. Press: Oxford, UK, 2013; pp. 134–144.
91. Engelbrecht, H.-J. A proposal for a 'national innovation system plus subjective well-being' approach and an evolutionary systemic normative theory of innovation. In *Foundations of Economic Change*; Pyka, A., Cantner, U., Eds.; Springer: Cham, Switzerland, 2017; pp. 207–231.
92. Lahsen, M. The social status of climate change knowledge: An editorial essay. *WIREs Clim. Chang.* **2010**, *1*, 162–171. [CrossRef]
93. Grunwald, A. Working towards sustainable development in the face of uncertainty and incomplete knowledge. *J. Environ. Policy Plan.* **2007**, *9*, 245–262. [CrossRef]
94. Wiek, A.; Binder, C. Solution spaces for decision-making—A sustainability assessment tool for city-regions. *Environ. Impact Assess. Rev.* **2005**, *25*, 589–608. [CrossRef]
95. Rydin, Y. Re-examining the role of knowledge within planning theory. *Plan. Theory* **2007**, *6*, 52–68. [CrossRef]
96. Pyka, A. Transformation of economic systems: The bio-economy case. In *Knowledge-Driven Developments in the Bioeconomy*; Dabbert, S., Lewandowski, I., Weiss, J., Pyka, A., Eds.; Springer: Cham, Switzerland, 2017; pp. 3–16.
97. Steward, F. Breaking the Boundaries: Transformative Innovation for the Global Good, NESTA Provocation 07. 2008. Available online: http://www.nesta.org.uk/publications/breaking-boundaries (accessed on 16 April 2018).
98. Steward, F. Transformative innovation policy to meet the challenge of climate change: Sociotechnical networks aligned with consumption and end-use as new transition arenas for a low-carbon society or green economy. *Technol. Anal. Strat. Manag.* **2012**, *24*, 331–343. [CrossRef]
99. Grunwald, A. Strategic knowledge for sustainable development: The need for reflexivity and learning at the interface between science and society. *IJFIP* **2004**, *1*, 150–167. [CrossRef]
100. Schellnhuber, H.-J.; Crutzen, P.J.; Clark, W.C.; Claussen, M.; Held, H. (Eds.) *Earth System Analysis for Sustainability*; MIT Press in Cooperation with Dahlem University Press: Cambridge, MA, USA, 2004.
101. Biermann, F.; Betsill, M.M.; Gupta, J.; Kanie, N.; Lebel, L.; Liverman, D.; Schroeder, H.; Siebenhüner, B.; Zondervan, R. Earth system governance: A research framework. *Int. Environ. Agreem.* **2010**, *10*, 277–298. [CrossRef]
102. Boulding, K.E. *The World as a Total System*; SAGE: Beverly Hills, CA, USA, 1985.
103. Schramm, M. *Der Geldwert der Schöpfung: Theologie, Ökologie, Ökonomie*; Schöningh: Paderborn, Germany, 1994.
104. Seidler, R.; Bawa, K.S. Dimensions of sustainable development. In *Dimensions of Sustainable Development*; Bawa, K.S., Seidler, R., Eds.; Eolss Publishers Co Ltd.: Oxford, UK, 2009; Volume 1, pp. 1–20.
105. Colander, D.C.; Kupers, R. *Complexity and the Art of Public Policy: Solving Society's Problems from the Bottom Up*; Princeton University Press: Princeton, NJ, USA, 2014.
106. Almudi, I.; Fatas-Villafranca, F. Promotion and coevolutionary dynamics in contemporary capitalism. *J. Econ. Issues* **2018**, *52*, 80–102. [CrossRef]
107. Young, O.R. *Governing Complex Systems: Social Capital for the Anthropocene*; The MIT Press: Cambridge, MA, USA, 2017.
108. Lamberson, P.J.; Page, S.E. Tipping points. *QJPS* **2012**, *7*, 175–208. [CrossRef]
109. Wassmann, P.; Lenton, T.M. Arctic tipping points in an Earth system perspective. *Ambio* **2012**, *41*, 1–9. [CrossRef] [PubMed]
110. Gladwell, M. *The Tipping Point: How Little Things Can Make a Big Difference*; Little, Brown and Company: Boston, MA, USA, 2000.
111. Morone, P.; Tartiu, V.E.; Falcone, P. Assessing the potential of biowaste for bioplastics production through social network analysis. *J. Clean. Prod.* **2015**, *90*, 43–54. [CrossRef]
112. Scheiterle, L.; Ulmer, A.; Birner, R.; Pyka, A. From commodity-based value chains to biomass-based value webs: The case of sugarcane in Brazil's bioeconomy. *J. Clean. Prod.* **2018**, *172*, 3851–3863. [CrossRef]
113. Boulding, K.E. The economics of knowledge and the knowledge of economics. *Am. Econ. Rev.* **1966**, *56*, 1–13.
114. Leemans, R.; van Amstel, A.; Battjes, C.; Kreileman, E.; Toet, S. The land cover and carbon cycle consequences of large-scale utilizations of biomass as an energy source. *Glob. Environ. Chang.* **1996**, *6*, 335–357. [CrossRef]

115. Searchinger, T.; Heimlich, R.; Houghton, R.A.; Dong, F.; Elobeid, A.; Fabiosa, J.; Tokgoz, S.; Hayes, D.; Yu, T.-H. Use of U.S. croplands for biofuels increases greenhouse gases through emissions from land-use change. *Science* **2008**, *319*, 1238–1240. [CrossRef] [PubMed]

116. Dopfer, K. Evolutionary economics: A theoretical framework. In *The Evolutionary Foundations of Economics*; Dopfer, K., Ed.; Cambridge University Press: Cambridge, UK, 2005; pp. 3–55.

117. Dopfer, K. Economics in a cultural key: Complexity and evolution revisited. In *The Elgar Companion to Recent Economic Methodology*; Davis, J.B., Hands, D.W., Eds.; Edward Elgar: Cheltenham, UK, 2011; pp. 319–340.

118. Dopfer, K. Evolutionary economics. In *Handbook on the History of Economic Analysis*; Faccarello, G., Kurz, H.-D., Eds.; Edward Elgar: Cheltenham, UK; Northampton, MA, USA, 2016; pp. 175–193.

119. Dopfer, K.; Potts, J. On the theory of economic evolution. *Evol. Inst. Econ. Rev.* **2009**, *6*, 23–44. [CrossRef]

120. Dopfer, K. The economic agent as rule maker and rule user: Homo Sapiens Oeconomicus. *J. Evol. Econ.* **2004**, *14*, 177–195. [CrossRef]

121. Dopfer, K.; Foster, J.; Potts, J. Micro-meso-macro. *J. Evol. Econ.* **2004**, *14*, 263–279. [CrossRef]

122. Blind, G.; Pyka, A. The rule approach in evolutionary economics: A methodological template for empirical research. *J. Evol. Econ.* **2014**, *24*, 1085–1105. [CrossRef]

123. Nelson, R.R.; Winter, S.G. *An Evolutionary Theory of Economic Change*; The Belknap Press of Harvard University Press: Cambridge, MA, USA, 1982.

124. Ostrom, E. *Understanding Institutional Diversity*; Princeton University Press: Princeton, NJ, USA, 2005.

125. Ostrom, E. The complexity of rules and how they may evolve over time. In *Evolution and Design of Institutions*; Schubert, C., von Wangenheim, G., Eds.; Routledge: London, UK; New York, NY, USA, 2006; pp. 100–122.

126. Ostrom, E.; Basurto, X. Crafting analytical tools to study institutional change. *J. Inst. Econ.* **2011**, *7*, 317–343. [CrossRef]

127. Meadows, D.H. *Thinking in Systems: A Primer*; Wright, D., Ed.; Earthscan: London, UK, 2008.

128. Capra, F.; Luisi, P.L. *The Systems View of Life: A Unifying Vision*; Cambridge University Press: Cambridge, UK, 2014.

129. Daimer, S.; Hufnagl, M.; Warnke, P. Challenge-oriented policy-making and innovation systems theory: Reconsidering systemic instruments. In *Innovation System Revisited: Experiences from 40 Years of Fraunhofer ISI Research*; Koschatzky, K., Ed.; Fraunhofer Verlag: Stuttgart, Germany, 2012; pp. 217–234.

130. Lindner, R.; Daimer, S.; Beckert, B.; Heyen, N.; Koehler, J.; Teufel, B.; Warnke, P.; Wydra, S. Addressing Directionality: Orientation Failure and the Systems of Innovation Heuristic. Towards Reflexive Governance. In *Fraunhofer ISI Discussion Papers Innovation and Policy Analysis*; Fraunhofer Institute for Systems and Innovation Research: Karlsruhe, Germany, 2016; Volume 52.

131. Almudi, I.; Fatas-Villafranca, F.; Potts, J. Utopia competition: A new approach to the micro-foundations of sustainability transitions. *J. Bioecon.* **2017**, *19*, 165–185. [CrossRef]

132. Miller, T.R.; Wiek, A.; Sarewitz, D.; Robinson, J.; Olsson, L.; Kriebel, D.; Loorbach, D. The future of sustainability science: A solutions-oriented research agenda. *Sustain. Sci.* **2014**, *9*, 239–246. [CrossRef]

133. Beddoe, R.; Costanza, R.; Farley, J.; Garza, E.; Kent, J.; Kubiszewski, I.; Martinez, L.; McCowen, T.; Murphy, K.; Myers, N.; et al. Overcoming systemic roadblocks to sustainability: The evolutionary redesign of worldviews, institutions, and technologies. *Proc. Natl. Acad. Sci. USA* **2009**, *106*, 2483–2489. [CrossRef] [PubMed]

134. Matutinović, I. Worldviews, institutions and sustainability: An introduction to a co-evolutionary perspective. *Int. J. Sustain. Dev. World Ecol.* **2007**, *14*, 92–102. [CrossRef]

135. Brewer, J.; Karafiath, L. Why Global Warming Won't Go Viral: A Research Report Prepared by DarwinSF 2013. Available online: https://www.slideshare.net/joebrewer31/why-global-warming-wont-go-viral (accessed on 14 April 2018).

136. Leach, M.; Stirling, A.; Scoones, I. *Dynamic Sustainabilities: Technology, Environment, Social Justice*; Earthscan: Abingdon, UK; New York, NY, USA, 2010.

137. Van Opstal, M.; Hugé, J. Knowledge for sustainable development: A worldviews perspective. *Environ. Dev. Sustain.* **2013**, *15*, 687–709. [CrossRef]

138. Breslin, D. Towards a generalized Darwinist view of sustainability. In *Beyond Sustainability*; Scholz, C., Zentes, J., Eds.; Nomos: Baden-Baden, Germany, 2014; pp. 13–35.

139. Zwier, J.; Blok, V.; Lemmens, P.; Geerts, R.-J. The ideal of a zero-waste humanity: Philosophical reflections on the demand for a bio-based economy. *J. Agric. Environ. Ethics* **2015**, *28*, 353–374. [CrossRef]

140. Blok, V.; Gremmen, B.; Wesselink, R. Dealing with the wicked problem of sustainability in advance. *Bus. Prof. Ethics J.* **2016**. [CrossRef]
141. Urmetzer, S.; Pyka, A. Varieties of knowledge-based bioeconomies. In *Knowledge-Driven Developments in the Bioeconomy*; Dabbert, S., Lewandowski, I., Weiss, J., Pyka, A., Eds.; Springer: Cham, Switzerland, 2017; pp. 57–82.
142. Hodgson, G.M. The evolution of morality and the end of economic man. *J. Evol. Econ.* **2014**, *24*, 83–106. [CrossRef]
143. Wuketits, F.M. Moral systems as evolutionary systems: Taking evolutionary ethics seriously. *J. Soc. Evol. Syst.* **1993**, *16*, 251–271. [CrossRef]
144. Boyd, R.; Richerson, P.J. *Culture and the Evolutionary Process*; University of Chicago Press: Chicago, IL, USA, 1985.
145. Boyd, R.; Richerson, P.J. The evolution of norms: An anthropological view. *J. Inst. Theor. Econ.* **1994**, *150*, 72–87.
146. Boyd, R.; Richerson, P.J. *The Origin and Evolution of Cultures*; Oxford University Press: Oxford, UK; New York, NY, USA, 2005.
147. Richerson, P.J.; Boyd, R. *Not by Genes Alone: How Culture Transformed Human Evolution*; University of Chicago Press: Chicago, IL, USA, 2005.
148. Ayala, F.J. The biological roots of morality. *Biol. Philos.* **1987**, *2*, 235–252. [CrossRef]
149. Waring, T.M.; Kline, M.A.; Brooks, J.S.; Goff, S.H.; Gowdy, J.; Janssen, M.A.; Smaldino, P.E.; Jacquet, J. A multilevel evolutionary framework for sustainability analysis. *E&S* **2015**, *20*. [CrossRef]
150. Markey-Towler, B. The competition and evolution of ideas in the public sphere: A new foundation for institutional theory. *J. Inst. Econ.* **2018**, *43*, 1–22. [CrossRef]
151. Almudi, I.; Fatas-Villafranca, F.; Izquierdo, L.R.; Potts, J. The economics of utopia: A co-evolutionary model of ideas, citizenship and socio-political change. *J. Evol. Econ.* **2017**, *27*, 629–662. [CrossRef]
152. Johnson, B. Institutional learning. In *National Systems of Innovation: Toward a Theory of Innovation and Interactive Learning*; Lundvall, B.-Å., Ed.; Anthem Press: London, UK, 2010; pp. 23–45.
153. Göpel, M. *The Great Mindshift*; Springer International Publishing: Cham, Switzerland, 2016.
154. Schaper-Rinkel, P. Bio-politische Ökonomie: Zur Zukunft des Regierens von Biotechnologien. In *Bioökonomie: Die Lebenswissenschaften und die Bewirtschaftung der Körper*; Lettow, S., Ed.; Transcript: Bielefeld, Germany, 2012; pp. 155–179.
155. Banks, J.A. The canon debate, knowledge construction, and multicultural education. *Educ. Res.* **1993**, *22*, 4–14. [CrossRef]
156. Sterling, S. Transformative learning and sustainability: Sketching the conceptual ground. *Learn. Teach. High. Educ.* **2011**, 17–33.
157. Mezirow, J. *Transformative Dimensions of Adult Learning*; Jossey-Bass: San Francisco, CA, USA, 1991.
158. Dirkx, J.M. Transformative learning theory in the practice of adult education: An overview. *PAACE J. Lifelong Learn.* **1998**, *7*, 1–14.
159. Focus. Kretschmann kauft sich neues Dieselauto—Fahrverbote für alte Diesel bleiben. *Focus.* 21 May 2017. Available online: https://www.focus.de/auto/news/abgas-skandal/mache-was-ich-fuer-richtig-halte-gruener-ministerpraesidentin-kauft-sich-neues-dieselauto-fahrverbote-fuer-alte-diesel-bleiben_id_7160275.html (accessed on 7 April 2018).
160. Brewer, J. Tools for culture design: Toward a science of social change? *Spanda J.* **2015**, *6*, 67–73.
161. Wilson, D.S. Intentional cultural change. *Curr. Opin. Psychol.* **2016**, *8*, 190–193. [CrossRef] [PubMed]
162. Wilson, D.S.; Hayes, S.C.; Biglan, A.; Embry, D.D. Evolving the future: Toward a science of intentional change. *Behav. Brain Sci.* **2014**, *37*, 395–416. [CrossRef] [PubMed]
163. Biglan, A.; Barnes-Holmes, Y. Acting in light of the future: How do future-oriented cultural practices evolve and how can we accelerate their evolution? *J. Context. Behav. Sci.* **2015**, *4*, 184–195. [CrossRef] [PubMed]
164. Ramcilovic-Suominen, S.; Pülzl, H. Sustainable development—A 'selling point' of the emerging EU bioeconomy policy framework? *J. Clean. Prod.* **2018**, *172*, 4170–4180. [CrossRef]
165. Schmid, O.; Padel, S.; Levidov, L. The bio-economy concept and knowledge base in a public goods and farmer perspective. *Bio-Based Appl. Econ.* **2012**, *1*, 47–63.
166. Hilgartner, S. Making the bioeconomy measurable: Politics of an emerging anticipatory machinery. *BioSocieties* **2007**, *2*, 382–386. [CrossRef]

167. Birch, K.; Levidow, L.; Papaioannou, T. Sustainable capital? The neoliberalization of nature and knowledge in the European "knowledge-based bio-economy". *Sustainability* **2010**, *2*, 2898–2918. [CrossRef]

168. McCormick, K.; Kautto, N. The bioeconomy in Europe: An overview. *Sustainability* **2013**, *5*, 2589–2608. [CrossRef]

169. Fatheuer, T.; Fuhr, L.; Unmüßig, B. *Kritik der Grünen Ökonomie*; Oekom Verlag: München, Germany, 2015.

170. Albrecht, S.; Gottschick, M.; Schorling, M.; Stirn, S. Bio-Ökonomie—Gesellschaftliche Transformation ohne Verständigung über Ziele und Wege? Biogum Univ: Hamburg, Germany, 2012.

171. Raghu, S.; Spencer, J.L.; Davis, A.S.; Wiedenmann, R.N. Ecological considerations in the sustainable development of terrestrial biofuel crops. *Curr. Opin. Environ. Sustain.* **2011**, *3*, 15–23. [CrossRef]

172. ten Bos, R.; van Dam, J.E.G. Sustainability, polysaccharide science, and bio-economy. *Carbohydr. Polym.* **2013**, *93*, 3–8. [CrossRef] [PubMed]

173. Schütte, G. What kind of innovation policy does the bioeconomy need? *New Biotechnol.* **2018**, *40*, 82–86. [CrossRef] [PubMed]

174. European Commission. *Innovating for Sustainable Growth: A Bioeconomy for Europe*; European Commission: Brussels, Belgium, 2012.

175. European Commission. *Review of the 2012 European Bioeconomy Strategy*; European Commission: Brussels, Belgium, 2017.

176. The European Bioeconomy Stakeholders Panel. European Bioeconomy Stakeholders Manifesto. Available online: https://ec.europa.eu/research/bioeconomy/pdf/european_bioeconomy_stakeholders_manifesto.pdf (accessed on 16 April 2018).

177. Meyer, M. The rise of the knowledge broker. *Sci. Commun.* **2010**, *32*, 118–127. [CrossRef]

178. Castells, M. *The Rise of the Network Society*, 2nd ed.; Wiley-Blackwell: Chichester, UK, 2010.

179. Castells, M. *The Power of Identity*, 2nd ed.; Wiley-Blackwell: Chichester, UK, 2010.

180. Castells, M. *End of Millennium*, 2nd ed.; Wiley-Blackwell: Chichester, UK, 2010.

181. Copagnon, D.; Chan, S.; Mert, A. The changing role of the state. In *Global Environmental Governance Reconsidered*; Biermann, F., Pattberg, P., Eds.; MIT Press: Cambridge, MA, USA, 2012; pp. 237–264.

182. Wyborn, C.A. Connecting knowledge with action through coproductive capacities: Adaptive governance and connectivity conservation. *E&S* **2015**, *20*. [CrossRef]

183. Folke, C.; Hahn, T.; Olsson, P.; Norberg, J. Adaptive governance of social-ecological systems. *Annu. Rev. Environ. Resour.* **2005**, *30*, 441–473. [CrossRef]

184. Boyd, E.; Folke, C. (Eds.) *Adapting Institutions: Governance, Complexity and Social-Ecological Resilience*; Cambridge University Press: Cambridge, UK, 2012.

185. Voß, J.-P.; Bauknecht, D.; Kemp, R. (Eds.) *Reflexive Governance for Sustainable Development*; Edward Elgar: Cheltenham, UK, 2006.

186. Biermann, F. *Earth System Governance: World Politics in the Anthropocene*; The MIT Press: Cambridge, MA, USA, 2014.

187. von Schomberg, R. A vision of responsible research and innovation. In *Responsible Innovation: Managing the Responsible Emergence of Science and Innovation in Society*; Owen, R., Bessant, J.R., Heintz, M., Eds.; John Wiley & Sons Inc.: Chichester, UK, 2013; pp. 51–74.

188. Scoones, I.; Leach, M.; Newell, P. (Eds.) *The Politics of Green Transformations*; Earthscan: London, UK, 2015.

189. Milkoreit, M. *Mindmade Politics: The Cognitive Roots of International Climate Governance*; The MIT Press: Cambridge, MA, USA, 2017.

190. Bugge, M.M.; Coenen, L.; Branstad, A. Governing socio-technical change: Orchestrating demand for assisted living in ageing societies. *Sci. Public Policy* **2018**, *38*, 1235. [CrossRef]

191. Evans, J.; Jones, R.; Karvonen, A.; Millard, L.; Wendler, J. Living labs and co-production: University campuses as platforms for sustainability science. *Curr. Opin. Environ. Sustain.* **2015**, *16*, 1–6. [CrossRef]

192. Frantzeskaki, N.; Kabisch, N. Designing a knowledge co-production operating space for urban environmental governance—Lessons from Rotterdam, Netherlands and Berlin, Germany. *Environ. Sci. Policy* **2016**, *62*, 90–98. [CrossRef]

193. Kahane, A. *Transformative Scenario Planning: Working Together to Change the Future*; Berrett-Koehler Publishers: San Francisco, CA, USA, 2012.

194. Luederitz, C.; Schäpke, N.; Wiek, A.; Lang, D.J.; Bergmann, M.; Bos, J.J.; Burch, S.; Davies, A.; Evans, J.; König, A.; et al. Learning through evaluation—A tentative evaluative scheme for sustainability transition experiments. *J. Clean. Prod.* **2017**, *169*, 61–76. [CrossRef]

195. Mauser, W.; Klepper, G.; Rice, M.; Schmalzbauer, B.S.; Hackmann, H.; Leemans, R.; Moore, H. Transdisciplinary global change research: The co-creation of knowledge for sustainability. *Curr. Opin. Environ. Sustain.* **2013**, *5*, 420–431. [CrossRef]

196. Moser, S.C. Can science on transformation transform science? Lessons from co-design. *Curr. Opin. Environ. Sustain.* **2016**, *20*, 106–115. [CrossRef]

197. Wiek, A. Challenges of transdisciplinary research as interactive knowledge generation: Experiences from transdisciplinary case study research. *GAIA* **2007**, *16*, 52–57. [CrossRef]

198. Wiek, A.; Ness, B.; Schweizer-Ries, P.; Brand, F.S.; Farioli, F. From complex systems analysis to transformational change: A comparative appraisal of sustainability science projects. *Sustain. Sci.* **2012**, *7*, 5–24. [CrossRef]

199. Albrecht, S.; Gottschick, M.; Schorling, M.; Stirn, S. Bioökonomie am Scheideweg: Industrialisierung von Biomasse oder nachhaltige Produktion? *GAIA* **2012**, *21*, 33–37. [CrossRef]

200. Costanza, R. How do cultures evolve, and can we direct that change to create a better world? *Wildl. Aust.* **2016**, *53*, 46–47.

201. Mesoudi, A. Cultural evolution: Integrating psychology, evolution and culture. *Curr. Opin. Psychol.* **2016**, *7*, 17–22. [CrossRef]

202. Mesoudi, A. Cultural evolution: A review of theory, findings and controversies. *Evol. Biol.* **2016**, *43*, 481–497. [CrossRef]

203. Mesoudi, A. Pursuing Darwin's curious parallel: Prospects for a science of cultural evolution. *Proc. Natl. Acad. Sci. USA* **2017**. [CrossRef] [PubMed]

204. Voß, J.-P.; Kemp, R. Sustainability and reflexive governance: Introduction. In *Reflexive Governance for Sustainable Development*; Voß, J.-P., Bauknecht, D., Kemp, R., Eds.; Edward Elgar: Cheltenham, UK, 2006; pp. 3–28.

205. Fagerberg, J. Mission (Im)possible? The Role of Innovation (and Innovation Policy) in Supporting Structural Change & Sustainability Transitions; Centre for Technology, Innovation and Culture, University of Oslo, Oslo, Norway, 2017. Available online: https://www.sv.uio.no/tik/InnoWP/tik_working_paper_20180216.pdf (accessed on 16 April 2018).

206. Bioökonomierat. *Empfehlungen des Bioökonomierates: Weiterentwicklung der "Nationalen Forschungsstrategie Bioökonomie 2030"*; Bioökonomierat: Berlin, Germany, 2016.

207. BMEL. *Fortschrittsbericht zur Nationalen Politikstrategie Bioökonomie*; Bundesministerium für Ernährung und Landwirtschaft: Berlin, Germany, 2016.

208. Imbert, E.; Ladu, L.; Morone, P.; Quitzow, R. Comparing policy strategies for a transition to a bioeconomy in Europe: The case of Italy and Germany. *Energy Res. Soc. Sci.* **2017**, *33*, 70–81. [CrossRef]

209. Abson, D.J.; Fischer, J.; Leventon, J.; Newig, J.; Schomerus, T.; Vilsmaier, U.; von Wehrden, H.; Abernethy, P.; Ives, C.D.; Jager, N.W.; et al. Leverage points for sustainability transformation. *Ambio* **2017**, *46*, 30–39. [CrossRef] [PubMed]

sustainability

Article

Socioeconomic Indicators to Monitor the EU's Bioeconomy in Transition

Tévécia Ronzon * and Robert M'Barek

European Commission, Joint Research Centre (JRC), Directorate for Sustainable Resources,
Economics of Agriculture Unit, Edificio EXPO, C/ Inca Garcilaso 3, 41092 Seville, Spain;
Robert.M'BAREK@ec.europa.eu
* Correspondence: tevecia.ronzon@ec.europa.eu

Received: 16 April 2018; Accepted: 22 May 2018; Published: 26 May 2018

Abstract: The monitoring of the European bioeconomy is hampered by a lack of statistics on emergent and partially bio-based sectors. In this study, we complete the picture of the bioeconomy in the European Union (EU) by first estimating a set of socioeconomic indicators in missing sectors. Second, we identify four broad bioeconomy patterns within the EU that differ according to the specialisation of Member States' labour markets in the bioeconomy (location quotient) and according to the apparent labour productivity of their bioeconomies. The patterns are geographically distributed in (i) Eastern Member States and Greece and Portugal; (ii) Central and Baltic Member States; (iii) Western Member States; and (iv) Northern Member States. They are strongly related to the level of gross domestic product (GDP) per capita in Member States, and to their political histories (e.g., their year of accession to the EU, and the existence and maturity of their bioeconomy strategies). Within each group, diversity exists in terms of sectoral bioeconomy development. Third, we examine temporal dynamics over the period 2008–2015, stressing with the cases of Slovenia, Portugal, Greece and Finland that a transition from one group to another is possible. Finally, we take a closer look at the East–West bioeconomy disparities within Europe and suggest measures to promote EU bioeconomies.

Keywords: bioeconomy; EU Member States; apparent labour productivity; jobs and growth; typology

1. Introduction

Europe's bioeconomy encompasses the production of renewable biological resources and the conversion of these resources and waste streams into value-added products such as food, feed, bio-based products and bioenergy [1]. As a core principle, the European Bioeconomy Strategy aims to balance social, environmental and economic gains by linking the sustainable use of renewable resources with the protection and restoration of biodiversity, ecosystems and natural capital across land and water [2].

The 2017 review of the Bioeconomy Strategy created a major opportunity for a new political impetus and orientation [3]. The roadmap 'Update of the 2012 Bioeconomy Strategy,' published in early 2018, reinforced the main purpose of the strategy and provided an updated plan for concrete actions [2]. These actions relate to research and innovation, including education, special attention to the development of bio-based markets, the minimisation of harmful impacts and maximisation of co-benefits of the bioeconomy, and better exploitation of the potential in EU Member States and their regions [2].

Both the review and the roadmap stress the need to assess progress through better monitoring and assessment frameworks, in particular providing SMART (specific, measurable, attainable, relevant and timely) indicators across relevant sectors.

In this article we present the most recent socioeconomic key numbers, discuss and propose additional indicators, and illustrate the use of further derived indicators for the analysis of EU

Member States' bioeconomies in transition, as a contribution to the upcoming set-up of monitoring and assessment frameworks. The paper gives some insights into the situation of individual countries, as such contemplating also potentially existing path dependencies.

The majority of bioeconomy policy documents [1–3], but also scientific publications [4], describe the bioeconomy in terms of indicators, turnover and jobs as calculated by the Joint Research Centre (JRC) and the nova-Institute for Ecology and Innovation (www.nova-institute.eu) [5–7].

According to our most recent estimations, the EU-28 bioeconomy created 18 million full-time jobs and generated €2.3 trillion of turnover in 2015. To date, turnover has been the main economic indicator used to quantify the bioeconomy and its different sectors. In this article we argue that value added should receive more attention, primarily because it avoids double-counting and provides the additional value to the whole economy created by a sector. Furthermore, it is more in line with EU Member State calculations and allows comparisons with national accounts. Expressed in value added, the bioeconomy amounted to €620 billion in Europe in 2015. In particular, sectors with high inputs reduce its overall share in the bioeconomy.

In a second step, with the objective of analysing the state and potential pathways of individual EU Member States in the bioeconomy, this paper puts forward the use of the sectoral apparent labour productivity (or value added per person employed) and the so-called location quotient (see also [5]). Addressing economic performance through a productivity measure gives further insights into the growth potential of specific bioeconomy sectors in individual EU Member States, which is of particular importance from the perspective of the EU and its various policies that have implications for territorial coherence.

Accordingly, this paper proposes a simplified socioeconomic indicator framework, the positioning of EU Member States on a transition path to higher productivity and, finally, a grouping of Member States.

It should be stressed that this approach does not constitute a holistic assessment framework, in particular because it omits the environmental dimension of sustainability. Nonetheless, it gives insights into the state and the transition of EU Member States in this matrix of key economic and social indicators, and thus provides an initial contribution to the monitoring of the bioeconomy in the EU.

The article is structured as follows. Following the introduction, the second section describes the materials and methods, in particular the way the indicators are calculated. The third section provides the results, starting with an overview of the socioeconomic indicators and followed by a detailed analysis of the clustering of EU Member States. The fourth section discusses the choice of value added as the main economic indicator and then looks into the untapped potential and bioeconomy-related strategies for Eastern EU Member States.

2. Materials and Methods

2.1. Scope of the 'European Bioeconomy'

This paper follows the official definition of the bioeconomy as published in communication COM (2012) 60 of the European Commission mentioned in the introduction [1]. Briefly put, the bioeconomy incorporates all the economic activities related to the production and manufacturing of biomass. According to the official statistical classification of economic activities of the European Community (NACE rev. 2) [8], these economic activities correspond to the list of sectors presented in

Table 1. Note that the NACE classification does not differentiate bio-based and non-bio-based activities. This is the case, for instance, in the manufacture of textiles that can use biomass as a feedstock (cotton, ,wool, silk, etc.), or synthetic fibres, or both. Sectors making use of biomass and other kinds of feedstock are called 'hybrid' sectors. They are marked with a * in Table 1.

Table 1. NACE sectors considered part of the bioeconomy.

NACE Code	Bioeconomy Sector (Parent Categories in Bold)
A01	**Agriculture**
A02	**Forestry**
A03	**Fishing and aquaculture**
A032	Aquaculture
A031	Fishing
-	**Manufacture of food, beverages and tobacco**
C10	Manufacture of food
C11	Manufacture of beverages
C12	Manufacture of tobacco
-	**Manufacture of bio-based textiles**
C13 *	Manufacture of bio-based textiles
C14 *	Manufacture of bio-based wearing apparel
C15	Manufacture of leather
-	**Manufacture of wood products and furniture**
C16	Manufacture of wood products
C31 *	Manufacture of wooden furniture
C17	**Manufacture of paper**
-	**Manufacture of bio-based chemicals, pharmaceuticals, plastics and rubber (excluding biofuels)**
C20 *	Manufacture of bio-based chemicals (excluding biofuels)
C21 *	Manufacture of bio-based pharmaceuticals
C22 *	Manufacture of bio-based plastics and rubber
-	**Manufacture of liquid biofuels**
C2014 *	Manufacture of bioethanol
C2059 *	Manufacture of biodiesel
D3511 *	**Production of bioelectricity**

* hybrid sector.

2.2. Determination of the Bio-Based Proportion of Hybrid Sectors

The extent to which a given hybrid sector is bio-based is determined following the approach set out by Ronzon et al. [5]: experts estimate the proportion of biomass incorporated in each product produced by the hybrid sector; and, at sector level, the proportion of biomass incorporated in all products from this sector makes up the sectoral bio-based share.

In the present study, the quantification of sectoral bio-based shares strictly follows the same methodology as described by Ronzon et al. [5], except for the case of the manufacture of bio-based chemicals and pharmaceuticals (NACE rev. 2 sectors C20 and C21). In order to track possible developments, the nova-Institute for Ecology and Innovation has updated the product bio-based shares of these two sectors after a new round of expert interviews in which experts have worked on the statistical classification of products by category (CPA) list of products belonging to sectors C20 and C21. Consequently, the sectoral bio-based shares of the two aforementioned sectors were estimated using the production value from the EUROSTAT-Prodcom dataset (see Equation (1)).

$$\text{BBS}_{i,k,l} = \frac{\sum_{j=1}^{n} \text{bbs}_j \times \text{Production value}_{j,k,l}}{\sum_{j=1}^{n} \text{Production value}_{j,k,l}}, \tag{1}$$

where:

- $\text{BBS}_{i,k,l}$ is the bio-based share of sector i (NACE Rev. 2), in EU Member State k and for year l;
- bbs_j is the bio-based share of product j, given that sector i manufactures $j = n$ products. Bio-based shares vary from 0 for products that do not incorporate biomass (e.g., Prodcom code 20.12.23.30, Synthetic organic tanning substances) to 1 for those that are made entirely of biomass (e.g., Prodcom code 20.12.22.50, Tanning extracts of vegetable origin);
- Production value$_{j,k,l}$ is the production value of product j, by EU Member State k and for year l.

Finally, the sectoral bio-based share for the production of bio-electricity is derived from the EUROSTAT—energy balances, using the nrg_105a dataset, which decomposes the gross electricity generation by source. As for the precedent sectoral bio-based shares, it is calculated per Member State and year and then applied to EUROSTAT—Structural Business Statistics data (sbs_na_ind_r2).

2.3. Calculation of Monitoring Indicators

This study is based on the analysis of the number of persons employed in bioeconomy sectors, their turnover, value added and derived indicators (see details in official definitions [9]):

- The number of people employed (code V16110 in EUROSTAT—Structural Business Statistics) is the total number of persons who work in the observation unit, as well as persons who work outside the unit who belong to it and are paid by it.
- The turnover (code V12110 in EUROSTAT—Structural Business Statistics) comprises the totals invoiced by the observation unit.
- The value added at factor cost (code V12150 in EUROSTAT—Structural Business Statistics) is the gross income from operating activities after adjusting for operating subsidies and indirect taxes.

In the case of fully bio-based sectors, data for these indicators are retrieved from different EUROSTAT datasets and from Scientific, Technical and Economic Committee for Fisheries (STECF) reports (see Table 2).

Table 2. Data source for fully bio-based sectors.

Bioeconomy Sector	NACE Code Used for Calculations	Number of Persons Employed	Turnover	Value Added
Agriculture	A01	EUROSTAT—Labour Force Survey (lfsa_egan22d)	EUROSTAT—Economic accounts for agriculture (aact_eaa01)	EUROSTAT—National accounts (nama_10_a64)
Forestry	A02	EUROSTAT—Forestry Employment (for_emp_lfs)	EUROSTAT—Forestry economic accounts (for_eco_cp)	
Fishing	A03	STECF 2014	STECF 2016	
Manufacturing sectors	C10; C11; C12; C15; C16; C17	EUROSTAT—Structural Business Statistics (sbs_na_ind_r2)		

Sources: [10–12].

Indicators are estimated for hybrid sectors by applying their sectoral bio-based share to EUROSTAT—Structural Business Statistics data (sbs_na_ind_r2; see Ronzon et al. for more details [5]).

The derived indicators used in this study include apparent labour productivity (code V91110 in EUROSTAT—Structural Business Statistics) and location quotient. Apparent labour productivity refers to the ratio of value added to persons employed. Location quotient refers to the proportion of persons employed in a particular sector and in a given Member State compared with the European proportion (see Equation (2)). A location quotient of sector i in Member State k greater than 1 means the labour market of Member State k is more 'concentrated' in sector i than the EU-28 labour market.

$$LQ_{i,k,l} = \frac{\% \text{ people employed}_{i,k,l}}{\% \text{ people employed}_{i,EU28,l}}, \tag{2}$$

where:

- $LQ_{i,k,l}$ is the location quotient of sector i (NACE Rev. 2), in EU Member State k and for year l;
- % people employed$_{i,k,l}$ is the proportion of people employed in sector i (the bioeconomy or a NACE Rev. 2 sector), in EU Member State k and for year l; and
- % people employed$_{i,EU-28,l}$ is the proportion of people employed in sector i (the bioeconomy or a NACE Rev. 2 sector), in the EU-28 and for year l.

For example, the bioeconomy location quotient of 3.8 in Romania means that in the Romanian labour market the proportion of persons employed in bioeconomy sectors is nearly four times higher than the proportion of bioeconomy workers on the EU28 labour market.

Data are compiled within the JRC-Bioeconomics dataset and they can be gathered at https://datam.jrc.ec.europa.eu/datam/perm/od/7d7d5481-2d02-4b36-8e79-697b04fa4278 (see also the QR codes in the Supplementary Materials).

3. Results

3.1. Key Socioeconomic Indicators

According to our estimations, the EU-28 bioeconomy employed 18 million people and generated €2.3 trillion of turnover or €620 million of value added in 2015. In other words, this sector employed 8.2% of the EU-28 labour force and generated 4.2% of the EU-28 GDP. Agriculture and the manufacture of food, beverages and tobacco accounted for about two thirds of the value added and turnover of the bioeconomy and three quarters of bioeconomy employment. These sectors generated €174 billion and €233 billion of value added respectively in 2015.

Even though these two sectors dominate the bioeconomy when measured with the three indicators—employment, turnover and value added—sectoral contributions vary according to the degree of labour intensiveness of the sector (see Table 3 and Appendix A Table A1). Agriculture, being a low labour productive sector, employs 51% of bioeconomy workers but generates only 28% of the bioeconomy value added. The other sectors tend to increase their contribution to the bioeconomy when measured in value terms rather than employment terms. For example, the manufacture of food, beverages and tobacco employs only 25% of the bioeconomy workers, but generates 37% of the bioeconomy value added. The manufacture of bio-based chemicals, pharmaceuticals, plastics and rubber (excluding biofuels) employs fewer than 3% of the workers in the bioeconomy, but generates more than 9% of its value added.

Table 3. Contribution of bioeconomy sectors to the total bioeconomy labour market, turnover and value added (%), EU-28, 2015.

Sector	Workers	Turnover	Value Added
Agriculture	51.0	16.8	28.0
Forestry	3.0	2.2	3.8
Fishing	1.2	0.5	1.1
Manufacture of food, beverages and tobacco	25.1	51.0	37.6
Manufacture of bio-based textiles	5.6	4.6	4.6
Manufacture of wood products and furniture	7.8	7.7	7.6
Manufacture of paper	3.6	8.3	7.3
Manufacture of bio-based chemicals, pharmaceuticals, plastics and rubber (excluding biofuels)	2.5	7.8	9.1
Manufacture of liquid biofuels	0.1	0.5	0.4
Production of bioelectricity	0.1	0.5	0.5

Differences in sectoral contribution to the bioeconomy turnover versus value added arise from differences in cost structure. Costs of bought-in goods and services are relatively higher in the manufacture of food, beverages and tobacco than in agriculture. As a result, its proportion of the bioeconomy's turnover (51%) is far larger than its proportion of value added (38%), while the opposite is true in agriculture (17% of turnover versus 28% of value added).

Over time, bioeconomy employment tends to reduce while value added increases. The number of people working in the bioeconomy was 2.5 million fewer in 2015 than in 2008, mainly because of the ongoing restructuring of the agricultural sector, which lost 1.5 million people during the same period (which equates to 63% of the reduction in jobs in the EU-28 bioeconomy). In contrast, the value added generated by the bioeconomy has increased by €45 million and the apparent labour productivity has also improved from €28,000 of value added per person employed in 2008 to €34,400 in 2015.

We can observe five levels of sectoral apparent labour productivity in the EU bioeconomy, which more or less follow sectoral levels of capitalisation (see bars on Figure 1): (i) €19,000 of value added per person employed reached in agriculture; (ii) around €30,000 of value added per person employed in the manufacture of bio-based textiles, in fishing and aquaculture and in the manufacture of wood products and furniture; (iii) between €40,000 and €50,000 per person employed in forestry, the manufacture of bio-plastics and the manufacture of food, beverages and tobacco; (iv) €70,000 per person employed in the manufacture of paper and paper products; and (v) more than €120,000 per person employed in the manufacture of bio-based chemicals and bio-based pharmaceuticals as well as in the production of bioelectricity. Nevertheless, behind these EU averages, sectoral levels of apparent labour productivity show very wide ranges of variation at Member State level (see points on Figure 1).

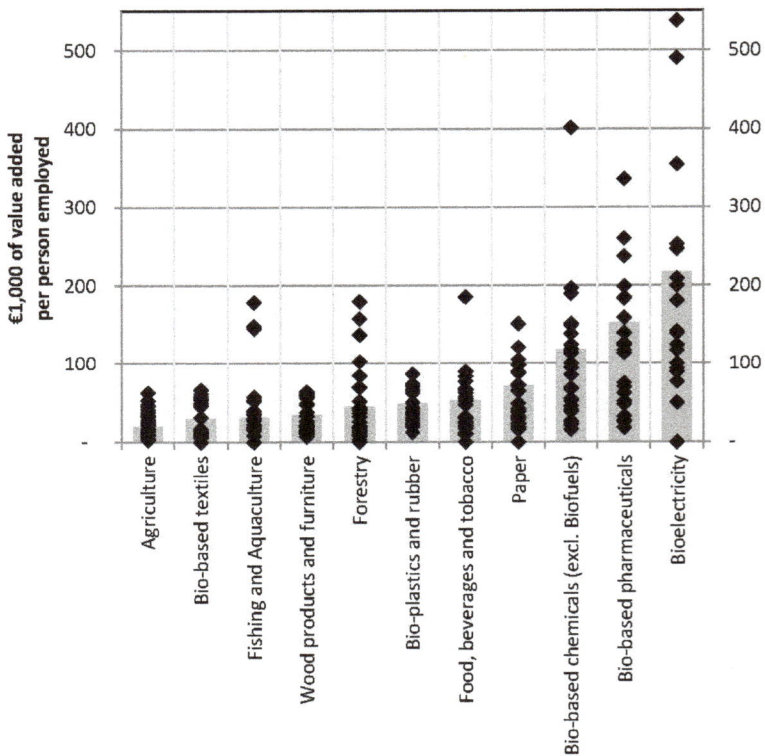

Figure 1. Sectoral apparent labour productivity in the EU-28 bioeconomy, 2015: bars show EU-28 apparent labour productivity; points show Member States' apparent labour productivity. Note that apparent labour productivity in Ireland for the manufacture of bio-based pharmaceutical is out of scale on this graph, reaching 784 k€ per person employed.

3.2. Clustering of EU Member States

The indicators shown in Section 2 illustrate considerable heterogeneity among EU Member States. As a first step to identify the state and potential evolution of Europe's bioeconomy, we propose a clustering or typology for the EU Member States based on the socioeconomic indicators selected. The clustering reduces the complexity of the analysis of 28 countries and can therefore support monitoring and policy decision-making.

Policy initiatives already exist at the supra-national level. The Scandinavian countries of Denmark, Finland and Sweden have founded the Nordic Bioeconomy [13]. The BIOEAST initiative comprises the Central and Eastern European countries of Bulgaria, the Czech Republic, Hungary, Poland, Romania and Slovakia [14]. This initiative argues that an east-west divide exists in the EU, with eastern countries serving only as raw material providers for big companies in the west and having limited access to research. Developing bioeconomy strategies for these countries can contribute to overcome existing or perceived geographical imbalances and better exploit untapped potential.

Literature related to bioeconomy typologies is still in its infancy. Philippidis et al. [15], and Mainar Causapé et al. [16] employed disaggregated social accounting matrices and multipliers to analyse the bioeconomy of the EU. Employing the same tools of analysis, a similar case study was conducted for the Spanish economy (Cardenete et al.) [17]

A typology of European regions according to their bioeconomy profile and approach is proposed by Spatial Foresight et al. [18] with the aim of gathering stakeholders around specific bioeconomy sectors/products.

In this article we focus on a more aggregated and macro-economic analysis, taking into account two criteria for a typology: the concentration of national labour markets into the bioeconomy, and apparent labour productivity of the bioeconomy. The first criterion acts as a proxy for the employment situation, while the second reflects economic growth potential. This approach enables comparisons between countries, insights into the complex interactions of job and growth creation, and, finally, the identification of potential future pathways of countries that exhibit similar dynamic patterns.

3.2.1. Eastern Member States, Portugal and Greece (Group 1.1)

This group is defined by a strong specialisation of national labour markets in the bioeconomy (location quotient higher than 1.6 in 2015) but a level of apparent labour productivity of the bioeconomy below half the EU-28 level (i.e., less than €18,000 of value added per person employed in 2015). It comprises Romania, Greece, Lithuania, Poland, Croatia, Portugal, Latvia and Bulgaria (see Figure 2).

These Member States joined the European Union after 2004 and show the lowest levels of GDP per capita of the EU-28 (below €11,600 per capita), with the exception of Greece and Portugal, which entered the EU in 1981 and 1986 respectively and reached around €17,000 of GDP per capita in 2015 (far below the EU-28 level of €26,600 per capita (Eurostat sdg_08_10, http://ec.europa.eu/eurostat/web/products-datasets/-/sdg_08_10)).

Over half of the bioeconomy labour force in this group is concentrated in biomass-producing sectors (i.e., agriculture, forestry and the fishing sector), which generate 33–63% of the bioeconomy value added. The agriculture sector alone contributes between 38% and 81% of bioeconomy jobs and 23–55% of the value added. The agricultural focus is extremely strong in Romania and Greece, where it provides more than 70% of bioeconomy jobs and around 55% of the bioeconomy value added. In Latvia, Bulgaria and Lithuania, the relatively low contribution of agriculture to bioeconomy jobs (38–48%) is compensated for by a strong contribution of the forestry sector (6–14% versus 3% on average in the EU-28). The contribution of agriculture and forestry to the bioeconomy value added is also higher in this group than the average in the EU-28. The fishing sector contributes nearly 5% of the bioeconomy jobs in Greece, which is not negligible compared with other EU-28 Member States (1.2% on average across the EU-28).

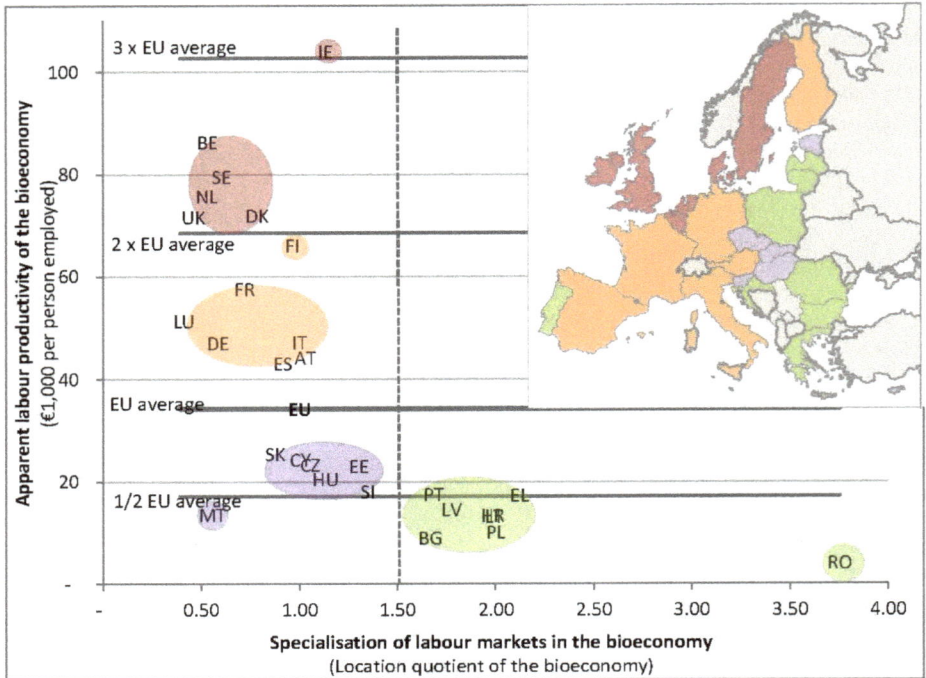

Figure 2. Bioeconomy patterns across EU-28 Member States.

As regards downstream bio-based value chains, the manufacture of food, beverage and tobacco is the second largest employing sector (7–24% of bioeconomy employment), after agriculture, and the second largest contributor to the bioeconomy value added (15–38%). Nevertheless, its relative contribution is below the EU-28 average in both employment and value added terms (see Figure 3). The strong forestry basis of Latvia and Lithuania has triggered the development of the manufacture of wood products and furniture (respectively 22% and 16% of bioeconomy jobs; 27% and 16% of the bioeconomy value added) but not the manufacture of paper sector (which is less than the EU average). The manufacture of bio-based textiles contributes more to the bioeconomy labour market in Portugal (17%), Bulgaria (15%), Croatia (7%) and Lithuania (6%) than on average in the EU (5.5%). Finally, the manufacture of bio-based chemicals, pharmaceuticals, plastic and rubber is under-represented in the bioeconomy labour market (less than 1.7% of bioeconomy employment) compared with the EU average (2.5%) as well as in value added terms (less than 4% of the bioeconomy value added versus 9% in the EU). The case of Portugal is remarkable: it reaches high levels of productivity compared with the other Member States of this group in forestry, the manufacture of wood products and furniture and the manufacture of paper sectors. Thus, these three sectors generate one-quarter of the bioeconomy value added in Portugal while employing only 11% of the bioeconomy labour force.

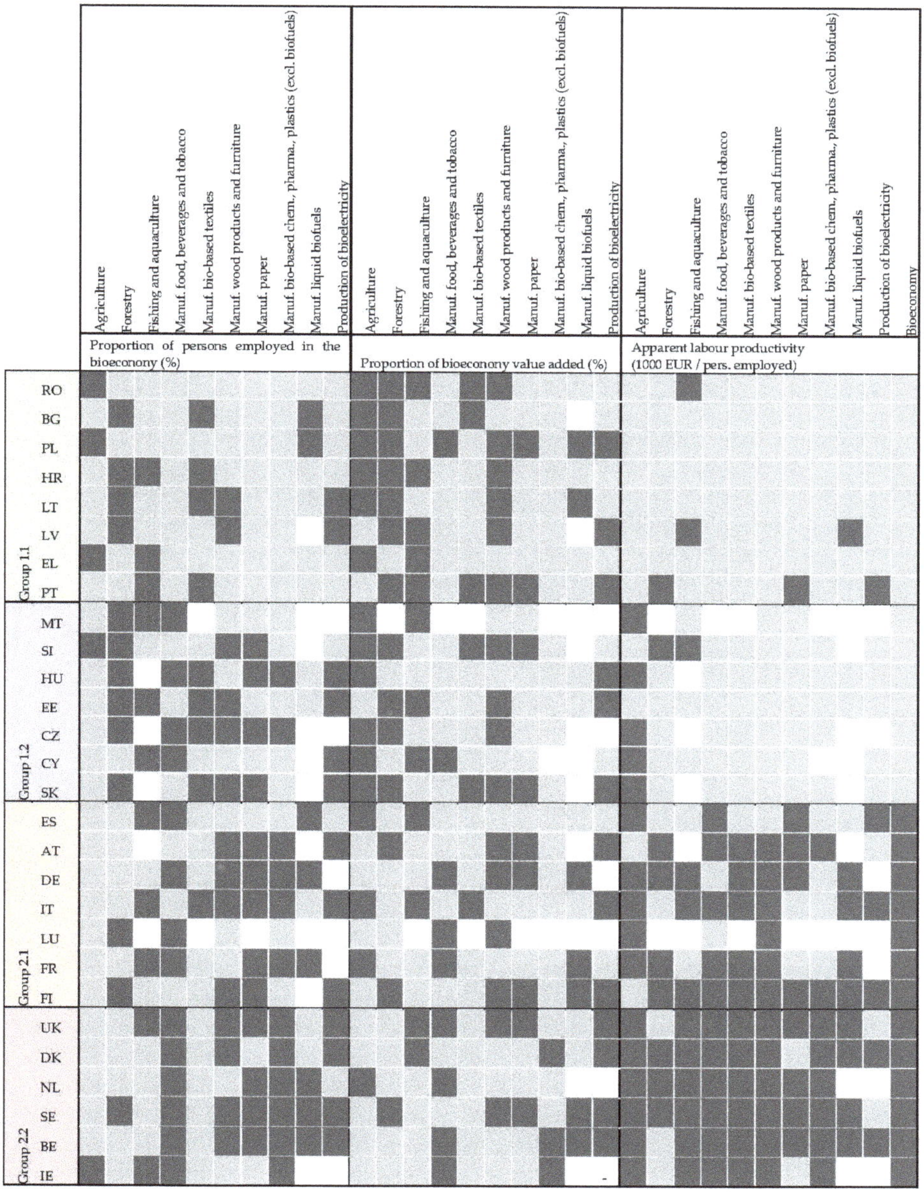

Figure 3. Heat map of the sectoral contribution to bioeconomy jobs and value added in EU Member States, 2015: light grey shows a contribution below the EU average, dark grey shows a contribution above the EU average, white indicates missing data.

In summary, the bioeconomy pattern of this group is geared towards biomass-producing sectors and the manufacture of food, beverage and tobacco sector. Other low-productive manufacturing sectors can play a significant role according to historical sectoral specialisation (e.g., the manufacture of bio-based textiles in Portugal) or to biomass endowment (e.g., the manufacture of wood products and furniture in Latvia and Lithuania). More labour-productive sectors, such as the manufacture of paper and the manufacture of bio-based chemicals, pharmaceuticals, plastics and rubber, are not very developed. The apparent labour productivity in all bioeconomy sectors is lower than the EU-28 average, except in the Portuguese forestry sector. Therefore, there remains a large potential for the development of the bioeconomy in these Member States. This is in line with the research and innovation (R&I) bioeconomy index calculated in 2017. This index refers to the bioeconomy R&I maturity level of a given country or region. It derives from four variables: (i) the innovation capacity and activity, (ii) the existence of a specific bioeconomy strategy, (iii) the existence of bioeconomy-related clusters and (iv) the intensity level of bioeconomy-related activities [18]. In the case of Member States of group 1.1, the index does not exceed 5 on a scale from 1 to 10 in any of these Member States except Portugal (which has a level of 6/10) [18]. It is also important to note that, among these countries, only Latvia has its own bioeconomy strategy, the LIBRA. Drawn up in 2017, the strategy runs until 2030 [19].

3.2.2. Baltic and Central Member States (Group 1.2)

This group is defined by a medium specialisation of national labour markets in the bioeconomy on the EU-28 scale (location quotient from 0.9 to 1.3) and a level of apparent labour productivity of the bioeconomy of between half the EU-28 level and the EU-28 average level (i.e., €18,000 to €26,000 of value added per person employed in 2015). It comprises Slovakia, Cyprus, the Czech Republic, Estonia, Hungary, Malta and Slovenia (see Figure 2). Note that Malta presents a lower location quotient (1.6) and a lower bioeconomy value added per capita (€14,000) than the other Member States in this group. Nevertheless, we classify it in group 1.2 because this group is the closest to Malta's characteristics.

These Member States joined the European Union after 2004. In 2015, their GDP per capita varied between €11,000 and €21,000, below the EU-28 average level of €26,600 per capita (Eurostat sdg_08_10, http://ec.europa.eu/eurostat/web/products-datasets/-/sdg_08_10) but in general higher than levels reached in group 1.1.

Agriculture and the manufacture of food, beverage and tobacco are the main sources of bioeconomy jobs and value added in group 1.2. This group differs from group 1.1 in that the highest levels of apparent labour productivity reached were in the agricultural sector. As a result, agriculture in group 1.2 contributes a similar proportion to the whole bioeconomy value added (18–48% according to the Member States) while employing a lower proportion of bioeconomy workers (24–46%) than in group 1.1. The very low proportion of agricultural jobs within the bioeconomy labour market reflects a sectoral development towards other biomass-producing sectors, in which the Member States are better endowed (e.g., in Malta the fishing sector provides 19% of bioeconomy jobs and in Estonia, Slovakia and the Czech Republic forestry provides 8–11% of bioeconomy jobs). The manufacture of food, beverage and tobacco contributes to the whole bioeconomy value added in a similar proportion to the Member States of group 1.1 (17–30%) and has an apparent labour productivity below the EU-28 average level (see Figure 3). However, the Czech Republic is an exception, with 47% of the bioeconomy value added generated by the manufacture of food, beverage and tobacco. The manufacture of food, beverage and tobacco is the top employing sector of the Maltese and Czech bioeconomies and the second in the other group 1.2 Member States (except Estonia).

The manufacture of wood and furniture is another important manufacturing sector of this group, in particular in Estonia, where it represents the main source of bioeconomy jobs (31%) and value added (34%). It is the third source of bioeconomy jobs in the other Member States of this group (except in Malta where the fishing sector contributes to 19% of bioeconomy jobs) and it generates 11% and 15% of the bioeconomy value added in the Czech Republic and Slovenia, respectively. The paper industry contributes 7–9% to the bioeconomy value added in these countries. Finally, the most labour productive sectors of the bioeconomy—i.e., the manufacture of bio-based chemicals, pharmaceuticals, plastics and rubber—are less developed than the average in the EU, Hungary's and Czech's manufacture of bio-based pharmaceuticals being an exception (see Figure 3).

This group of Member States could illustrate the initial stage of a bioeconomy transition characterised by the intermediate levels of apparent labour productivity achieved in low productive sectors. Intermediate levels of productivity are indeed observed in agriculture and forestry. Labour productivity remains low in the other bioeconomy sectors. Therefore, there is still potential for (i) improving the apparent labour productivity in bio-based manufacturing sectors and (ii) developing the bio-based industry in general. R&I bioeconomy indices confirm a low to medium maturity in bioeconomy R&I in this group of Member States (the index ranges from 1/10 to 6/10). The transition could be assisted in future years by national R&I orientations: the biochemical and biopharmaceutical sectors are a R&I priority in Slovenia, the Czech Republic and Slovakia; the biochemical sector alone is a priority in Estonia; and Malta and Slovenia focus their bioeconomy R&I on biorefineries [18]. Hungary has a 2011–2020 'National Environmental Technology Innovation Strategy' in place [20] that addresses some sectors of the bioeconomy: renewable energy, agriculture and soil protection, the construction industry, and waste. R&I priorities in these Member States demonstrate efforts to develop the bio-based sectors.

3.2.3. Western Member further States (Group 2.1)

This group is defined by a low-to-medium specialisation of national labour markets into the bioeconomy on the EU-28 scale (location quotient from 0.4 to 1) and a level of apparent labour productivity of the bioeconomy between the EU-28 level and twice the EU-28 average level (i.e., €43,000 to €67,000 of value added per person employed in 2015). It comprises Austria, Spain, Italy, Germany, Luxembourg, France and Finland (see Figure 2). However, the classification of Finland in group 2.1 ('Western Member States') might be temporary, as it is close to meeting all the characteristics of the bioeconomy in group 2.2 ('Northern Member States'); See Section 3.3). Note that Luxembourg will not be included in the following comments because of a lack of data for many sectors, which distorts the overall picture. Indeed, data for Luxembourg are available for only a few sectors: agriculture, the manufacture of food, beverages and tobacco, and the manufacture of wood products.

This group includes Member States that joined the EU between 1952 and 1995. They reach a medium level of GDP per capita (€23,100 to €36,200, Luxembourg being an outlier with €81,300 of GDP per capita).

Although agriculture and the food industry dominate the bioeconomy in this group of Member States, these states tend to be more diversified across sectors than those in group 1.2. A slightly higher proportion of bioeconomy workers is employed in the manufacture of paper and bio-based chemicals and pharmaceuticals. Higher apparent labour productivity levels are reached in all the bioeconomy sectors (see Figure 3), so high labour productive sectors contribute even more to the total bioeconomy value added. The bioeconomy profiles of France and Spain are squeezed into the agriculture and food industry sectors, which generate more than the three quarters of their bioeconomy value added. In contrast, the contribution of agriculture to the bioeconomy value added is low in Germany (16%) and Austria (17%). The food industry sector in these countries' bioeconomies is pivotal (respectively 44% and 36% in value added) but they have also developed woody biomass value chains (respectively 25% and 36% of the bioeconomy value added from forestry and the manufacture of wood and paper) and the manufacture of bio-based chemicals, pharmaceuticals, plastics and rubber (12% for Germany and 7% for Austria).

Within this group, Finland presents a special profile with (i) the lowest representation of agriculture and the manufacture of food, beverages and tobacco in its bioeconomy (in terms of both jobs and value added) and (ii) among the highest levels of labour productivity in the EU-28 in forestry and the manufacture of wood and paper. Consequently, woody biomass value chains provide 61% of the bioeconomy value added in Finland (versus 10–36% in the other Member States of this group).

In more general terms, the higher sectoral diversification and productivity in this group suggest a higher maturity of the bioeconomy manufacturing sectors than in groups 1.1 and 1.2. Biorefineries and biochemicals are among the R&I priorities of all Member States of this group (except Luxembourg) as are biopharmaceuticals for all Member States except Italy [18]. In addition, most of these Member States have developed a national bioeconomy strategy, which reflects a certain political support to this sector. Indeed, national bioeconomy strategies were already in place in 2014 in Finland [21] and Germany [22]. National strategies were released more recently, in 2016 and 2017, in Mediterranean Member States: Spain [23], France [24] and Italy [25]. A bioeconomy policy paper is under development in Austria. Luxembourg is the only Member State of this group with no strong strategic orientation towards the bioeconomy.

3.2.4. Northern Member States (Group 2.2)

This group is defined by a low degree of specialisation of national labour markets in the bioeconomy on the EU-28 scale (location quotient from 0.4 to 0.8, except Ireland with 1.1) and the highest levels of apparent labour productivity of the bioeconomy in the EU-28 (i.e., €72,000 to €104,000 of value added per person employed in 2015, which is above twice the EU-28 average level). It comprises Denmark, the Netherlands, the United Kingdom, Sweden, Belgium and Ireland (see Figure 2).

This group comprises Member States that joined the EU between 1952 and 1995. Their GDP per capita is above the EU average (€31,300 to €51,400).

The bioeconomy of these Member States is close to that of group 2.1 but it differs in the even greater significance of the manufacture of bio-based chemicals and pharmaceuticals. Another characteristic of this group is that the manufacture of bio-based textiles is not well developed (fewer than 3% of bioeconomy jobs and less than 2% of the bioeconomy value added) and there is only one Member State, Sweden, that exhibits an orientation towards the woody biomass value chains. As in group 2.1, sectoral apparent labour productivities tend to be higher than in groups 1.1 and 1.2 except in agriculture and fishing. However, they are even higher than in group 2.1 in the manufacture of food, beverages and tobacco and in the manufacture of bio-based pharmaceuticals. Bioeconomy labour markets are dominated by jobs in agriculture and the food industry, but, because of the very high levels of labour productivity achieved in other sectors, the distribution among sectors of the bioeconomy value added is less concentrated. The major contributors are, in descending order, the manufacture of food, beverages and tobacco, the manufacture of bio-based pharmaceuticals, agriculture, the manufacture of paper and the manufacture of wood products and furniture.

R&I bioeconomy indices confirm a medium to high maturity in bioeconomy R&I in this group of Member States (the index ranges from 5/10 to 10/10). Biorefineries and the biochemical and biopharmaceutical sectors are R&I priorities in these Member States (except for the biochemical sector in Denmark, and the biopharmaceutical sector in Denmark and Ireland) [18]. All states in this group put a bioeconomy or a bioeconomy-related strategy in place between 2012 and 2015 (Denmark [26], the Netherlands [27], the United Kingdom [28], Sweden [29], Belgium (Flanders [30]) and Ireland [31]). Their bioeconomy strategies and visions tend to cover a wide variety of sectors, considering market and research and innovation aspects as well as more efficient and environmentally friendly processes.

3.3. EU Member States in Transition

Over the period 2008–2015, the four bioeconomy patterns observed have evolved in a context of agricultural restructuring and a reduction of the agricultural labour force (1.5 million fewer persons employed in the agriculture sector in the EU-28 during this period), which was not compensated for with job creation in other bioeconomy sectors. During this same period, the apparent labour productivity has improved, with major gains in high labour productive sectors (growth of more than €13,000/person employed in the manufacture of paper, of bio-based chemicals, pharmaceuticals and plastics and in the production of bioelectricity) than in low labour productive sectors (growth of less than €6,000/person employed in agriculture and the manufacture of wood and bio-based textiles). A gradient is also observed alongside the four bioeconomy patterns identified: the apparent labour productivity increases by €10,000 to €33,000/person in group 2.2, by €5000 to €16,000/person in group 2.1, by −€2000 to €8000/person in group 1.2 and by 0 to €6000/person in group 1.1. Overall, bigger increases in bioeconomy value added are observed in Member States of groups 2.1 and 2.2 (except Spain and Luxembourg).

Although the four bioeconomy patterns continued between 2008 and 2015, a few Member States have experienced individual developments: Slovenia, Portugal, Greece and Finland.

Slovenia is a very interesting case, as it shows a type 1.2 profile in 2015 with some characteristics of type 1.1 that could be the mark of an uncompleted transition from group 1.1 to group 1.2. Slovenia's characteristics of type 1.1 in 2015 are the relatively high specialisation of its labour market in agriculture (7% of total jobs, 59% of bioeconomy jobs), the low apparent productivity of agriculture and the low diversification of its bioeconomy (in particular the manufacture of food, beverages and tobacco, which accounts for only 16% of bioeconomy jobs).

Looking at Slovenia in 2008, its bioeconomy presented all the characteristics of a member of group 1.1, that is to say a location quotient higher than 1.5, an apparent labour productivity lower than half the EU-28 average and a high specialisation in agriculture with a low level of labour productivity (see Figure 4). Nevertheless, it had already achieved levels of apparent labour productivity at the top end of the group 1.2 range in many bioeconomy sectors (i.e., forestry, fishing, the manufacture of bio-based textiles, the manufacture of bio-based plastics and the manufacture of food, beverages and tobacco). It also already generated a GDP per capita comparable to those of Member States in group 1.2. Between 2008 and 2015, the apparent labour productivity of the Slovenian bioeconomy improved by €5000 of value added per person employed. Major improvements occurred in fishing, the manufacture of paper, forestry and the manufacture of bio-based chemicals, bio-based pharmaceuticals and bio-based plastics. Therefore, in 2015 the apparent labour productivity in Slovenia attained the first or second rank of group 1.2 in all bioeconomy sectors except agriculture and the production of bioelectricity.

Assuming that the Slovenian bioeconomy has been evolving towards a bioeconomy pattern of type 1.2 from a 1.1 pattern, the last step would be a rise in the level of apparent labour productivity in agriculture. This would also raise the overall apparent labour productivity of the bioeconomy. It would certainly be concomitant to a reduction in agricultural jobs, entailing a further reduction of the location quotient of the Slovenian bioeconomy.

Within group 1.1, the Portuguese bioeconomy has also evolved substantially so that, by 2015, its characteristics were closer to those of group 1.2. First, the Portuguese bioeconomy labour market was less concentrated in agriculture in 2015 (51% of bioeconomy workers) than it was in 2008 (62% of bioeconomy workers). The number of agricultural workers in Portugal has shrunk drastically (reducing by 234,000 workers during this period). This has been accompanied by a stronger specialisation in the manufacture of food, beverages and tobacco and in the manufacture of bio-based textiles (both of which employed 17% of bioeconomy workers in 2015 versus 13% in 2008). Second, the apparent labour productivity of the Portuguese bioeconomy has grown more (+€6000 of value added per person employed) than in other Member States of group 1.1. The production of bioelectricity generated €40,000 per person employed more in 2015 than it did in 2008. The manufacture of paper sector and the forestry sector recorded progress of around €30,000 per person employed, and the production of bio-based chemicals, pharmaceuticals and plastics sector was €10,000 more productive per person employed in 2015 than it was in 2008. Consequently, apart from the agriculture and fishing sectors, where labour productivity remained very low throughout the period, the Portuguese bioeconomy sectors attained the highest levels of apparent labour productivity within group 1.1. The labour productivity of forestry and the manufacture of paper is even higher in Portugal than in group 1.2. As a result, in 2015, the apparent labour productivity of the Portuguese bioeconomy was as high as half the EU-28 average, i.e., at the threshold between groups 1.1 and 1.2. Nevertheless, its location quotient remained high (1.7) compared with group 1.2. It seems that the non-bioeconomy sectors did not develop as fast as in the other EU Member States between 2008 and 2015, a period that was marked by a deep economic crisis in Portugal (the GDP per capita contracted from €17,200 to €16,600 in this period). Although the proportion of bioeconomy workers in the whole economy reduced from 17% in 2008 to 14% in 2015, this remains high compared with Member States of group 1.2 (where it was at most 11%). Finally, if the trends of 2008–2015 continue in the coming years, the Portuguese bioeconomy could complete the last steps of a transition and present a bioeconomy pattern of type 1.2.

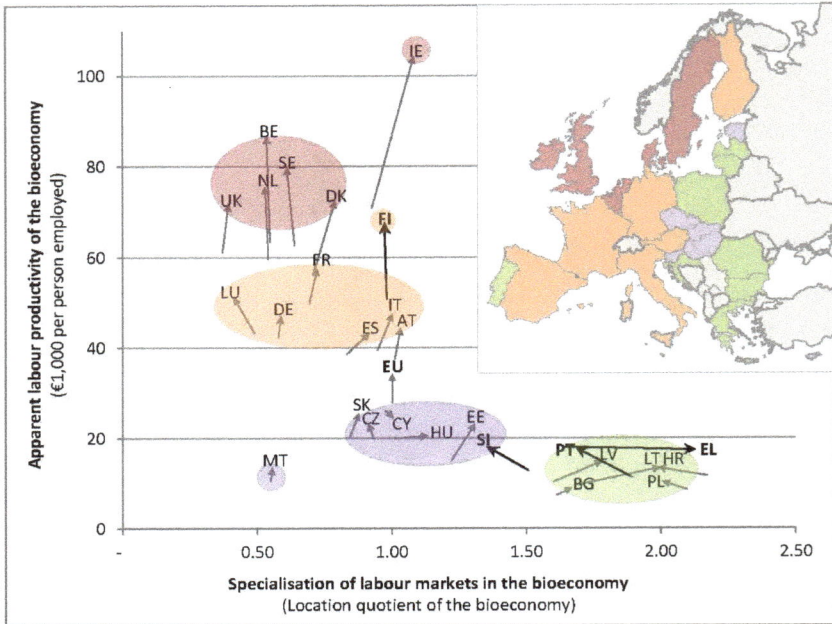

Figure 4. Evolution of the location quotient and apparent labour productivity in the bioeconomy of the 28 EU Member States, 2008–2015 (the cases commented on in Section 3.3 are shown in bold). Note that Romania is missing since it is out of scale (location quotient of 3.76). Its position on the graph is almost unchanged in 2015 compared to 2008.

The Greek case illustrates the opposite trend: in 2008, it displayed a hybrid bioeconomy pattern that positioned it between groups 1.1 and 1.2, but by 2015 Greece had consolidated its position within group 1.1. Indeed, over the 2008–2015 period its location quotient increased from 1.7 to 2.1 and its apparent labour productivity did not stay above half the EU-28 average threshold. The change in location quotient is linked with the very strong impacts of the 2008 economic crisis on the employment structure of Greece. Total employment in Greece decreased by 22% in this period, affecting the bioeconomy sectors to a lesser extent (11% of job losses). Within the bioeconomy, only the manufacture of bio-based textiles and the manufacture of wood and furniture reduced their labour market share. All other bioeconomy sectors either maintained or increased their standing in the national labour market, especially the agriculture sector, which contributed 10.69% of Greek jobs in 2008 and 12.37% in 2015.

The evolution of the apparent labour productivity of the Greek bioeconomy has to be put in perspective with the European dynamic: the bioeconomy labour productivity remained stable in Greece (around €18,000 per person employed) while it was improving in all the other EU-28 Member States (except Cyprus and Romania). As a result, Greece's performance deteriorated compared with the EU-28 average: from above half the EU average in 2008, to just half the average in 2015.

The Greek case emphasises the relationship between the bioeconomy and the rest of the economy. In the same way, Greece has moved fast in a context of recent economic crisis, suggesting that it can catch up once the economic environment recovers.

The Slovenian, Portuguese and Greek examples are illustrations of ongoing developments within the Member States of groups 1.1 and 1.2. However, we have not observed any case of transformation of a bioeconomy pattern of type 1 (including 1.1 and 1.2) to type 2. This suggests that the necessary leap in productivity might be more difficult to achieve than a transition from group 1.1 to 1.2, or that it requires longer than the time period here observed, or that the enabling conditions for such a transition were not met between 2008 and 2015. In contrast, it seems that an evolution from a bioeconomy pattern of type 2.1 towards one of type 2.2 is unfolding in Finland. The Finnish bioeconomy has shown a leap in productivity from €51,000 per person employed in 2008 to €67,000 per person employed in 2015 and it is approaching the threshold of twice the EU average level. Labour productivity has grown faster in Finland than in the other bioeconomies of group 2.1 with major improvements in the manufacturing sectors (mainly biofuels, paper, bio-based chemicals, bio-based pharmaceuticals and wood) and the forestry sector. In 2015, the apparent labour productivity in woody biomass value chains reached levels equivalent to that of Member States of group 2.2. The bioeconomy strategy adopted in Finland in 2014 may reinforce this trend as it focuses on the sectors that have progressed more: the wood value chain (with actions in forestry, timber construction and bioenergy) and the manufacture of bio-based chemicals and pharmaceuticals as part of the strategic focus on the health sector [21]. In conclusion, the conditions seem to be met that will enable the Finnish bioeconomy to acquire the full characteristics of group 2.2, the group of Northern European Member States.

4. Discussion

4.1. Reflections on Turnover and Added Value as Economic Indicators

The 2012 Communication from the Commission to the European Parliament, the Council, the European Economic and Social Committee and the Committee of the Regions on 'Innovating for Sustainable Growth: A Bioeconomy for Europe' [1] presented the economic significance of the bioeconomy in terms of turnover. The JRC [5,6,32] and the nova-Institute for Ecology and Innovation [5,7] followed this approach and presented results at EU and Member State levels in terms of turnover.

Turnover corresponds to the total value of market sales of goods and services to third parties and therefore is, in most cases, higher than value added or the share of GDP [9].

Sectors with a high proportion of inputs (or high costs of bought-in goods and services) have a particularly high turnover. This becomes evident when directly comparing turnover and value added by bioeconomy sector, (gross) value added being defined by EUROSTAT as output minus intermediate consumption. As shown in Section 3.1, the turnover share of the manufacture of food, beverage and tobacco sector in the bioeconomy (51%) is far more important than its value-added share (38%), while the opposite is true in agriculture (17% of turnover share versus 28% of value-added share).

According to our investigations, ongoing research by the Food and Agricultural Organization of the United Nations (FAO) [33] and the MontBioeco study [34], most countries provide value-added data, for example Germany [35] or the Netherlands [36].

The authors of this article propose that, in the future, value added should be considered the main economic indicator instead of turnover, primarily because it avoids double-counting and provides the additional value created by a sector. Furthermore, it is more in line with EU Member States' own calculations and allows for comparisons with national accounts.

The use of value added or GDP has several limitations, in particular in the context of sustainable development (see, for example, the World Economic Forum [37]). However, as one of several indicators describing the economic, social and environmental status of the bioeconomy, both provide key information on economic growth compared with the rest of the economy.

4.2. The Untapped Potential in Central and Eastern Europe

Many regions in Central and Eastern Europe have a strong agricultural tradition and have selected the bioeconomy as a smart specialisation strategy [38]. As already outlined in Section 3.2, the BIOEAST initiative assists Central and Eastern European countries (CEEC, this term is used throughout the rest of this chapter and also when the literature refers to new Member States) to operationalise their bioeconomy visions for 2030, drawing on their biomass potential to develop a sustainable increase in biomass production and circular processing of the available biomass and, in viable rural areas, also to develop an innovative, inclusive, climate-ready and inclusive growth model [14].

This article focuses on socioeconomic indicators to explain the development and potential of the bioeconomy in individual EU Member States and, eventually, in a cluster of states within the EU. It is not within the scope of this analysis to prove that much more biomass can be produced sustainably within CEEC to drive the development of their bioeconomies. Two main aspects are discussed below, which could support the argument of the untapped potential in CEEC: land abandonment and the increase of productivity.

Agricultural land abandonment, or outflow, in the EU is expected to continue at a rate of a decrease of 0.2% in utilised agricultural area (UAA) per year until 2030. This is, however, much lower than the average observed between 2011 and 2016, when UAA decreased by 0.7% per year. By 2030, arable land is projected to have decreased by 3% to reach 104 million ha [39]. For the CEEC, depending on the sources, different developments are plausible. According to the Agricultural Member States Modelling (AGMEMOD) model results, the sown area has not changed significantly in the past and is not expected to do so in the future [40]. On the other hand, Stürck et al. [41] came to the conclusion, in different modelled scenarios, that land abandonment will occur, particularly in Eastern and Southern Europe.

The EU in total shows only marginal growth of yield development for major crops in the EU, particularly because of the high yield levels already achieved, mainly in the EU-15 (Member States that joined the EU before 2014: Austria, Belgium, Denmark, Finland, France, Germany, Greece, Ireland, Italy, Luxembourg, the Netherlands, Portugal, Spain, Sweden and the United Kingdom) [39]. In contrast, the production of major crops is expected to increase significantly in CEEC, almost entirely through increased yields (e.g., for wheat and maize, increases of 15% and 50% respectively are projected for 2026). Notwithstanding, according to the AGMEMOD projections for 2026, the EU-15's yields will still be around 40% higher than for CEEC [40]. The Global yield gap atlas (http://www.yieldgap.org/) explains the difference between actual yields and agro-climatically achievable yields in the same region. For CEEC, several examples underpin the existing gap from the northwestern EU countries [42]. Similarly, the potential from forestry and agroforestry residues could be further exploited [43].

The potential to provide more biomass from agriculture for different bio-based activities could be further enhanced through a development push for rural areas of CEEC, where, in some regions, small semi-subsistence farms still dominate [40]. Furthermore, double-cropping could substantially increase biomass output [42]. The same opportunity to increase productivity also applies, in principle, to animal production [40]. Higher productivity through an improved input/output ratio would therefore require less feedstock, which would then be available for other uses in the bioeconomy. In this context, the much smaller proportion of the processing industry in CEEC to date has to be stressed. Country-specific mapping of the potential for bio-based industries is provided by the Bio-based Industries Consortium for Poland [44] and Romania [45].

These considerations suggest the existence of an untapped potential in CEEC for further biomass production. Closing the yield gap would allow a higher level of labour productivity, which, all other things being equal, would reduce the location quotient and therefore further enable a transition path similar to that taken by northwestern EU countries.

4.3. Possible Strategies

The Member States typology developed in this paper suggests that, in spite of national specificities inherent to natural endowment and historical political and economic choices, including path dependencies, similarities in economic structures can help to identify type-specific needs and tailor bioeconomy development strategies.

The bioeconomy of group 1.1 (Eastern Member States) presents a strong specialisation—in terms of both employment and value added—in agriculture and the manufacture of food, beverages and tobacco, followed by low productive sectors. The bioeconomy of these Member States shows many similarities with that of group 1.2 (Baltic and Central Member States). In addition, the data from Slovenia indicate that a transition from a bioeconomy of type 1.1 to type 1.2 is possible. We might therefore expect future sectoral changes from the primary sector to the secondary sector as a first step of a transition towards a bioeconomy pattern of type 1.2. Such a transition can, of course, be supported by industrial and innovation policies, but the Greek case shows that it is also strongly dependent on the macro-economic environment: opposite sectoral changes occurred all over the Greek economic and financial crisis period. However, the agriculture sector in particular and the bioeconomy sectors more generally were less affected by job losses than the rest of the economy. The role agriculture plays in rural areas and as an economic and social buffer is not to be underestimated. The valorisation of biomass's untapped potential could ease this transition process, starting with measures aimed at reducing the yield gap in these countries while preserving their natural capital.

The bioeconomy labour market of Member States of group 1.2 suggests that outflows from primary sectors have already occurred. Nevertheless, levels of labour productivity remain low compared with Northern and Western Member States. The productivity divide widened between 2008 and 2015 because of lower bioeconomy productivity growth in Member States of groups 1.1 and 1.2 than in Member States of groups 2.1 and 2.2. No Member State of group 1.1 or 1.2 has effected such a productivity leap to reach the levels attained in group 2.1. Kuusk et al. also concluded that the sectoral reallocation at stake in Central and Eastern European countries (CEEC) from low productive sectors to more productive ones did not lead to substantial productivity gains over the period 2001–2012: 'In most of the CEE countries there has been some labour transfer into sectors with relatively higher initial productivity, whereas sectors with faster productivity growth over the year have on average seen lower employment shares. Nevertheless, given the relatively small size of the sectoral change effects, the contributions from the structural bonus and the structural burden are indeed very modest' [46]. They quantify higher productivity gains within sectors than between sectors (i.e., gains arising from sectoral reallocation). Looking at industrialisation processes in CEEC, Stojčić and Aralica describe two opposite trends during the period 2000–2015 [47]: (i) restructuring of manufacturing sectors in the Czech Republic, Hungary, Poland, Bulgaria, Lithuania and Romania in terms of labour productivity versus (ii) deindustrialisation due to declining competitiveness in Slovakia, Estonia, Latvia, Croatia and Slovenia. They argue that export competitiveness was the main driver of the restructuring of the manufacturing sectors (trend (i)) and it is most likely the result of quality upgrading and of integration into higher segments of global value chains. So, in order to foster industrial processes in the manufacturing sector, they recommend targeting the industries with the largest market potential, supporting their integration into global value chains and supporting their segmentation into high value-added segments. They also stress that past industrial policies that centred around international technology transfers have fallen short in creating growth in productivity. Innovation in firms and industries in CEEC is most likely to emerge from a 'doing-using-interacting' (DUI) mode of innovation, i.e., experience-based innovation emerging within firms or within firm networks [48] (The DUI mode is opposed to the STI (science, technology and innovation) mode, which accompanies technology transfers). On the one hand, these recommendations align with Smart Specialisation Strategy (S3) EU policies in the sense that they are rooted in a regional assessment, in the targeting of regional sectoral advantages and in regional networks of industries and stakeholders [48]. On the other hand, Smart Specialisation Strategies emphasise the role of research, development and innovation (RDI) in territorial development, while Stojčić and Aralica expect more productivity growth and spillovers to come from market competitiveness than from innovation [47].

Labour productivity in the Member States of groups 2.1 and 2.2 already reaches very high levels. The development of the bioeconomy in these Member States is more oriented towards shifting production and manufacturing processes towards more resource-efficient and environmentally friendly processes. Note that such orientations are also relevant to the Member States of groups 1.1 and 1.2, in combination with the measures described above. They include realising the transition to a low-carbon economy by replacing fossil fuels with bio-based drop-ins and dedicated bio-based chemicals [49]. With this aim, the bioeconomy strategy of the European Union, as designed in 2012, gave a strong emphasis to investments in research, innovation and skills as one of the three pillars of the overall strategy. €975 million of EU funds are allocated to the Bio-Based Industries (BBI) Public–Private Partnership (PPP) for the period 2012–2020, complementing €2.7 billion of private investments. Annual calls finance development activities and fund the construction of pilot plants in strategic value chains. Optimisation of biomass flows is another component of the bioeconomy shift. It implies enhancing process efficiency, valorising byproducts and waste streams in the production of more diverse final products from initial feedstock (e.g., the biorefinery concept) as well as promoting the cascading use of bio-based products until the end of their lives. These concepts are already included in the Circular Economy Action Plan [50]. The European Commission's Communication on the Circular Economy [51] called for the inclusion of these concepts in the Bioeconomy Strategy of the European Union after its revision in 2017. The updated bioeconomy strategy should be released in 2018.

Supplementary Materials: All the data used in this study are compiled in the JRC Bioeconomics dataset. Please find the bulk download at https://datam.jrc.ec.europa.eu/datam/perm/od/jrc-datam-biomass-estimates/download/dataset.zip. The data can also be browsed at https://datam.jrc.ec.europa.eu/datam/perm/od/7d7d5481-2d02-4b36-8e79-697b04fa4278. Pre-visualisations of the data are accessible as interactive infographics at https://datam.jrc.ec.europa.eu/datam/mashup/BIOECONOMICS.

Author Contributions: T.R. and R.M. conceptualised the original idea. R.M. wrote Section 1. T.R. wrote Sections 2 and 3. Both authors contributed to Section 4.

Acknowledgments: This study has been financed by European Commission funding in the framework of an Administrative Arrangement between the Directorate-General for Research and Innovation and the Joint Research Centre (JRC). The authors would like to thank the nova-Institute for Ecology and Innovation, which prepared the methodology for indicator quantification jointly with the JRC in previous years and continues to collaborate on methodological updates. The authors are very grateful to Saulius Tamosiunas, who automated the calculations and made them available online at https://datam.jrc.ec.europa.eu/datam/public/pages/index.xhtml. The views expressed in this article are purely those of the authors and may not in any circumstances be regarded as stating an official position of the European Commission.

Conflicts of Interest: The authors declare no conflict of interest. The views expressed are those solely of the authors and should not in any circumstances be regarded as stating an official position of the European Commission.

Appendix A

Table A1. Quantified Socioeconomic Indicators of the EU Bioeconomy in 2015 (number of persons employed, turnover, value added and apparent labour productivity).

Sector	Workers	Turnover	Value Added	Apparent Labour Productivity
	Number of persons employed	(€ million)	(€ million)	(€000 per person employed)
Agriculture	9,227,200	380,164	173,597	19
Forestry	539,000	50,101	23,834	44
Fishing	222,392	11,650	6957	31
Manufacture of food, beverages and tobacco	4,544,452	1,153,006	233,408	51
Manufacture of bio-based textiles	999,235	103,497	28,341	28
Manufacture of wood products and furniture	1,407,184	173,724	47,165	34
Manufacture of paper	643,104	186,616	45,590	71
Manufacture of bio-based chemicals, pharmaceuticals, plastics and rubber (excluding biofuels)	444,967	177,044	56,314	127
Manufacture of liquid biofuels	26,271	12,194	2560	97
Production of bioelectricity	13,844	10,831	3138	217
Bioeconomy	**18,067,648**	**2,258,827**	**620,903**	**34**

References

1. European Commission. *Innovating for Sustainable Growth: A Bioeconomy for Europe*; COM(2012) 60 Final; European Commission: Brussels, Belgium, 2012; p. 9. Available online: http://ec.europa.eu/research/bioeconomy/pdf/official-strategy_en.pdf (accessed on 24 May 2018).
2. European Commission. *Roadmap. Update of the 2012 Bioeconomy Strategy*; European Commission: Brussels, Belgium, 2018; p. 3. Available online: https://ec.europa.eu/info/law/better-regulation/initiatives/ares-2018-975361_en (accessed on 24 May 2018).
3. European Commission-DG RTD. *Review of the 2012 European Bioeconomy Strategy*; European Commission: Luxembourg, 2017; p. 86. Available online: https://ec.europa.eu/research/bioeconomy/pdf/review_of_2012_eu_bes.pdf (accessed on 24 May 2018).
4. El-Chichakli, B.; von Braun, J.; Lang, C.; Barben, D.; Philp, J. Five cornerstones of a global bioeconomy. *Nature* **2016**, *535*, 221–223. [CrossRef] [PubMed]
5. Ronzon, T.; Piotrowski, S.; M'Barek, R.; Carus, M. A systematic approach to understanding and quantifying the EU's bioeconomy. *Bio-Based Appl. Econ.* **2017**, *1*. [CrossRef]
6. Ronzon, T.; Lusser, M.; Klinkenberg, M.; Landa, L.; Sanchez Lopez, J.; M'Barek, R.; Hadjamu, G.; Belward, A.; Camia, A.; Giuntoli, J.; et al. (Eds.) *Bioeconomy Report 2016*; JRC Science for Policy Report, EUR 28468 EN; Publications Office of the European Union: Luxembourg, 2017; p. 124. [CrossRef]
7. Piotrowski, S.; Carus, M.; Carrez, D. *European Bioeconomy in Figures 2008–2015*; Nova-Institute for Ecology and Innovation: Hürth, Germany, 2018; p. 16. Available online: http://biconsortium.eu/sites/biconsortium.eu/files/documents/Bioeconomy_data_2015_20150218.pdf (accessed on 24 May 2018).
8. Eurostat. *NACE Rev. 2 Statistical Classification of Economic Activities in the European Community*; Eurostat Methodologies and Working Papers: Luxembourg, 2008; p. 367. Available online: http://ec.europa.eu/eurostat/web/nace-rev2 (accessed on 24 May 2018).
9. European Commission. Commission Regulation (EC) No 250/2009 of 11 March 2009 implementing Regulation (EC) No 295/2008 of the European Parliament and of the Council as regards the definitions of characteristics, the technical format for the transmission of data, the double reporting requirements for NACE Rev.1.1 and NACE Rev.2 and derogations to be granted for structural business statistics. *Off. J. Eur. Comm.* **2009**, 169.
10. Eurostat. Databases. Available online: http://ec.europa.eu/eurostat/data/database (accessed on 24 May 2018).
11. Scientific, Technical and Economic Committee for Fisheries (STECF). *The 2017 Annual Economic Report on the EU Fishing Fleet (STECF 17-12)*; European Commission-Joint Research Center: Luxembourg, 2017; p. 493. [CrossRef]
12. Scientific, Technical and Economic Committee for Fisheries (STECF). *Economic Report of EU Aquaculture Sector (STECF-16-19)*; EUR 28356 EN, JRC104210; Publications Office of the European Union: Luxembourg, 2016; p. 451. [CrossRef]
13. Refsgaard, K.; Teräs, J.; Kull, M.; Oddsson, G.; Jóhannesson, T.; Kristensen, I. *The Rapidly Developing Nordic Bioeconomy: Excerpt from State of the Nordic Region*; Nordic Council of Ministers: Copenhagen, Denmark, 2018; p. 18. ISBN 978-92-893-5443-1.
14. BIOEAST. BIOEAST Vision Paper. BIOEAST-Central and Eastern European Initiative for Knowledge-Based Agriculture, Aquaculture and Forestry in the Bioeconomy. 2018, p. 15. Available online: http://www.bioeast.eu/article/bioeastvisionpaper23022018 (accessed on 24 May 2018).
15. Philippidis, G.; Sanjuán, A.I.; Ferrari, E.; M'barek, R. Employing social accounting matrix multipliers to profile the bioeconomy in the EU member states: Is there a structural pattern? *Span. J. Agric. Res.* **2014**, *12*, 14. [CrossRef]
16. Mainar Causapé, A.J.; Philippidis, G.; Sanjuán López, A.I. *Analysis of Structural Patterns in Highly Disaggregated Bioeconomy Sectors by EU Member States Using SAM/IO Multipliers*; JRC Technical Reports, EUR 28591; Publications Office of the European Union: Luxembourg, 2017; p. 36. [CrossRef]
17. Cardenete, M.A.; Boulanger, P.; Del Carmen Delgado, M.; Ferrari, E.; M'Barek, R. Agri-food and biobased analysis in the Spanish economy using a key sector approach. *Rev. Urban Reg. Dev. Stud.* **2014**, *26*, 112–134. [CrossRef]
18. Berman Group, Infyde. *Bioeconomy Development in EU Regions. Mapping of EU Member States'/Regions' Research and Innovation Plans & Strategies for Smart Specialisation (RIS3) on Bioeconomy for 2014–2020*; Spatial

Foresight, SWECO, ÖIR, t33, Nordregio; European Commission: Brussels, Belgium, 2017; p. 93. Available online: https://ec.europa.eu/research/bioeconomy/pdf/publications/bioeconomy_development_in_eu_regions.pdf (accessed on 24 May 2018). [CrossRef]

19. The Latvian Ministry of Agriculture. Latvijas Bioekonomikas Stratēģija 2030. Riga, 2017; p. 29. Available online: http://tap.mk.gov.lv/doc/2017_12/ZMZin_071217_VSS831.2014.docx (accessed on 24 may 2018).

20. Hungarian Ministry of Rural Development. *National Environmental Technology Innovation Strategy 2010–2020*; Hungarian Ministry of Rural Development: Budapest, Hungary, 2012; p. 16. Available online: http://kornyezettechnologia.kormany.hu/admin/download/b/4f/50000/NETIS_English.pdf (accessed on 24 May 2018).

21. Finnish Ministry of Employment and the Economy. *Sustainable Growth from Bioeconomy. The Finnish Bioeconomy Strategy*; Finnish Ministry of Employment and the Economy: Helsinki, Finland, 2014; p. 15. Available online: http://biotalous.fi/wp-content/uploads/2014/08/The_Finnish_Bioeconomy_Strategy_110620141.pdf (accessed on 24 May 2018).

22. Federal Ministry of Food and Agriculture (BMEL). *National Policy Strategy on Bioeconomy*; Federal Ministry of Food and Agriculture (BMEL): Berlin, Germany, 2014; p. 78. Available online: http://www.bmel.de/SharedDocs/Downloads/EN/Publications/NatPolicyStrategyBioeconomy.pdf?__blob=publicationFile (accessed on 24 May 2018).

23. Ministerio De Económia Y Competitividad, Secretaría De Estado De Investigación Desarollo E Innovación. *The Spanish Bioeconomy Strategy 2030 Horizon*; Ministerio De Económia Y Competitividad, Secretaría De Estado De Investigación Desarollo E Innovación: Madrid, Spain, 2016; p. 44. Available online: http://bioeconomia.agripa.org/download-doc/102159 (accessed on 24 May 2018).

24. Le Ministère de l'Écologie du Développement Durable et de l'Énergie, Le Ministère de l'Éducation Nationale de l'Enseignement Supérieur et de la Recherche, Le Ministère de l'Économie de l'Industrie et du Numérique, Le Ministère de l'Agriculture de l'Agroalimentaire et de la Forêt. Une Stratégie Bioéconomie Pour La France. Paris, France, 2017; p. 34. Available online: http://agriculture.gouv.fr/telecharger/83595?token=4b2095fafe14f075309cc193dda53d70 (accessed on 24 May 2018).

25. Ministry for Economic Development (Co-Coordinator), Ministry of Agriculture Food and Forestry, Ministry of Education University and Research, Ministry of the Environment Land and Sea, Committee of Italian Regions, Agency for Territorial Cohesion, Italian Technology Clusters for Green Chemistry Agri-Food and Bluegrowth. BIT Bioeconomy in Italy. 2017; p. 74. Available online: http://www.agenziacoesione.gov.it/opencms/export/sites/dps/it/documentazione/S3/Bioeconomy/BIT_v4_ENG_LUGLIO_2017.pdf (accessed on 24 May 2018).

26. Government, T.D. Denmark At. Work. Plan. For. Growth For. Water, Bio and Environmental Solutions. 2013, p. 12. Available online: http://em.dk/english/~/media/files/2013/12-03-13-summary-plan-for-growth-for-water-bio-etc.ashx (accessed on 24 May 2018).

27. Netherlands Office for Science and Technology. *The Bio-Based Economy in the Netherlands*; Netherlands Office for Science and Technology: The Hague, Netherlands, 2013; p. 11. Available online: https://www.rvo.nl/sites/default/files/Bio%20Based%20Economy.pdf (accessed on 24 May 2018).

28. HM Government. Building a High Value Bioeconomy. Opportunities from Waste. 2015; p. 38. Available online: https://www.gov.uk/government/uploads/system/uploads/attachment_data/file/408940/BIS-15-146_Bioeconomy_report_-_opportunities_from_waste.pdf (accessed on 24 May 2018).

29. Formas. Swedish Research and Innovation Strategy for a Bio-based Economy. 2015, p. 34. Available online: http://www.formas.se/PageFiles/5074/Strategy_Biobased_Ekonomy_hela.pdf (accessed on 24 May 2018).

30. The Interdepartmental Working Group for the Bioeconomy. *Bioeconomy in Flanders. The Vision and Strategy of the Government of Flanders for a Sustainable and Competitive Bioeconomy in 2030*; Flemish Government-Environment, Nature and Energy Department: Brussels, Belgium, 2014; p. 21. Available online: http://ebl.vlaanderen.be/publications/documents/55157 (accessed on 24 May 2018).

31. Government of Ireland. Delivering our Green Potential, Government Policy Statement on Growth and Employment in the Green Economy. 2012; p. 45. Available online: https://dbei.gov.ie/en/Publications/Publication-files/Delivering-Our-Green-Potential.pdf (accessed on 24 May 2018).

32. Ronzon, T.; Santini, F.; M'Barek, R. *The Bioeconomy in the European Union in Numbers. Facts and Figures on Biomass, Turnover and Employment. Data Based on a Collaboration with Nova-Institute*; European Commission Joint Research Centre: Brussels, Belgium, 2015; p. 4. Available online: https://ec.europa.eu/jrc/sites/jrcsh/files/JRC97789%20Factsheet_Bioeconomy_final.pdf (accessed on 24 May 2018).

33. Bracco, S.; Calicioglu, O.; Gomez San Juan, M.; Flammini, A. Assessing the Contribution of Bioeconomy to the Total Economy: A Review of National Frameworks. *Sustainability* **2018**, *10*, 1698. [CrossRef]

34. MontBioeco. *Monitoring Bioeconomy–Current Approaches in EU Member States and at EU Level*; MontBioeco: Helsinki, Finland; Available online: https://www.luke.fi/en/projects/montbioeco/ (accessed on 24 May 2018).

35. Efken, J.; Dirksmeyer, W.; Kreins, P.; Knecht, M. Measuring the importance of the bioeconomy in Germany: Concept and illustration. *NJAS-Wagen. J. Life Sci.* **2016**, *77*, 9–17. [CrossRef]

36. Heijman, W. How big is the bio-business? Notes on measuring the size of the Dutch bio-economy. *NJAS-Wagen. J. Life Sci.* **2016**, *77*, 5–8. [CrossRef]

37. World Economic Forum. Agenda in Focus: Beyond GDP. Available online: https://www.weforum.org/focus/beyond-gdp (accessed on 24 May 2018).

38. Central and Eastern European Regions and Stakeholders. *Lodz Declaration of Bioregions*; Central and Eastern European Regions and Stakeholders: Lodz, Poland, 2016; Available online: http://bioeconomy.lodzkie.pl/en/deklaracja-lodzka-16/ (accessed on 24 May 2018).

39. Domínguez, I.P.; Fellmann, T.; Chatzopoulos, T.; Pieralli, S.; Jensen, H.; Barreiro-Hurle, J.; Micale, F. *EU Commodity Market Development: Medium-Term Agricultural Outlook. Proceedings of the October 2017 workshop*; JRC Conference and Workshop Reports, JRC109451; Publications Office of the European Union: Luxembourg, 2017; p. 133. [CrossRef]

40. Salamon, P.; Banse, M.; Barreiro-Hurlé, J.; Chaloupka, O.; Donnellan, T.; Erjavec, E.; Fellmann, T.; Hanrahan, K.; Hass, M.; Jongeneel, R. *Unveiling Diversity in Agricultural Markets Projections: From EU to Member States. A Medium-Term Outlook with the AGMEMOD Model*; JRC Technical Report, 29025 EUR; Publications Office of the European Union: Luxembourg, 2017; p. 90. [CrossRef]

41. Stürck, J.; Levers, C.; van der Zanden, E.H.; Schulp, C.J.E.; Verkerk, P.J.; Kuemmerle, T.; Helming, J.; Lotze-Campen, H.; Tabeau, A.; Popp, A.; et al. Simulating and delineating future land change trajectories across Europe. *Reg. Environ. Chang.* **2018**, *18*, 733–749. [CrossRef]

42. Szabó, Z. Europe's sustainable biomass potential is substantial. Euractiv, 2015. Available online: https://www.euractiv.com/section/central-europe/opinion/europe-s-sustainable-biomass-potential-is-substantial/ (accessed on 24 May 2018).

43. Thorenz, A.; Wietschel, L.; Stindt, D.; Tuma, A. Assessment of agroforestry residue potentials for the bioeconomy in the European Union. *J. Clean. Prod.* **2018**, *176*, 348–359. [CrossRef] [PubMed]

44. Bio-Based Industries Consortium. *Mapping the Potential of Poland for the Bio-Based Industry*; Bio-Based Industries Consortium: Brussels, Belgium, 2018; p. 40. Available online: http://biconsortium.eu/sites/biconsortium.eu/files/downloads/Country-Report-Poland.pdf (accessed on 24 May 2018).

45. Bio-Based Industries Consortium. *Mapping the Potential of Romania for the Bio-Based Industry*; Bio-Based Industries Consortium: Brussels, Belgium, 2018; p. 34. Available online: http://biconsortium.eu/sites/biconsortium.eu/files/downloads/Country-Report-Romania.pdf (accessed on 24 May 2018).

46. Kuusk, A.; Staehr, K.; Varblane, U. Sectoral change and labour productivity growth during boom, bust and recovery in Central and Eastern Europe. *Econ. Chang. Restruct.* **2017**, *50*, 21–43. [CrossRef]

47. Stojčić, N.; Aralica, Z. *Choosing Right from Wrong: Policy and (De) Industrialization in Central and Eastern Europe*; EIZ Working Papers EIZ-WP-1703; EIZ: Zagreb, Croatia, 2017; p. 36. Available online: http://www.eizg.hr/UserDocsImages/publikacije/serijske-publikacije/radni-materijali/Choosing_Right_from_Wrong-Industrial_Policy_and_(De)industrialization_in_Central_and_Eastern_Europe.pdf (accessed on 24 May 2018).

48. Jensen, M.B.; Johnson, B.; Lorenz, E.; Lundvall, B.Å. Forms of knowledge and modes of innovation. In *The Learning Economy and the Economics of Hope*; Anthem Press: London, UK; New York, NY, USA, 2016; Volume 155, pp. 155–182.

49. Carus, M.; Dammer, L.; Puente, Á.; Raschka, A.; Arendt, O. *Bio-Based Drop-in, Smart Drop-in and Dedicated Chemicals*; Nova-Institute for Ecology and Innovation: Hürth, Germany, 2017; p. 3. Available online: http://bio-based.eu/download/?did=107519&file=0 (accessed on 24 May 2018).

50. European Commission. *COM/2015/0614 Final. Communication from the Commissionto the European Parliament, the Council, the European Economic and Social Committee and the Committee of the Regions. Closing the Loop-An. EU Action Plan for the Circular Economy*; European Commission: Brussels, Belgium, 2015; p. 21. Available online: http://eur-lex.europa.eu/legal-content/EN/TXT/?uri=CELEX:52015DC0614 (accessed on 24 May 2018).
51. European Commission. *COM(2014) 398 Final. Communication from the Commission to the European Parliament, the Council, the European Economic and Social Committee and the Committee of the Regions. Towards a Circular Economy: A Zero Waste Programme for Europe*; European Commission: Brussels, Belgium, 2014; p. 14.   Available online: http://eur-lex.europa.eu/resource.html?uri=cellar:50edd1fd-01ec-11e4-831f-01aa75ed71a1.0001.01/DOC_1&format=PDF (accessed on 24 May 2018).

sustainability

MDPI

Article

Assessing the Contribution of Bioeconomy to the Total Economy: A Review of National Frameworks

Stefania Bracco [1,*], Ozgul Calicioglu [1,2], Marta Gomez San Juan [1] and Alessandro Flammini [1]

[1]   Food and Agriculture Organization of the United Nations, Viale delle Terme di Caracalla,
      00153 Roma RM, Italy; Ozgul.Calicioglu@fao.org (O.C.); Marta.GomezSanJuan@fao.org (M.G.S.J.);
      Alessandro.Flammini@fao.org (A.F.)
[2]   Department of Civil and Environmental Engineering, The Pennsylvania State University,
      212 Sackett Building, University Park, PA16802, USA
*   Correspondence: stefania.bracco@fao.org or stefania.bracco2@gmail.com; Tel.: +39-065-705-5955

Received: 13 April 2018; Accepted: 17 May 2018; Published: 23 May 2018

Abstract: Developments in technology have enabled envisioning the derivation of materials and products from renewable biomass as an alternative to finite fossil-based resource consumption. Therefore, bioeconomy is regarded as an opportunity for sustainable economic growth. Countries are formulating strategies in accordance with their goals to attain a bioeconomy. Proper measurement, monitoring, and reporting of the outcomes of these strategies are crucial for long-term success. This study aims to critically evaluate the national methods used for the measurement, monitoring, and reporting of bioeconomy contribution to the total economy. For this purpose, research and surveys have been conducted on selected countries (Argentina, Germany, Malaysia, the Netherlands, South Africa, and the United States). The results reveal that the bioeconomy targets set up in the strategies often reflect the country's priorities and comparative advantages. However, comprehensive approaches to measure and monitor bioeconomy progress are frequently lacking. Most countries only measure the contribution to gross domestic product (GDP), turnover, and employment of the sectors included in their bioeconomy definition, which may provide an incomplete picture. In addition, this study identifies the mismatch between the targets and measurement methods, as the environmental and social impacts of bioeconomy are often foreseen, but not measured. It is concluded that existing global efforts towards sustainable bioeconomy monitoring can be strengthened and leveraged to measure progress towards sustainable goals.

Keywords: bioeconomy; bio-based products; GDP; policy measures; sustainability assessment; sustainable development

---

1. Introduction and Background

Modern economies rely on resources which are finite in nature. On top of their long-term unsustainability, utilization of fossil-fuel resources and unsustainable consumption of derived products also pose risks to societies and the environment due to their negative impacts such as climate change and ecosystem degradation [1,2]. Nevertheless, advancements in industrial biotechnology have enabled the derivation of materials, chemicals, and energy from renewable biomass, which could provide substitutes for fossil-based and finite resources [3]. This substitution potential forms the core of a still-evolving bioeconomy concept.

The literature on bioeconomy vision has been evolving in parallel with the concept and has been clustered under three major perspectives: (1) the biotechnology vision, which emphasizes innovations and utilization of biotechnology at commercial scales; (2) the bioresource vision, which emphasizes the improvement of value chains on upstream biomass production; (3) and the bioecology vision, which emphasizes the positive impacts of energy and resource optimization on ecosystem health [4].

These perspectives underline the potential of bioeconomy in the opportunities it offers, such as low-carbon economic growth, preservation of natural resources, restoration of environmental and ecosystem health, and welfare of rural communities.

Due to its promising potential in addressing these global challenges, bioeconomy has been directly or indirectly included in policy agendas worldwide [5,6]. Country objectives and bioeconomy priorities encompass economic growth, employment, energy security, food security, fossil-fuel reduction, mitigation and adaptation to climate change, and rural development [7]. With respect to their potentials in deploying the bioeconomy vision, countries have different opportunities, which might also affect their policies. Countries can be classified as countries with: (1) an abundance of renewable biological resources, but a lack of downstream processing industries; (2) both high feedstock potential and advanced processing industries; and (3) low feedstock potential but advanced processing industries [8]. These variations in potential also create differences in countries' objectives for adopting a bioeconomy strategy and in the evaluation of success towards their achievement.

Measuring bioeconomy contribution to countries' overall economy can be an important indicator of development. No internationally agreed methodology exists today to measure progress in attaining the ambitions and targets set by bioeconomy policies and strategies. Moreover, given the differences among countries' constraints, opportunities, and priorities, the development of a uniform way to assess the contribution of bioeconomy to the national economy is challenging. In addition, incomprehensive measurement processes might lead to the omission of potential negative impacts of bioeconomy. This lack of a coherent methodology could also create confusion when trying to compare the importance of bioeconomy within and across countries. One first step towards a globally recognized methodology could be to assess the current efforts of individual countries to define bioeconomy and the frameworks for measuring, monitoring, and reporting its contribution. In fact, regional efforts for the harmonization of the measurement of bioeconomy exist, for example, in the European Union (EU). A European Commission (EC) strategy for the bioeconomy was launched in 2012, and the EC Joint Research Centre (JRC) has been assigned to monitor jobs and turnover in the EU bioeconomy for all the member states and sectors. However, for a global methodology, a more geographically balanced analysis would be useful.

Typical economic models that can be adopted to measure the bioeconomy contribution to a country's economy include the value added/GDP approach, the input-output (I-O) and social accounting matrix (SAM) analysis, the computable general equilibrium (CGE) model, the partial equilibrium (PE) model, and other economic models and tools [9]. However, these approaches do not systematically consider environmental and social aspects. In fact, the objective of this study is to analyze how the contribution of bioeconomy is measured in the overall national economy, using information from a geographically representative list of countries (Argentina, Germany, Malaysia, the Netherlands, South Africa, and the United States). Furthermore, the alignment of the country objectives and the parameters measured have been analyzed in order to assess whether social and environmental impacts of bioeconomy were captured through selected measurement, monitoring, and reporting frameworks adopted by the countries.

Within the scope of this study, the bioeconomy has been defined as "the knowledge-based production and utilization of biological resources, biological processes, and principles to sustainably provide goods and services across all economic sectors" [7]. It involves three elements: (1) the use of renewable biomass and efficient bioprocesses to achieve sustainable production; (2) the use of enabling and converging technologies, including biotechnology; (3) and integration across applications such as agriculture, health, and industry. In accordance with the Food and Agriculture Organization of the United Nations (FAO) development [7], the term "bio-based economy" excluded food and feed production. Instead, it was used to take the production of nonfood goods into consideration, i.e., bio-based materials, chemicals, and pharmaceuticals; pulp and paper; construction materials; textiles; and bioenergy. "Bio-based industry" refers to the industrial production of all possible bio-based goods. The strategies related to bioeconomy, bio-based economy, and bio-industries were all considered as "bioeconomy strategies". This assessment has been built upon previous and ongoing efforts to foster global bioeconomy. In this respect, the countries to be investigated have been selected among the members of the FAO International Sustainable Bioeconomy Working Group (ISBWG), which includes 23 members as of March 2018: 11 countries (Argentina, Brazil, China, Germany, Italy, Kazakhstan, Malaysia, the Netherlands, South Africa, Uruguay, and the United States), the German Bioeconomy Council, the European Union (EU) Commission, Organisation for Economic Co-operation and Development (OECD), International Center for Tropical Agriculture (CIAT), Stockholm Environment Institute (SEI), United Nations Environment Programme (UNEP), World Wide Fund (WWF), the Nordic Council of Ministers, the EU Bio-Based Industries Consortium, Wageningen University, the World Business Council for Development (WBCSD), and FAO.

The countries selected for the study are located in five continents, differ in terms of levels of economic development, and have different bioeconomy strategies and priorities. For instance, some countries have little land availability but advanced technologies, while others prioritize farmers and rural development and have greater land availability. For all the countries, the study reviews bioeconomy objectives and priorities as well as measurement, monitoring, and reporting frameworks. From the sample countries and a review of existing literature, a pathway towards a sustainable bioeconomy monitoring is then proposed.

2. Approach and Methods

The study was based on desk research of policy documents, strategies, and statements on the bioeconomy and its measurement for a selected number of countries (Argentina, Germany, Malaysia, the Netherlands, South Africa, and the United States). Because of their demonstrated interest in developing a global bioeconomy framework, the selection process was primarily performed among the ISBWG member countries, following the selection process illustrated in Figure 1. The structural organization of the analysis included an evaluation of the available information in terms of: (1) how countries define bioeconomy; (2) which are the objectives and/or priorities of their strategy; (3) and the methodology they use to measure, monitor, and report the contribution of bioeconomy to their economy or objectives (Figure 1).

The relevant information for the analysis was gathered from official bioeconomy strategies and documents, upon availability. When a government official document on how to measure the contribution of bioeconomy to the total economy was not available, commissioned studies, studies from research institutes, and/or non-profit organizations were used instead. Table 1 summarizes the sources of information and documents analyzed in order to understand the definition of bioeconomy, the objectives/priorities of the bioeconomy strategy, and the measurement, monitoring, and reporting frameworks established by each selected country. Whenever possible, the information was complemented and validated by a survey, which was distributed to government representatives (Table 2). The survey included the same questions to which answers were sought while scanning through the written materials.

Table 1. Sources of information for the selected countries.

	Bioeconomy Definition and Strategy	Objectives/Priorities of the Strategy	Measurement, Monitoring and Reporting Framework
Argentina	Ministry of Agroindustry (MINAGRO) [10]	MINAGRO [10]	Bolsa de Cereales [11]
Germany	Federal Ministry of Food and Agriculture (BMEL), and the Federal Ministry for Education and Research (BMBF) [12]	National Policy Strategy on Bioeconomy [13]	A comprehensive and system monitoring approach to measure the contribution of German BE to the overall economy is currently under development
Malaysia	National Biotechnology Policy (NBP); the BioNexus Status (BNX); the Bioeconomy Transformation Programme (BTP); the Bioeconomy Community Development Programme (BCDP); National Biomass Strategy 2020 ("NBS 2020") [14]	BTP and BCDP [14]	MOSTI and Bioeconomy Corporation ([14,15])
The Netherlands	Netherlands Enterprise Agency (RVO) [16]; CE Delft [17]; Agency Ministry of Economic Affairs (NOST) [18]; NNFCC [19]	NOST [18]; NNFCC [19]	Bio-based economy protocol monitor [16]; NOVA Institute [20]; CE Delft [17]; EC Bioeconomy Knowledge Centre [21]
South Africa	Public Understanding of Biotechnology [22]; National Biotechnology Strategy [23]	National Biotechnology Strategy [23]	National Biotechnology Strategy [23]; ongoing study to establish a framework to develop indicators to measure the growth of the BE in South Africa [24]
USA	National Bioeconomy Blueprint [25]; The Billion Ton Bioeconomy Vision [26]	National Bioeconomy Blueprint [25]; The Billion Ton Bioeconomy Vision [26]	USDA reports ([27–29]); Department of Energy [30]

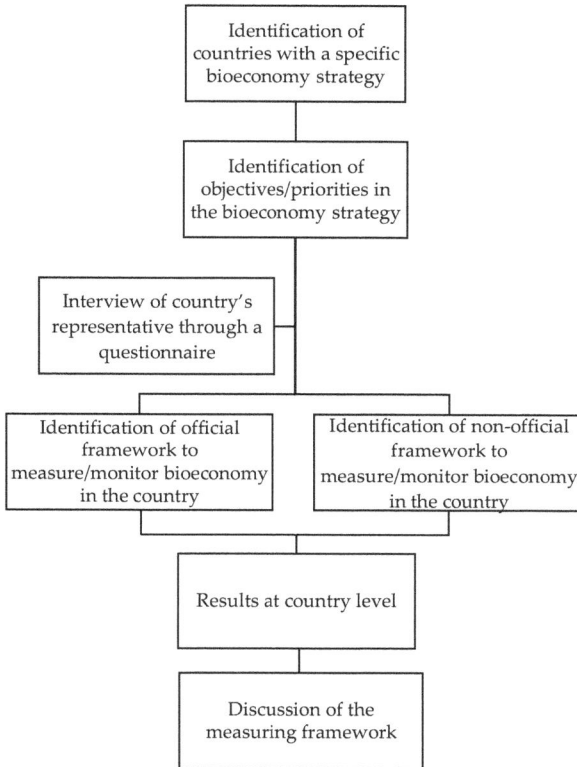

Figure 1. Selection process and structural organization of the analysis.

Information collected through desk research and surveys has been used for the analysis and has been complemented with an extensive literature review of objectives and priorities of bioeconomy strategies in other low-, middle-, and high-income countries (outside the study focus) in order to improve the quality of the discussion. For this purpose, the survey was distributed to all ISBWG members (not only to those in the study focus).

Table 2. Questionnaire submitted to ISBWG members on assessing bioeconomy contribution to countries' economy.

Bioeconomy Definition
How does your Country define bioeconomy?
Which sectors are included into your bioeconomy strategy?
(e.g., Agriculture; Automotive and mechanical engineering; Chemistry (incl. bioplastics); Biofuels/bioenergy; Biorefining; Construction/Building industry; Consumer goods such as cosmetics and cleaning products; Feed; Fisheries; Food and Beverage industry; Forestry; Health; Knowledge/Innovation; Mining; Pharmaceuticals industry; Pulp and paper; Textiles)
Objectives/Priorities
Which are the objectives/priorities of your country strategy (e.g., food security, energy security, fossil fuel reduction, rural development, economic growth, employment, mitigation and adaptation to climate change, etc.)?

Table 2. *Cont.*

Measurement, Monitoring and Reporting Framework
Does the country strategy include criteria to measure the contribution of bioeconomy to the overall economy? If yes, which ones?
Which approach does your country use to measure bioeconomy contribution? (e.g., GDP approach; Input-Output matrix; Computable General Equilibrium (CGE) Model; Partial Equilibrium (PE) Model)
Does your country measure the impact of bioeconomy on the following areas? (Turnover/sales; Value added; Job creation; Market development; Investments; Intellectual property; R&D spending; Trade balance; Poverty alleviation; Food security and sustainable agriculture; Health and well-being; Education; Gender equality; Availability and sustainable management of water; Access to affordable, reliable, sustainable and modern energy; Inclusive and sustainable industrialization and innovation; Inequality and inclusiveness; Inclusive, safe, resilient and sustainable cities; Ensure sustainable consumption and production patterns; Climate change; Oceans, seas and marine resources; Terrestrial ecosystems, forests, land degradation and biodiversity)
Are indicators to measure bioeconomy contribution defined? If so, which ones?

Short Discussion
Which is, in our opinion, the main limitation of your country approach to measure bioeconomy contribution?

3. Analysis and Results

3.1. Bioeconomy Definitions and Strategies

It is observed that the sectors and subsectors considered in bioeconomy were different among the analyzed countries, which is a reflection of the differences in their priorities and strategies. Table 3 summarizes the various sectors included in the bioeconomy definition by the six analyzed countries, and the sectors taken into account for the quantification of bioeconomy contribution to overall national economy. The first important point is the variation in the definition of bioeconomy. For instance, the Netherlands has a focus on bio-based economy, excluding the agriculture and food sectors [17] and still has not agreed on a which sectors are included in the bioeconomy [31]. The United States aims to analyze the bioeconomy, but USDA shows results limited to bio-based products industries, which exclude the energy, food, feed, livestock, and pharmaceutical industries.

Table 3. Sectors included into BE strategy and monitoring in the selected countries.

	Argentina	Germany	Malaysia	The Netherlands *	South Africa	USA *
Agriculture	■■	■■	■■		■	■■
Automotive and mechanical engineering		■■				
Chemistry (incl. bioplastics)	■■	■■	■■	■■	■	■■
Biofuels/bioenergy	■■	■■	■■	■■	■	
Biorefining		■■	■■		■	■■
Construction/Building industry		■■				
Consumer goods (e.g., cosmetics, cleaners)	■■	■■			■	
Feed	■■	■■	■■		■	
Fisheries	■■	■■	■■		■	
Food and Beverage industry	■■	■■	■■		■	
Forestry	■■	■■	■■	■■ **	■	■■
Health		■■			■	
Knowledge/Innovation		■■	■■	■■	■	
Mining					■	
Pharmaceutical industry	■■	■■	■■	■■	■	
Pulp and paper	■■	■■		■■	■	
Textiles	■■	■■		■■	■	■■
References:	[11]	[12]	[32]	[20]	[23,33]	[29]

* The monitoring system analysis for the Netherlands refer to bio-based economy and the results for the United States refer to bio-based products industries. ** Only forest-based industry. Legend: ■: included in bioeconomy strategy, ■■: included in the bioeconomy strategy and monitored or measured.

According to Argentina, the bioeconomy includes agriculture, forestry, fishing, food production, and pulp and paper production, as well as parts of textile, chemical, energy, and biotechnological industries (medical and pharmaceutical industry). The German bioeconomy includes agriculture, forestry, fishing, manufacturing, and trading of bio-based products. Similarly, the Malaysian bioeconomy includes agriculture, forestry, fisheries, food, feed, healthcare wellness products, chemicals, and renewable energy. South Africa's bioeconomy strategy focuses on agriculture, industrial, and environmental bio-innovation and health but has yet to develop metrics to monitor performance.

It is worthwhile to note that the decision on which sectors to include is also relevant when trying to measure the bioeconomy's contribution to countries' economies in terms of GDP and value added, since the calculations would only take the included sectors into account [34].

3.2. Bioeconomy Objectives and Priorities

The sectors included in the bioeconomy strategy often reflect the priorities identified by the country and comparative advantages linked, for instance, to endowment in biomass resources, historical economic specialization, labor productivity, and past investments in R&D [7,11,35–39].

For instance, in Argentina, bioeconomy is seen as a tool for sustainable development in the country. It is recognized as a positive alternative for the generation of new behaviors and sources of employment to face the double challenge of climate change and the continued need for economic progress indispensable for poverty reduction [10].

In Germany, the National Policy Strategy on Bioeconomy's priorities for advancing towards a knowledge-based bioeconomy are: the development of a secure supply of high-quality food; the transition from a fossil-based economy to an economy that is increasingly efficient in terms of raw materials and based on renewable resources; the supply of renewable resources; the sustainable use of renewable resources while conserving biodiversity and soil fertility; protection of the climate; the strengthening of Germany's innovative power and its international competitiveness in business and research; securing and creating employment and added value, particularly in rural areas; and sustainable consumption [13].

In Malaysia, bioeconomy is seen as a key contributor to economic growth, which can provide benefits to the society via breakthroughs in agricultural productivity, innovations in healthcare, and the adoption of sustainable industrial processes [15].

The objectives of the South African bioeconomy strategy are to make the country more competitive internationally (especially in the industrial and agriculture sectors), to create more sustainable jobs, to enhance food security, and to create a greener economy as the country shifts towards a low-carbon economy [23]. In particular, the strategic economic sectors identified are (i) agriculture, (ii) health, and (iii) industry and environment.

In the Netherlands, the bioeconomy strategy encourages knowledge development and innovation in nine top sectors: agriculture and food, water, chemicals, energy, life sciences and health, horticulture and propagation materials (seed stock), logistics, high-tech systems, and materials and creative industries [18]. However, the Dutch government focuses more often on the bio-based economy, defined as "economic activity based on biomass, with the exception of human food and feed", with the condition that it is based on recently captured carbon [40]. The most important drivers behind the adoption of a bio-based economy strategy were: striving for more sustainability (reduction of CO_2 emissions, circular economy); the awareness of the finite nature of fossil fuels; and the economic opportunities offered to Dutch businesses through the use of renewable biological resources and residues [18].

In the United States, the five strategic objectives introduced by the National Bioeconomy Blueprint aimed to generate economic growth and address societal needs. Some of these strategies include supporting R&D investments, facilitating the transition of bio-inventions from research lab to market, developing and reforming regulations, updating training programs, and aligning academic institution incentives with student training for national workforce needs. They also include identifying and supporting opportunities for the development of public–private partnerships and precompetitive collaborations in their objectives [25].

3.3. Measurement, Monitoring, and Reporting Frameworks

The countries selected for the study use various approaches for measuring the bioeconomy contribution to their economy and in attaining the objectives.

In the case of Argentina, a standard approach was adopted to measure the contributions (gross production value and value added) of bio-based products to the national GDP, referencing the general principles of the System of National Accounts (SNA) for the calculation of GDP and internationally comparable satellite accounts (e.g., for education, capital, productivity, and environment). A paper by an intermediate service provider *Bolsa de Cereales* [11] designs a general methodology for the criteria, procedures, and databases to be used in the measurement of bioeconomy and its contribution to GDP. The sectors included in the calculation of the bioeconomy's contribution to the GDP are "agriculture, forestry, fishing, food production, and pulp and paper production, as well as [bio-based] fractions of textile and chemical industry, and energy and biotechnological industries (health and pharmaceutical industry)" [11], which are in alignment with the sector included in the country's bioeconomy strategy. However, the study from *Bolsa de Cereales* considers only the economic variables, without taking into account the social and environmental aspects of the bioeconomy. For instance, the study does not report on regional and territorial development, employment, food security, energy security, sustainability, or climate change mitigation and adaptation, which are among the objectives of the bioeconomy vision. Not addressing these issues would not only give an incomplete picture of the achievement of these objectives, but it also might pose the risk of overlooking the potential negative impacts of bioeconomy on these dimensions.

In Germany, most of the areas contributing to bioeconomy are monitored by traditional statistical accounts [12]. However, in most of the cases, methodologies for data collection and assessment are not streamlined to assess the impact of the bioeconomy. This leads to sparse information on impacts, along with data gaps, uncertainties, lack of comparability of results, and the potential double counting of impacts. A comprehensive and system monitoring approach to measure the contribution of German bioeconomy to the overall economy is currently under development by a joint interministerial undertaking, consisting of three main projects: the monitoring of biomass flows, the Systemic Monitoring and Modelling of the Bioeconomy (SYMOBIO), and the identification of economic key performance indicators to monitor the bioeconomy [41].

Malaysia has developed a Bioeconomy Contribution Index (BCI) to measure the contribution of bioeconomy to the overall economy, which is a combination of five components/parameters: bioeconomy value added, bio-based exports, bioeconomy investments, bioeconomy employment, and productivity performance [14]. The BCI is a comparative tool designed to provide a holistic view, encompassing multiple aspects of the bioeconomy, and it is used to identify trends, patterns, and synergies within the industry. The index compares the performance of each specific component for a selected year against the adjusted (i.e., accounting for changes in variables such as inflation rates and import–export values) expected base performance (in the base year 2005), determined by a dynamic computable general equilibrium (DCGE) model [15]. The share of bioeconomy contribution to national development is estimated from the SAM, assuming that Malaysia is a price taker country [42]. Until now, the BCI has been primarily measuring revenues and economic flows, but it could be further improved to consider broader socioeconomic or environmental aspects. For instance, the BCI could incorporate social measures (e.g., poverty reduction and income inequality in the bioeconomy industry), and environmental measures (e.g., CO_2 emissions and level of local biodiversity), in order to evaluate whether the bioeconomy poses a risk or contributes to sustainability in all dimensions [42].

Since the Netherlands lacks a clear protocol on defining the boundaries of the bioeconomy, until now their focus has been on the bio-based economy (BBE). In 2013, the Netherlands established a BBE monitoring protocol to quantify its size and to monitor its development over time in order to make trends visible and comparable with developments abroad [16]. The protocol defines the system boundaries, the units to express the size of BBE, and utilization of available data or their collection, if missing. The protocol is built on existing statistical data on production and consumption, prevents double counting of bio-based raw materials, and accounts for raw material flows monitoring. However, it raises some problems linked to classification of business sectors, classification of product groups, and timely acquisition of data [16].

The South African bioeconomy strategy [23] contains some indicators to monitor progress in the bioeconomy in comparison with other high- and middle-income countries, broadly divided into "knowledge and skills" indicators (full-time equivalent researchers, scientific publications, and bioeconomy-related publications) and "financial support" indicators (gross domestic expenditure on R&D as percentage of GDP and funding and governmental support). Moreover, the methodology provides 18 output indicators (related to industry, market, knowledge transmission and application, and knowledge base and human resources) to be used to track and monitor the bioeconomy strategy [23]. However, systematic metrics to measure and monitor South Africa's bioeconomy have not yet been implemented. Moreover, most indicators in the strategy are derived from the measurement of a knowledge-based economy or biotechnology innovation policies. Therefore, they do not cover social and environmental issues, which would make it difficult to determine negative impacts, if they exist, on these aspects. Ongoing efforts are anticipated to result in detailed implementation plans and value propositions for specific sectors and initiatives that will help refine targets [33].

The USDA report [29] examines and quantifies the effect of the bio-based products industry from an economics and jobs perspective at the state level. It was preceded by a report analyzing the effect at national level [28] and by another report [43] which provides a snapshot of available information on the bioeconomy in the country and a platform upon which to build future efforts to measure the bioeconomy. The report adopts a three-pronged approach to gather information: interviews of representatives of government, industry, and trade associations involved in the bio-based products; data collection from government agencies and published literature on the bio-based products industry; and economic modelling. Despite being intended as a platform for understanding and tracking the progress of the bioeconomy in the United States, the USDA 2016 report does not provide a complete picture of the bioeconomy, as it does not report on the bioenergy sector, which is included in the US strategy. Instead, it only focuses on seven major sectors chosen to represent the bio-based industry's contribution to the US economy (agriculture and forestry, biorefining, bio-based chemicals, enzymes, bioplastic bottles and packaging, and forest products and textiles). In 2017, the Department of Energy (DOE) also provided some figures about the size of the bioeconomy [30], building up the Billion Ton Bioeconomy Vision, but without a systematic measurement approach. The DOE estimates are taken from a paper considering direct employment and revenues from biomass resources fed into a number of end-uses and products including heat and power generation, bio-based chemicals and products (including wood pellets), and biofuels and coproducts [44].

3.4. Limitations in Data Availability and Statistical Approaches

Most of the analyzed countries currently measure the contribution of bioeconomy to their GDP and other economic variables only. This economic approach, however, has some limitations in reflecting the contribution in the economic sphere, above all because no standard methodology has been established to enable international comparison of bioeconomy contribution to GDP. Additionally, as mentioned above, products and activities comprised within the bioeconomy greatly vary according to country's priorities and comparative advantages.

The most common classifiers of economic activity, trade and products at the international level (International Standard Industrial Classification (ISIC), North American Industry Classification System (NAICS), Classification of Economic Activities in the European Community (NACE), Nomenclature for External Trade (NET), and Classifier per Category (CPC)) are not compatible with the complexity of the bioeconomy [11,37] since they are not appropriate for the heterogeneous nature and variety of bio-based products. ISIC, NACE, and NAICS group production units according to the similarity of their productive processes, technology, inputs, and equipment. Their classification criteria make no distinction between bio- or nonbio-inputs [11]. Even the System of National Accounts (SNA 08) from the United Nations, which provides recommendations for measuring the national production, wellbeing, and other economic issues in an internationally comparable way, does not allow for the measurement of the bioeconomy [11]. Classifiers based on traditional industrial activities are not compatible with the bio-based industry. This can lead to under or overestimation of the size of the bioeconomy.

The high number of bio-based products and their heterogeneity make it very difficult to provide a full quantitative picture of the status and evolution of the bioeconomy [45]. Often, data on the bioeconomy are retrieved from surveys of the bio-based industry [11,27,46]. These surveys represent an important first step for a systematic approach to quantify the bioeconomy. However, they face difficulties in assembling the requested data and suffer from incomplete response rate [46]. These limitations are even more relevant in low- and middle-income countries, where statistical systems are not well developed. Under these circumstances, the surveys may not be updated and/or may include limited and biased samples (as shown for instance by the Argentinian and South Africa analyses, where the last company surveys were taken in 2003 [47]). Digitalization efforts as the ones undertaken in Malaysia to improve data collection can play an important role in the measurement and monitoring of bioeconomy.

4. Discussion

4.1. The Need for Defining the Bioeconomy Boundaries at National, Regional, and Global Levels

Due to the lack of a homogenous definition of bioeconomy and its sectors, a common ground which enables comparing the contribution of bioeconomy among countries is missing. Also, at national level, the definition of the bioeconomy boundaries is sometimes unclear. For instance, in the case of the Netherlands, the estimated impacts of bioeconomy were different among the official studies analyzed due to the variety of methodologies and input data [17,20,21]. Also in the United States, most of the sectors considered by the DOE to estimate the size of bioeconomy were excluded from the USDA 2016 report, leading to different estimates of the bioeconomy impacts in the country [29,30]. For this reason, the efforts of the US Biomass R&D Board to coordinate programs within and among departments and agencies of the federal government towards a single, harmonized bioeconomy vision should ideally produce a single comprehensive approach able to monitor and measure all the sectors included in the vision in a coherent way.

When the countries do not have a holistic bioeconomy strategy, they tend to adopt a fragmented approach by separately considering the different uses of biomass in each sector (e.g., agriculture, forestry, energy, and transport). This approach to governing the bioeconomy leads to different policies for different uses of biomass, different incentives for investment, and different regulations for the areas from which feedstocks are sourced [48]. In these countries, the efforts should aim at integrated approaches across different levels, sectors, landscapes, and end-uses in order to avoid boom and bust policies as it happened for first-generation biofuels in the EU and elsewhere.

Nevertheless, some regional efforts to harmonize the measurement of the bioeconomy's economic significance exist. For instance, since the launch of the European Commission (EC) strategy for the bioeconomy in 2012, the EC Joint Research Centre (JRC) is monitoring jobs and turnover in the European Union bioeconomy for all the member states and sectors [49]. More specifically, the EC Bioeconomy Knowledge Centre shows turnover, employment, and location quotient (i.e., the share of employment in the bioeconomy in a member state divided by the EU employment share in the bioeconomy) [49]. In order to enable achievement of the full potential of the bioeconomy, global guidelines on the measurement and regulation of the value chains could be beneficial [8]. FAO has been already coordinating the global efforts towards the development of international Bioeconomy Sustainability Guidelines. These could be used by the countries to measure sustainability aspects of their bioeconomy strategy and monitor the achievement of economic, social, and environmental targets and priorities.

4.2. Bioeconomy as a Means to Achieve Sustainable Development Goals

The bioeconomy has already been adopted by a significant number of countries as a new vision of development to decouple the economy from the fossil-fuel dependence and as a valid path towards the achievement of the Sustainable Development Goals (SDGs) and the commitments under the Paris Climate Agreement. For instance, efficient and sustainable natural resource management is directly tied to at least 12 of the 17 SDGs and can cut greenhouse gas (GHG) emissions by 60% by 2050 [50]. In addition, for lower-income countries, better management of natural resources is often a key component of poverty eradication, climate-change mitigation, and resilient economic growth [51].

In low- and middle-income countries with available biomass resources and/or well-developed primary sectors, a sustainable bioeconomy could unlock new opportunities for economic development and industrialization and support economic and social objectives, such as reducing unemployment and expanding access to energy. For instance, in Argentina (and similarly in other Latin American countries with high feedstock availability), the increase in the value added to agricultural production can create employment and improve the competitiveness of export-oriented sectors. The agriculture sector of the region has the potential to generate productivity gains, which could result in significant improvements in countries' inclusion in international trade [11]. Improvement in agricultural productivity can also play an important role in building resilience while increasing yields for farmers [35]. Countries with a low labor-productivity level in the bioeconomy sectors but abundant primary production and a sound manufacturing base could add value through bio-based methods of production [37].

The agriculture sector is also a key component of the bioeconomy strategy for middle-income countries such as Malaysia and South Africa. In Malaysia, the performance of the palm oil sector seems to somewhat determine the overall direction of Malaysian bioeconomy development [36]. In South Africa, enabling job creation through the expansion and intensification of sustainable agricultural production and processing is part of the three strategic objectives of the bioeconomy.

In the United States, with both high feedstock potential and advanced industries, the bioeconomy is based both on the expansion of biomass and on 'bio-inventions' [25,52]. However, a bioeconomy vision based on the expansion of biomass can face challenges, such as the reliable availability of raw materials due to the increased climate and severe weather impacts, water availability, and stability of the markets [43].

In contrast, some high-income countries such as the Netherlands have excluded the agriculture and food sector from their bio-based economy strategy. The main reasons for this is the limited domestic supply of ecologically sustainable biomass, which concerns several other EU countries as well. Estimates suggest that, for the EU, the sustainable biomass supply will be enough to meet about 10–20% of the final energy and feedstock consumption in 2030 [39]. Considering that land use is the most critical issue in sustainable biomass production, countries with limited land availability face relevant constraints. In countries with limited biomass availability, such as some Western European countries, the bioeconomy strategies focus more on biochemistry and bio-pharmacy benefiting from long-standing experience and R&D investments [49]. In countries focusing on high value-added bioeconomy sectors, the bioeconomy can generate higher turnover compared to the employment generated, whereas the less value-added sectors of the bioeconomy (mainly primary biomass production in agriculture, forestry, and fisheries) typically generate more employment.

Technical innovation and new business models associated with the bioeconomy should also potentially aim at decoupling economic growth from resource use in countries with available resources [51]. A sustainable bioeconomy would not foster depletion of resources, degradation of the environment, loss of biodiversity, and social injustice. As Germany recognizes in its bioeconomy strategy, the structural transition towards a bio-based economy can only be successful if it secures the supply of food, it protects the environment, the climate, and biodiversity, and it supports the development-policy objectives in developing countries and emerging economies [13].

4.3. Linking Goals and Measurement Frameworks

If a bioeconomy strategy aims to contribute to sustainable development and to environmental and social objectives (e.g., employment, food security, energy security, and mitigation and adaptation to climate change [10]), these should be clearly included in the strategy objectives and should be measurable (by means of quantitative, qualitative, or as aggregate indicators). Environmental and sustainability components in bioeconomy development approaches should be closely connected with supply and production of bio-resources, as well as with consumption patterns. In fact, the core of transformational strategies is not limited to the technological aspects but includes behavior change and institutional innovations for enabling settings and long-term incentives, both at the company and of international policy levels [38].

This study shows that a means to monitor progress in reaching the targets set in the bioeconomy policies and strategies is lacking in many countries, and the difficulty of measuring it can be a consequence of the lack of a clear definition of the bioeconomy concept and of concrete and measurable objectives. In fact, strategies often show nonmeasurable objectives and qualitative targets. In the case of South Africa and the United States, for example, the suggested output indicators of critical factors to monitor bioeconomy strategy [23] and the bio-based economy indicators and composite indicators [27] have not been measured in the practice yet due to the lack of sufficient data.

Most countries monitor bioeconomy progress just with economic values and shares of GDP, while other aspects of sustainability and resource availability are addressed only to a limited extent [53]. The GDP is a parameter which certainly gives information on the bioeconomy contribution to the economy. However, it is not ideal due to the inadequacy of the standard industrial classification systems to systematically monitor bio-based production, the lack of systematic data, and the often scattered information collected at national level. In addition, GDP is being increasingly criticized as an inappropriate indicator to measure sustainable development since it includes activities considered detrimental to humans and the environment and does not take into consideration social aspects that define human wellbeing nor the environmental aspects (which are all important information to assess the real contribution to the overall economy). Moreover, the GDP does not include transfer payments, such as subsidies for fossil fuels [54].

In addition to GDP, other economic indicators often used to measure bioeconomy are: turnover (revenue from sales); employment; resource use (crops, wood, waste, land, capital, etc.); primary production of biomass in the country (agriculture, forestry, residues, fisheries, and waste); import of biomass to the country; global land use for biomass based consumption in the country; production of bio-based products; price of biomass and bio-based products; consumption of bioeconomy products; and trade flows [18,27,34,49]. Further indicators focus on the drivers of innovation, such as investments and spending in R&D or intellectual property. However, it can be difficult to capture the impacts of a new innovation due to a time lag between investments and outcomes. These types of indicators could be used to compare country performance in the development of a bioeconomy (e.g., which countries have a bioeconomy strategy or have dedicated R&D funds).

Some countries currently measure only the effect of bio-based products on the GDP, although bioeconomy related services could be included in the measurement of the bioeconomy. For instance, Finland includes nature tourism, hunting, and fishing as bioeconomy services in its bioeconomy strategy [55]. Also, the Malaysian BCI currently measures primarily revenues and economic flows, but it could be improved to take bio-services into account as well as broader socioeconomic or environmental aspects. For instance, the BCI could incorporate measures of poverty reduction or income inequality in the bioeconomy industry, it could account for CO_2 emissions, or level of local biodiversity [42].

Nevertheless, some efforts to develop measurable social and environmental indicators to monitor the bioeconomy exist. For instance, Italy has developed a set of sustainability indicators with measurable impacts on food security, natural resources sustainability, dependence on nonrenewable resources, and climate change, in addition to economic growth [56]. These indicators on sustainability dimension are based on the results of Systems Analysis Tools Framework for the EU Bio-Based Economy Strategy (Sat-BBE) consortium [34]. In order to measure the environmental impact of the bioeconomy, the EC JRC has developed and integrated modelling framework (IMF) to implement the consequential life cycle assessment (C-LCA). This framework identifies the consequences that a decision in the foreground system has for other processes and systems of the economy, both in the analyzed background system and on other systems outside the boundaries, and allows policy impact assessment once it is fully implemented [46]. However, data gaps still need to be filled and concepts and methodology, including the IMF for the environmental impact assessment, need to be further developed and implemented.

Other environmental assessment and environmental management techniques include carbon footprinting, eco-audit, environmental and social impact assessment, and strategic environmental assessment [57]. An ongoing project, MontBioEco (from the Natural Resources Institute Finland (Luke), the Standing Committee on Agricultural Research (SCAR) Bioeconomy Strategic Working group (BSW), and CASA Ministry of Agriculture and Forestry Finland (MMM)) is developing a synthesis on bioeconomy monitoring systems in the EU Member States, including indicators and subindicators. The analysis has currently developed 22 indicators and 146 subindicators around 5 main objectives: creating jobs and maintaining competitiveness; reducing dependence on nonrenewable resources; mitigating and adapting to climate change; and ensuring food security and managing natural resources sustainably [58]. This assessment includes the development of measurable indicators and subindicators that go beyond economic monitoring.

In countries with an existing bioenergy or biofuels strategy, efforts towards monitoring the sustainability of bioeconomy can be linked with the previous efforts on biofuels, biomass, and bioenergy. Moreover, several standard, certification, and labelling initiatives already set some indications both on the "quality" of the bio-products and on their sustainability. For instance, the USDA has developed a Certified Biobased Product label which certifies the carbon content of a number of bio-products. Many existing bio-product certifications and standards give indications for monitoring environmental and social sustainability.

A further step in monitoring bioeconomy should ideally include the measurement of taxation and regulatory support. For instance, the Dutch RVO estimates also the investment in BBE R&D through tax credits and fiscal exemptions [20].

Other national studies are analyzing existing policy to assess how public finance, regulations, and capacity building can enable growth of the bioeconomy (see for instance [59] for Thailand).

Bioeconomy is also an opportunity for young people and next generations, and it is often linked to improving science, technology, engineering, and mathematics (STEM) education and training programs to meet the workforce needs. For instance, these aspects come as a key priority in the US bioeconomy strategy and the Finnish bioeconomy strategy, for which developing the bioeconomy competence base by upgrading education, training, and research is a key objective [55]. Also, the South African bioeconomy strategy and the indicators suggested in the strategy, being driven by the Department of Science and Technology, have an "innovation" bias. In fact, most indicators in the strategy are related to science, technology, and innovation, as they are derived from the measurement of a knowledge-based economy or biotechnology innovation policies [23].

Finally, in Argentina, Malaysia, and South Africa, one of the bioeconomy objectives is to strengthen infrastructure to support economic growth and increase access to national and international markets [10,15,23]. Therefore, this aspect should ideally be included in a pathway towards a sustainable bioeconomy monitoring including socioeconomic and environmental impacts. Other aspects, such as the inclusion of gender (not mentioned as priority in any of the strategy analyzed), may be included in the measurement framework, in order to reflect country's priorities and strategy. Consideration of social and environmental aspects would not only minimize the risks associated with bioeconomy transition but would also enable evaluation of the real picture of the bioeconomy impacts. Similarly, in countries heavily depending on the import of biomass, a broader perspective on the evaluation of bioeconomy would enable internalization of the potential negative impacts on societies and the environment at a global scale [60].

5. Conclusions

This study underlines the lack of a homogenous definition of bioeconomy across the countries analyzed, which does not allow any straightforward comparison of the relevance of bioeconomy in the different economies.

The sectors comprised mostly reflect priorities identified by the country and comparative advantages linked, for instance, to availability of natural resources, traditional industries, labor productivity, and past investments in R&D. For instance, the agri-food sector is identified as a priority for Argentina, Malaysia, and South Africa, while the Netherlands and the Unites States focus more on nonfood sectors.

Most countries measure bioeconomy progress just with economic values and shares of GDP. On top of the lack of international consensus on which products and activities are comprised within the bioeconomy, the GDP approach has several limitations due to the inadequacy of the standard industrial classification systems to systematically monitor bio-based production, the lack of systematic data, and the often scattered information collected at national level.

Some ongoing efforts aim to harmonize the definition and measurement of bioeconomy, at least across macroregions (such as the EU) will allow development of structured and comparable measurement and monitoring methodology of the trends in bioeconomy, at least for some sets of countries (as the EC already does for few economic indicators).

Usually, bioeconomy strategies also consider intangible aspects, such as institutional set-up, policies, governance, regulations, incentives, and financial instruments, which create an enabling environment for the bioeconomy as well as social and environmental issues. Several countries emphasize the role that bioeconomy plays in their development strategy, which is an important aspect to reflect in their measurement efforts. This could allow monitoring, for instance, progress in meeting the SDGs or environmental targets. In fact, important synergies between countries commitments towards the measurement of SDGs and bioeconomy can be leveraged.

In order to facilitate the measurement and monitoring of bioeconomy at a national level, the governments could enhance and coordinate communication between different domestic agencies and entities and establish protocols for sharing data, formalize bio-based industry measurement standards, develop a comprehensive survey for bio-based industry and commodity usage, and review and revise industry classification systems. Ongoing efforts aim to harmonize the definition and measurement of bioeconomy, at least across macroregions such as the EU. These efforts will allow for the structured and comparable measurement and monitoring of the trends in bioeconomy. These efforts should go hand in hand with the development of relevant and comprehensive guidelines on how to measure the sustainability of the bioeconomy, possibly agreed at an international level. These sets of indicators should also consider the social and environmental dimensions of sustainability, since their omission or exclusion might lead to an overlook of the potential stresses caused by bioeconomy on the social wellbeing of communities and the environmental viability of ecosystems.

Author Contributions: M.G.S.J., S.B. and A.F. conceived and designed the paper; S.B., O.C. and M.G.S.J. conducted scientific literature review and collected the data; S.B. and A.F. analyzed and interpreted the data; S.B. and O.C. wrote the paper; A.F. supervised the work.

Acknowledgments: This study was funded in the framework of FAO project GCP/GLO/724/GER "Towards Sustainable Bioeconomy Guidelines (SBG)" supported by the German Federal Ministry of Food and Agriculture (BMEL). Inputs and comments were provided by several members of the International Sustainable Bioeconomy Working Group (ISBWG). The authors are particularly grateful to Olivier Dubois, FAO Senior Natural Resources Officer; Zurina Che Dir, Bioeconomy Corporation Malaysia; Ben Durham, Department of Science and Technology, South Africa; Mariano Lechardoy, Argentine Ministry of Agroindustry; Marte Mathisen, Nordic Council of Ministers; Elina Nikkola, Finnish Ministry of Agriculture and Forestry; Timothy Ong, Malaysia's National Innovation Agency; Enrico Prezio, European Commission; Tilman Schachtsiek, German Ministry for Agriculture and Food; Cinzia Tonci, Italian Ministry of Economic Development; Jan W.J. van Esch, Dutch Ministry of

Economic Affairs and Climate Policy; Francis X. Johnson, Rocio A. Diaz-Chavez and Matthew Fielding, Stockholm Environment Institute, for their valuable inputs.

Conflicts of Interest: The authors declare no conflict of interest. The founding sponsors had no role in the design of the study; in the collection, analyses, or interpretation of data; in the writing of the manuscript, and in the decision to publish the results. The views expressed in this report reflect those of the authors and do not necessarily reflect the views of the Food and Agriculture Organization of the United Nations.

References

1. McCormick, K.; Kautto, N. The Bioeconomy in Europe: An Overview. *Sustainability* **2013**, *5*, 2589–2608. [CrossRef]
2. De Besi, M.; McCormick, K. Towards a bioeconomy in Europe: National, regional and industrial strategies. *Sustainability* **2015**, *7*, 10461–10478. [CrossRef]
3. Richardson, B. From a fossil-fuel to a biobased economy: The politics of industrial biotechnology. *Environ. Plan. C Gov. Policy* **2012**, *30*, 282–296. [CrossRef]
4. Bugge, M.M.; Hansen, T.; Klitkou, A. What is the bioeconomy? A review of the literature. *Sustainability* **2016**, *8*. [CrossRef]
5. German Bioeconomy Council. *Bioeconomy Policy (Part II): Synopsis of National Strategies around the World*; German Bioeconomy Council: Berlin, Germany, 2015.
6. Viaggi, D. Towards an economics of the bioeconomy: Four years later. *Bio-based Appl. Econ.* **2016**, *5*, 101–112. [CrossRef]
7. Food and Agriculture Organization of United Nations (FAO). *How Sustainability Is Addressed in Official Bioeconomy Strategies at International, National, and Regional Leveles—An Overview*; Food and Agriculture Organization of United Nations: Rome, Italy, 2016; ISBN 978-92-5-109364-1.
8. Axelsson, L.; Franzén, M.; Ostwald, M.; Berndes, G.; Lakshmi, G.; Ravindranath, N.H. Perspective: Jatropha cultivation in southern India: Assessing farmers' experiences. *Biofuels Bioprod. Biorefin.* **2012**, *6*, 246–256. [CrossRef]
9. SAT-BBE. *Annotated Bibliography on Qualitative and Quantitative Models for Analysing the Bio-Based Economy*; Systems Analysis Tools Framework for the EU Bio-Based Economy Strategy: The Hague, The Netherlands, 2014; Deliverable 2.3.
10. MINAGRO. *BioEconomía Argentina Visión desde Agroindustria*; MINAGRO: Buenos Aires, Argentina, 2016.
11. Wierny, M.; Coremberg, A.; Costa, R.; Trigo, E.; Regúnaga, M. *Measuring the Bioeconomy: Quantifying the Argentine Case*; Grupo Bioeconomia: Buenos Aires, Argentina, 2015.
12. Bundesministerium für Ernährung und Landwirtschaft (BMEL). *Bioeconomy in Germany: Opportunities for a Bio-Based and Sustainable Future*; Bundesministerium für Ernährung und Landwirtschaft (BMEL): Berlin, Germany, 2015.
13. Bundesministerium für Ernährung und Landwirtschaft (BMEL). *National Policy Strategy on Bioeconomy. Renewable Resources and Biotechnological Processes as a Bais for Food, Industry and Energy*; Bundesministerium für Ernährung und Landwirtschaft (BMEL): Berlin, Germany, 2014.
14. Bioeconomy Corporation. *Bioeconomy Transformation Program: Enriching the Nation, Securing the Future*; Bioeconomy Corporation: Kuala Lumpur, Malaysia, 2016.
15. Bioeconomy Corporation. *Bioeconomy Transformation Programme: Enriching the Nation*; Bioeconomy Corporation: Kuala Lumpur, Malaysia, 2007.
16. Meesters, K.P.H.; van Dam, J.E.G.; Bos, H.L. *Protocol Monitoring Materiaalstromen Biobased Economie*; Rijksdienst voor Ondernemend Nederland (RVO): Wageningen, The Netherlands, 2013.
17. CE Delft. *Sustainable Biomass and Bioenergy in The Netherlands: Report 2015*; CE Delft: Delft, The Netherlands, 2016.
18. The Netherlands Offices for Science and Technology (NOST). *The Bio-Based Economy in the Netherlands*; The Netherlands Offices for Science and Technology (NOST): The Hague, The Netherlands, 2013.
19. NNFCC-The Bioeconomy Consultants. *Bioeconomy Factsheet—Netherlands*; NNFCC-The Bioeconomy Consultants: York, UK, 2015.
20. Kwant, K.; Hamer, A.; Siemers, W.; Both, D. *Monitoring Biobased Economy in Nederland 2016*; Rijksdienst voor Ondernemend Nederland (RVO): Wageningen, The Netherlands, 2017.

21. European Commission. Bioeconomy Knowledge Centre. Bioeconomy Data Catalogue. Available online: https://data-bioeconomy.jrc.ec.europa.eu/ (accessed on 4 April 2018).
22. Public Understanding of Biotechnology (PUB). *South Africa Launches its Bio-Economy Strategy*; The South African Agency for Science and Technology Advancement: Pretoria, South Africa, 2014.
23. Department of Science and Technology. *The Bio-Economy Strategy*; Department of Science and Technology: Pretoria: South Africa, 2013.
24. NACI Council. *NACI Annual Report 2016/17*; NACI Council: Pretoria, South Africa, 2017.
25. The White House. *National Bioeconomy Blueprint*; The White House: Washington, DC, USA, 2012.
26. U.S. Department of Energy. *2016 Billion-Ton Report: Advancing Domestic Resources for a Thriving Bioeconomy, Volume 1: Economic Availability of Feedstocks*; U.S. Department of Energy: Washington, DC, USA, 2016; Volume 1160.
27. United States Department of Agriculture (USDA). *Biobased Economy Indicators*; United States Department of Agriculture (USDA): Washington, DC, USA, 2011.
28. United States Department of Agriculture (USDA). *An Economic Impact Analysis of the U.S. Biobased Products Industry: A Report to the Congress of the United States of America*; United States Department of Agriculture (USDA): Washington, DC, USA, 2015; Volume 11.
29. United States Department of Agriculture (USDA). *An Economic Impact Analysis of the U.S. Biobased Products Industry: 2016 Update*; United States Department of Agriculture (USDA): Washington, DC, USA, 2016.
30. U.S. Department of Energy. *The U.S. Bioeconomy by the Numbers 2017*; U.S. Department of Energy: Washington, DC, USA, 2017; Volume 1.
31. Van Esch, J.W.J. *Personal Communication. Questionnarie: Assessing Bioeconomy Contribution to Countries' Economy*; Dutch Ministry of Economic Affairs and Climate Policy: The Hague, The Netherlands, 2018.
32. Che Dir, Z. *Personal Communication. Questionnarie: Assessing Bioeconomy Contribution to Countries' Economy*; Bioeconomy Corporation: Kuala Lumpur, Malaysia, 2018.
33. Durham, B. *Personal Communication. Questionnarie: Assessing Bioeconomy Contribution to Countries' Economy*; Department of Science and Technology: Pretoria, South Africa, 2018.
34. SAT-BBE. *Tools for Evaluating and Monitoring the EU Bioeconomy: Indicators*; Systems Analysis Tools Framework for the EU Bio-Based Economy Strategy: The Hague, The Netherlands, 2013; Deliverable 2.2.
35. The European Innovation Partnership 'Agricultural Productivity and Sustainability' (EIP-AGRI). *EIP-AGRI Workshop "Opportunities for Agriculture and Forestry in the Circular Economy"*; Workshop Report; European Commission: Brussels, Belgium, 2015.
36. Bioeconomy Corporation. *Analysing the Contribution of Malaysian Bioeconomy Using the GDP Approach*; Bioeconomy Corporation: Kuala Lumpur, Malaysia, 2015.
37. Ronzon, T.; Piotrowski, S.; M'Barek, R.; Carus, M. A systematic approach to understanding and quantifying the EU's bioeconomy. *Bio-based Appl. Econ.* **2017**, *6*, 1–17. [CrossRef]
38. Von Braun, J. Bioeconomy and sustainable development—Dimensions. *Rural* **2014**, *21*, 6–9.
39. PBL. *Sustainability of Biomass in a Bio-Based Economy*; PBL Netherlands Environmental Assessment Agency: The Hague, The Netherlands, 2012.
40. CE Delft. *Sustainable Biomass and Bioenergy in The Netherlands: Report 2016*; CE Delft: Delft, The Netherlands, 2017.
41. Schachtsiek, T. *Personal Communication. Questionnarie: Assessing Bioeconomy Contribution to Countries' Economy*; Federal Ministry for Agriculture and Food (BMEL): Bonn, Germany, 2018.
42. Al-Amin, A.Q. *Developing a Measure for Quantifying Economic Impacts: The Bioeconomy Contribution Index*; Bioeconomy Corporation: Kuala Lumpur, Malaysia, 2015.
43. Golden, J.S.; Handfield, R.B. *Why Biobased? Opportunities in the Emerging Bioeconomy*; US Department of Agriculture, Office of Procurement and Property Management: Washington, DC, USA, 2014.
44. Rogers, J.N.; Stokes, B.; Dunn, J.; Cai, H.; Wu, M.; Haq, Z.; Baumes, H. An assessment of the potential products and economic and environmental impacts resulting from a billion ton bioeconomy. *Biofuels Bioprod. Biorefin.* **2017**, *11*, 110–128. [CrossRef]
45. Nattrass, L.; Biggs, C.; Bauen, A.; Parisi, C.; Rodríguez-Cerezo, E.; Gómez-Barbero, M. *The EU Bio-Based Industry: Results from a Survey*; Joint Research Centre (JRC) Technical Report; Joint Research Centre: Brussels, Belgium, 2016; Volume EUR 27736 EN. [CrossRef]

46. Ronzon, T.; Lusser, M.; Landa, L.; M'Barek, R.; Giuntoli, J.; Cristobal, J.; Parisi, C.; Ferrari, E.; Marelli, C.; Torres de Matos, C.; et al. *Bioeconomy Report 2016*; Klinkenberg, M., Sanchez Lopez, J., Hadjamu, G., Belward, A., Camia, A., Eds.; Joint Research Centre (JRC) Scientific and Policy Report; Joint Research Centre: Brussels, Belgium, 2017; Volume EUR 28468.

47. Mulder, M. *National Biotech Survey*; Egolibio: Pretoria, South Africa, 2003.

48. Johnson, F.X. Biofuels, Bioenergy and the Bioeconomy in North and South. *Ind. Biotechnol.* **2017**, *13*, 289–291. [CrossRef] [PubMed]

49. European Commission. Jobs and Turnover in the European Union Bioeconomy. European Commission: EU Sciente Hub, DataM. 2018. Available online: https://datam.jrc.ec.europa.eu/datam/mashup/BIOECONOMICS/index.html (accessed on 21 May 2018).

50. Ekins, P.; Hughes, N.; Bringezu, S.; Clarke, C.A.; Fischer-Kowalski, M.; Graedel, T.; Hajer, M.; Hashimoto, S.; Hatfield-Dodds, S.; Havlik, P.; et al. *Resource Efficiency: Potential and Economic Implications. A Report of the International Resource Panel*; United Nations Environment Programme: Nairobi, Kenya, 2016; ISBN 9789280736458.

51. Preston, F.; Lehne, J. *A Wider Circle? The Circular Economy in Developing Countries*; Chatham House, The Royal Institute of International Affairs: London, UK, 2017; ISBN 978 1 78413 256 9.

52. US Biomass R&D Board. *Federal Activities Report on the Bioeconomy*; US Biomass R&D Board: Washington, DC, USA, 2016.

53. Staffas, L.; Gustavsson, M.; McCormick, K. Strategies and policies for the bioeconomy and bio-based economy: An analysis of official national approaches. *Sustainablity* **2013**, *5*, 2751–2769. [CrossRef]

54. Van den Bergh, J.C.J.M. Abolishing GDP. *SSRN Electron. J.* **2007**. [CrossRef]

55. Finnish Ministry of Employment and the Economy. *The Finnish Bioeconomy Strategy*; Finnish Ministry of Employment and the Economy: Helsinki, Finland, 2014.

56. Presidency of Council of Ministers. *Bioeconomy in Italy*; Presidency of Council of Ministers: Roma, Italy, 2017.

57. Stockholm Environment Institute. *Re-linking Objectives and Potentials*; Stockholm Environment Institute: Stockholm, Sweden, 2018.

58. Natural Resources Institute Finland (LUKE). *MontBioeco—Synthesis on Bioeconomy Monitoring Systems in the EU Member States*; LUKE: Helsinki, Finland, 2018.

59. Fielding, M.; Aung, M.T. *Bioeconomy in Thailand: A Case Study*; Stockholm Environment Institute: Stockholm, Sweden, 2018.

60. Lewandowski, I.; Faaij, A.P.C. Steps towards the development of a certification system for sustainable bio-energy trade. *Biomass Bioenergy* **2006**, *30*, 83–104. [CrossRef]

![sustainability logo] *sustainability*

MDPI

Article

Bridging the Gaps for a 'Circular' Bioeconomy: Selection Criteria, Bio-Based Value Chain and Stakeholder Mapping

Kadambari Lokesh [1,*], Luana Ladu [2] and Louise Summerton [1]

[1]   Green Chemistry Centre of Excellence, University of York, York YO10 5DD, UK;
      louise.summerton@york.ac.uk
[2]   Faculty of Economics and Management, Technische Universität Berlin, Sekr. MAR 2-5, Marchstraße 23,
      10587 Berlin, Germany; luana.ladu@tu-berlin.de
*   Correspondence: kadambari.lokesh@york.ac.uk

Received: 20 April 2018; Accepted: 14 May 2018; Published: 23 May 2018

Abstract: Bio-products and bio-based value chains have been identified as one of the most promising pathways to attaining a resource-efficient circular economy. Such a "valorization and value-addition" approach incorporates an intricate network of processes and actors, contributing to socio-economic growth, environmental benefits and technological advances. In the present age of limited time and funding models to achieve ambitious sustainable development targets, whilst mitigating climate change, a systematic approach employing two-tier multi-criteria decision analysis (MCDA) can be useful in supporting the identification of promising bio-based value chains, that are significant to the EU plans for the bio-economy. Their identification is followed by an elaborate mapping of their value chains to visualize/foresee the strengths, weaknesses, opportunities and challenges attributable to those bio-based value chains. To demonstrate this methodology, a systematic review of 12 bio-based value chains, prevalent in the EU, sourcing their starting material from biomass and bio-waste, has been undertaken. The selected value chains are mapped to visualize the linkages and interactions between the different stages, chain actors, employed conversion routes, product application and existing/potential end-of-life options. This approach will help chain-actors, particularly investors and policy-makers, understand the complexities of such multi-actor systems and make informed decisions.

Keywords: value chain; multi-criteria decision analysis; circular economy; value chain-mapping; bioeconomy

---

1. Introduction

Escalating environmental and economic pressure to use our resources responsibly and add value to the used material/products in the commercial sphere has helped the development of technology routes and material circularity in nearly every global sector. According to the EU Circular Economy Strategy, the aim of such systems thinking is to "close the loop by becoming resource efficient through development and establishment of industrial symbiosis, to reduce the pressure on EU's natural capital" [1].

The approach to attaining/creating a circular economy is cascading of material, which may be virgin raw materials, by-products or wastes resulting from any given sector. The concept of cascading and its significance to the establishment and growth of a resource/energy efficient, green and low-carbon economy has been a recurring theme in EU policies since 2012, particularly in the EU Forest Strategy, EU Bioeconomy Strategy and EU Circular Economy package [2]. To understand the state of our transition, transparency on the EU-level biomass potential is essential. Bioeconomy, according to the European Commission, is a part of the economy that utilises bio-based renewable resources sourced both from land and sea, processed to produce materials and energy for consumption [3]. A fully-functional bioeconomy is one of the many pathways that have been identified to attaining a circular economy, both at micro-level (local rural development) and macro-level (nation-wide) [4]. The principles of both circular economy and bioeconomy are in synergy, in terms of the ultimate goal of attaining a sustainable technological and socio-economic development by decoupling economic growth from resource exhaustion and subsequent environmental degradation [5]. The two economies share a growing overlap, however, there is currently a need to push circularity down to the consumer level, which is where the largest share of waste is found with no end-of-life valorisation. This has been stated as one of the biggest challenges of the bioeconomy. However it is acknowledged that some sectors of the bioeconomy cannot satisfy the principles of circular economy (e.g., bioenergy and biofuel) because they are considered a dead end route for biomass [6]. In terms of current targets for sustainable growth, bioeconomy, in combination with circular economy, has the potential to directly contribute to 11 out of the 17 UN's Sustainability Development Goals (SDGs) (Figure 1). The direct contribution of circular and bio-based economy to sustainable consumption and production (SDG 12), reducing our pressure on the environment, air, water and land (SDG 13, 14, 15) is the ultimate aim of the concept. By working in partnership with rural communities and local bio-based infrastructure [7,8] (SDG17), utilising the rural knowledge pool, alleviation of poverty (SDG 1 and 2), forging skills among communities (SDG 4) to take an interest in guarding the local ecosystem services encourages the development of sustainable communities (SDG11), in addition to creating jobs and socio-economic opportunities (SDG 8). Use of bioenergy, devising smart strategies and value-chain pathways to lock the chain's GHG emissions, either via carbon capture or soil incorporation of high quality biochar, have been identified as potential means of achieving the ambitious Paris climate target [9].

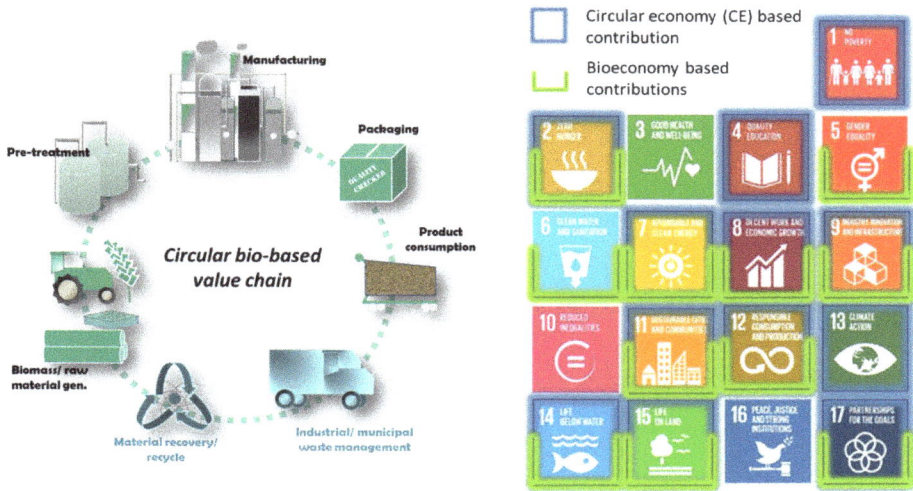

Figure 1. Potential for circular bio-based value chain to contribute to achieving UN's SDGs and the potential of value chain mapping and analysis in quantifying these goals.

Having comprehended the synergies between a circular economy and a bioeconomy and its potential for contributing to the global sustainability targets from a number of earlier studies [5,10], it is crucial to understand where we stand to take appropriate and smarter "next steps". Prior to exploring the opportunities that bioeconomy could offer, the probability of it being successfully established must also be systematically investigated.

A methodical approach for the selection and mapping of the most promising circular bio-based value chain is presented in this paper. The purpose of the methodology expressed in this paper is to encourage innovative thought processes, within the chain-actors and other external stakeholders, on how to identify and focus precious resources and efforts in the development of multi-functional value chains and business models using a set of selection criteria, and to be able to foresee the complexities and performance needs of the value chain, thereby minimising risks and shocks to the conceptual process system. This approach is clearly missing, not only within the bio-based sector but also within industrial sectors that are gearing-up towards a transition to circularising their supply chains. The suggested methodology can, therefore, be adopted for application within non-bio-based value/supply chains as well. To demonstrate this methodology, some exemplary bio-based value chains have been selected from a pool of bio-based value chains that are prevalent in the EU. This study employs a two-tier multi-criteria decision analysis to identify these most promising value chains.

1.1. Background

A value-chain is defined as a set of interlinked activities that deliver products/services by adding value to bulk material (feedstock). In a bio-based value chain, the feedstocks tend to be biomass drawn from an existing primary production route (e.g., agriculture, forestry and livestock), or of a novel (e.g., microalgae) or secondary origin (e.g., sludge, industrial wastewater and household organic waste). A generalised schematic for an ideally circular bio-based value chain has been presented in Figure 2.

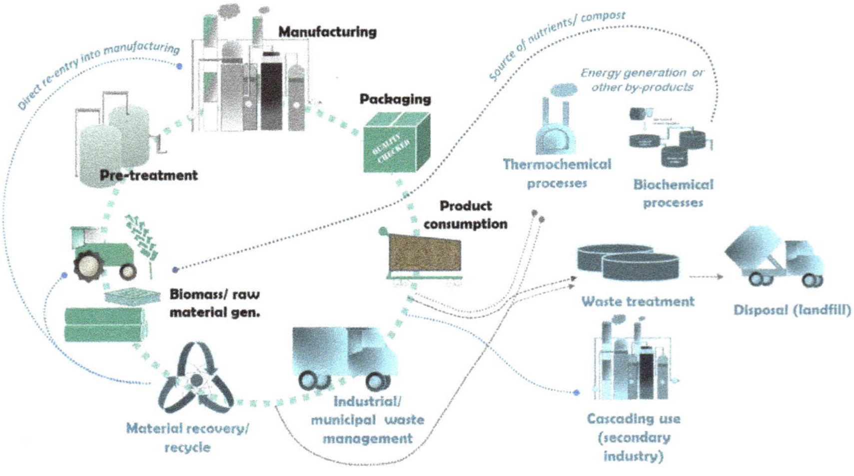

Figure 2. Figure 2. A generalised map of a bio-based value chain.

Value chains, in particular those that valorise secondary resources are designed to turn available organic material into different valuable product, ranging from high-value chemicals to secondary-use by-products and renewable energy [11]. Pathways that are capable of transforming waste/secondary feedstock into an array of high value products are called integrated biorefineries [12]. Integrated biorefineries contain a "pre-treatment plant" that prepares the feedstock for upcoming transformation and refining technologies within the supply chains, before packaging and distribution.

The first stage of a complex bio-based value chain is biomass availability. Brosowski et al. (2016) define the quantity of biomass, generated by a confined area of land (country) that is currently used or has the potential to be used as the "biomass potential" [13]. Biomass potential may be measured from a number of relevant sustainability-based angles: theoretical, environmental, economic and sustainable. Theoretical potential provides an estimation of potential biomass productivity based on the physical characteristics of all available arable land. Environmental and economic biomass potential, based on the fraction of the theoretical biomass potential, gradually adds land exclusions such as legally protected area, slopes, biodiversity richness and take into account the technical capability of the pre-existing biomass processing framework. Sustainability potential is the final filter that takes the technical, economic and environmental restrictions (and associated biomass capacities) into account providing the net estimated biomass production capacity for a given geographical location [14]. Focusing on EU biomass supply, the agricultural and forestry feedstock within the EU constitutes 1.13 billion tonnes of dry biomass [15]. From a stakeholder perspective, which draws data via engagement (interviews and survey questionnaire) with EU-based value chain actors, more than 40% of renewable material is invested in non-conventional industrial applications in EU-28. In such successful bio-based industries there are in-house developed frameworks for value-chain actors to communicate and synchronise their operations [15]. From the perspectives of bio-based industries, according to a 2016 study undertaken by the Joint Research Centre (JRC), European Commission, EU-28 has been determined to be home to 133 bio-based industries, excluding the relevant industrial research and development (R&D) institutions [16]. Nevertheless, a survey undertaken by the Bio-based Industries Consortium (BIC) and Nova Institute [17], revealed 224 bio-based industries with biorefineries processing different types of feedstock, mapped across EU demonstrating a sharp growth, as presented in Figure 3.

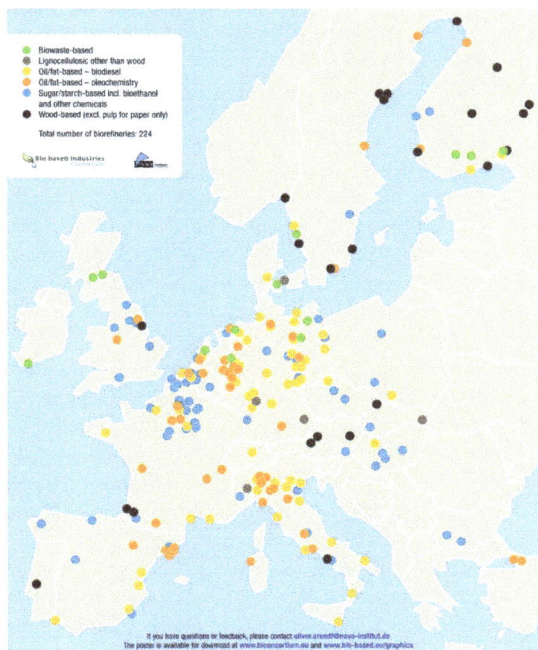

Figure 3. Biorefinery map in Europe (Source: [17]).

Long-term, innovative systems thinking which encourages the exploitation of organic waste/residues from agriculture, animal husbandry, domestic, industrial and commercial industries are exploited, is essential. Such solutions not only help gain access to and expand the innovation boundaries but also enable a systematic and feasible transition to a bioeconomy. Such a transition has been identified to benefit the economy through the creation of SMEs and skilled employment opportunities, in addition to reaching EU's climate change mitigation targets and to reduce dependence on fossil-derived resources [18]. With mounting pressure from a number of factors including limited time, finance, resources and impending environmental targets, the suggested method helps envisage the potential performance of any bio-based value chains. This is crucial to not only policy-makers, but also to any decision-making stakeholder in the value chain, in terms of material, financial, human resource allocation and operation [19,20]. From the socio-economic perspective, creation of a multi-regional/local value chain, networks, growth of SME's and other employment opportunities, development of waste-management infrastructure (where lacking), local skill-forging and knowledge dissemination are some of the practical benefits of a fully-functional bio-based value chain [21].

From a sectoral perspective, the EU biorefinery map (Figure 3) represents the prevalence of high numbers of oil and fat based biorefineries dedicated to the production of biofuel and oleochemical products.

1.2. Bioeconomy Strategies Initiatives

The primary policy framework of the European bioeconomy is represented by the European Bioeconomy Strategy and Action plan adopted by the EU in 2012 [22]. It provides policy framework conditions for developing new technologies and processes for the production and commercialization of renewable biological resources and their conversion into bio-based products. An ongoing revision of the bioeconomy strategy aims at establishing a more coherent and holistic policy and financial framework for the European bio-based economy, supporting access to sustainably produced biomass, fostering investments and further developing the commercialization of bio-based products. One important recommendation is to further develop the synergies and complementarities between EU policies and funding instruments with interconnected objectives with the bioeconomy. Among these policies, an important role is played by the circular economy strategy and by other sectorial policies that govern traditional sectors of the bioeconomy (e.g., the Common Agricultural Policy [CAP]).

However, from a value chain and bio-product perspective, a majority of current bioeconomy strategies are dedicated to bioenergy and biofuel-based value chains, followed by food and beverage chains. Prevalence of many such value chains may be attributed to the EU's race towards renewable energy consumption targets set in the Energy Strategy for 2030 [23]. However, a surge in integrated biorefineries that synthesise bio-based products other than biofuels is evident from the most recent report compiled and analysed from a stakeholder engagement approach undertaken by the Bio-based Industrial Consortium (BIC) and Nova Institute [17]. The growth of biomass-cascading biorefineries is also supplemented by the keenness of European bio-based industries to valorise organic-rich bio-waste (mainly agricultural residue and sludge). Besides, seizing an opportunity to synthesise value-added products from low-cost feedstock, an unhindered supply of starting material (one of the key barriers to bio-product synthesis), is a promising start for a bio-based business model. According to a study undertaken by Meyer-Kohlstock et al., (2015), the supply of biomass and waste for consumption within other value chains was predominantly sourced from the agri-food and livestock sector [24]. Though bio-based value chains create opportunities to circularise, some chain-level dynamics influence the environmental, commercial and social practicability of the same by varying degrees. These dynamic factors include biomass supply logistics, feedstock costs (influenced by whether the feedstock is primary or secondary), feedstock treatment requirements and ethical compliance requirements. Firstly, it is essential to identify a bio-based value chain with techno-economic potential, and that is environmentally and socio-economically sensible. Secondly, when developing a bio-based business model around the identified value chain, it is imperative to anticipate the various interactions among processes, stakeholders and related process-dynamics on the overall performance of the chain. The aim of this paper is to provide a two-fold methodology that helps identification of promising bio-based value chains and to demonstrate how value chain mapping can help stakeholders visualise the various interactions embedded in a value chain. Escalating environmental and economic pressure to use our resources responsibly and add value to the used material/products in the commercial sphere has helped the development of technology routes and material circularity, in nearly every global sector.

According to the EU Circular Economy Strategy, the aim of such systems thinking is to "close the loop by becoming resource efficient through development and establishment of industrial symbiosis, to reduce the pressure on EU's natural capital" [1].

2. Methodology

2.1. Value Chain Selection Criteria

Multi-criteria decision analysis (MCDA) is a valuable tool for decision making in complex process systems using multiple parameters that influence the embedded processes within a value chain. These parameters can be differently weighed as "significant factors" by the various chain-actors. Also, incorporating this flexibility into the scope of this assessment, MCDA in decision making enables a systematic investigation and transparency in analysis [20]. The goal of MCDA, in general, is to provide an opportunity to explore the knowledge and concerns put forward by the chain-actors, weigh them from an unbiased viewpoint, systematically analyse, identify the most important criteria, and subsequently, make decisions within a complex multi-actor process systems. This study employs two-tier MCDA to rank bio-based value chains based on a set of selection criteria, highlighting the significance of their adherence to the principles of circular economy. The outcomes of this analysis highlight the importance of these selection criteria dedicated to highlighting the circularity characteristics of any bio-based value-chain/business model. An elaborated mapping methodology to understand the strengths, weaknesses, opportunities and challenges embedded in a bio-based value chain, attributable to the synthesis of a variety of bio-products, also demonstrating the significance of upstream processes and material use on the downstream activities (mainly post-consumption and end-of-life management) has been presented as a part of this paper. Please see Figure 4 for a flow diagram that elaborates the MCDA methodology employed in this study for value chain selection and mapping.

A variety of bio-based value chains have been identified to be prevalent in the EU and are presented in Table 1. This list of preliminary value chains was drawn from the literature review earlier, focusing on bio-based value chains/products covered by EU certification schemes and market demand [25]. The list comprises bio-based value chains with (but not limited to) diverse characteristics covering:

- From virgin food-based feedstock to bio-waste cascading;
- 100% bio-based to partially bio-based, value chains;
- Those with a fully-functional waste management infrastructure to those that lack one;
- Diverse product functionality.

Table 1. List of EU-based value chains considered for selection, analysis and mapping exercise.

Sector	Value Chain
Chemicals	Cellulose to bio solvents
Disposable food packaging	Starch to bioplastic food packaging
Agriculture	Starch to bio-based mulch films
Fabrication	Starch to bioplastics for fabrication
Automotive	Vegetable fats to bio lubricants
Agriculture/waste management	Solid biomass to fine chemicals
Textiles	Cellulose to fabric
Food packaging	Cellulose to plastic paper cups
Construction	Waste biomass to insulation material
Construction	Waste biomass to wood-plastic composites
Agriculture	Polysaccharides to crop health inducers
Animal husbandry	Plant-based chemicals to fine chemicals

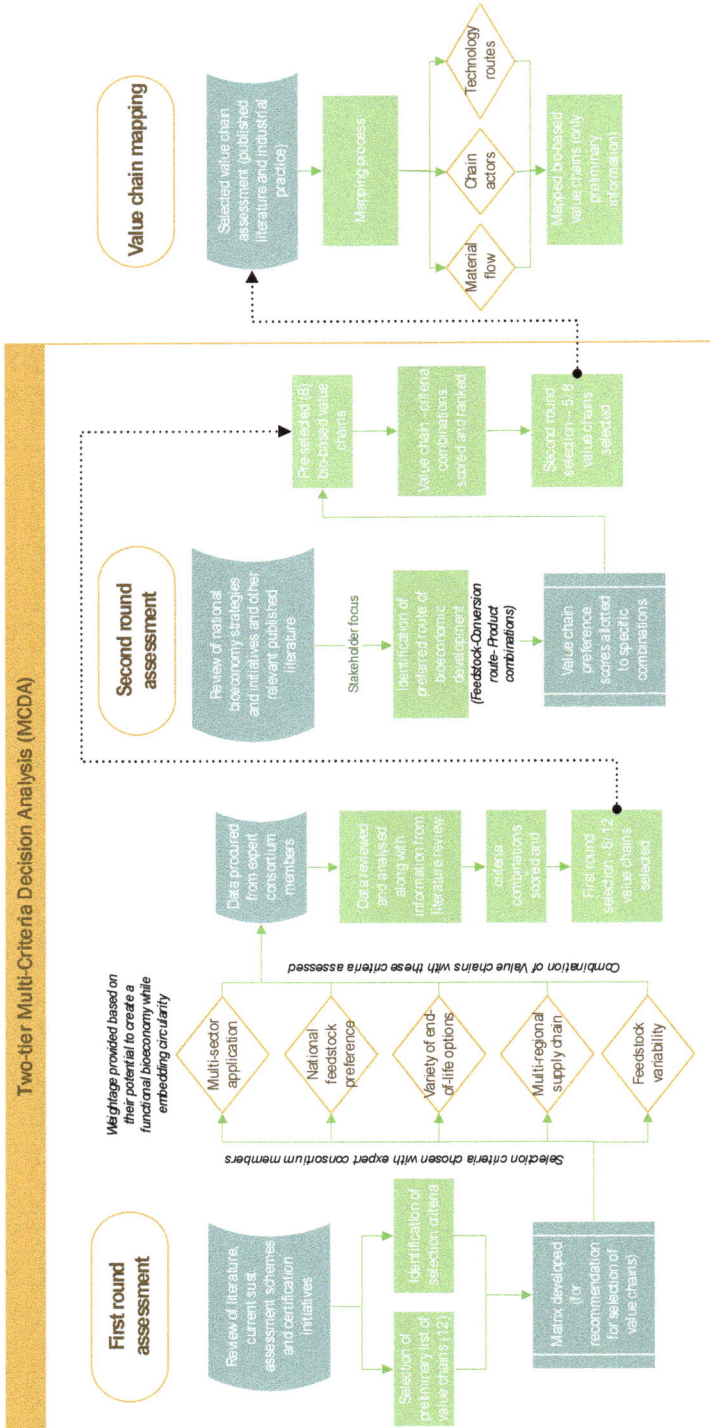

Figure 4. Flow diagram of the methodology involved in the two-tier multi-criteria decision analysis and mapping exercise adopted for the selection of the most promising bio-based value chains, in the context of a circular economy.

For systematic identification of promising value chains, a multi-criteria selection approach was used. A set of selection criteria were chosen based on the gaps that were identified from review of published literature and policy. There are a number of different bio-based value chains prevalent in the EU. To identify those multi-functional value chains (perhaps under-represented), for example, those that integrate material circularity, utilising agricultural by-products/waste, capability to create better and wider socio-economic growth/employment opportunities and needing less or no further investment in the form of a dedicated infrastructure, it is essential to use a few key criteria. The performance of candidate value chains from the viewpoint of these selected criteria must be assessed in detail. This approach will help this study weigh the potential and resilience of these candidates in the commercial market, against the backdrop of the EU's bioeconomy policies and in our journey towards "closing the loop". The selection criteria chosen for the first-round assessment of bio-based value chains are presented in Figure 5. The rationale for choosing these selection criteria may be found under the respective descriptions in the upcoming segments.

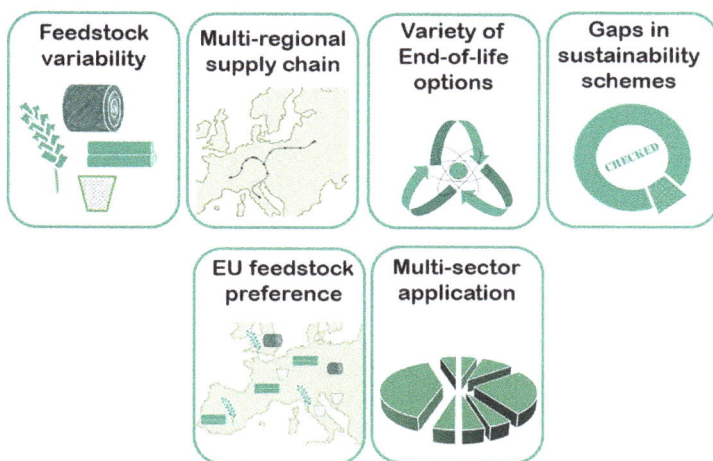

Figure 5. Selection criteria for the identification of the most promising bio-based value chains.

Feedstock variability: The flexibility of bio-based value chains to produce products and by-products from a variety of feedstocks is crucial. Biomass, the starting material of any bio-based value chain, is cost-susceptible to both market volatility and also seasonal in nature. Dependence on a seasonal feedstock will lead to a seasonal value chain, thereby resulting in seasonal products (and byproducts) which gives bio-based products a less attractive perspective among consumers [26].

Multi-regional supply chain: A multi-regional value chain, besides adding value to a low-value feedstock, also contributes to the economic growth of dependent communities via creation of jobs, development of skills and the knowledge pool of the local communities, leading to improved community wellbeing and social equity. Such approaches require a harmonised approach to reporting and communication of information among the embedded chain actors for transparency on practices and traceability of materials. However, such an approach could facilitate EU states with transition economies to establish bioeconomy models, with the needed investment from national funding initiatives [27].

Variety of end-of-life: End-of–life characteristics play a prominent role at any given stage of a value chain. From a top-down approach, a process that is capable of utilising waste biomass for raw feedstock, also called "cascading use" is a valuable, sustainable business model as there will a regular influx of low-cost feedstock, promising a continuous product supply to the market. From a "bottom-up" approach, strategic management and utilisation of waste (post-product consumption) is capable of delivering three-fold benefits: environmentally through reduction of waste for treatment and disposal; economically by enabling resource efficiency and through transformation of waste (as low-cost raw material for a secondary industry); and socially through creation of jobs, new value chains and social equity [28]. To be able to "catch up" with the 2014 EU Landfill Directive (which aims to phase out landfilling recyclable waste, e.g., bioplastics, paper, glass and bio-waste), we need to identify candidate value chains that generate products that can potentially circularise the value chain. Selection of value chains based on the capabilities of the products to demonstrate a variety of end-of life characteristics would be valuable to report via this study.

Gaps in sustainability schemes: From assessing the outcomes of the literature review, it is evident that sustainability schemes for bio-based products (e.g., bioplastics, bio-solvents, bio-based adhesives and binders, enzymes and cosmetics, etc. but not bioenergy) are either still in their infancy or have variable levels of maturity with major sustainability related gaps to cover. Some major gaps and limitations include a lack of clear criteria for sustainability/circularity assessment of bio-based products on one hand and on the other hand, an overlap in the existing certification schemes. For example, for the CEN standards for bioplastics (CEN/TC/249), some of the sustainability criteria such as the determination, declaration and reporting of the bio-based carbon content [29–32] are required via the following standards:

- CEN/TS 16137:2011: Plastics—Determination of bio-based carbon content
- CEN/TS 16295:2012: Plastics—Declaration of the bio-based carbon content
- CEN/TS 16398:2012: Plastics—Template for reporting and communication of bio-based carbon content and recovery options of biopolymers and bioplastics—Data sheet

However, these standards do not explicitly direct the economic operator to take further responsibility to address/quantify the sustainability criteria associated with bioplastics including production derived emissions to air, water and soil or economic and social impacts. A discrete set of standards is under development by the technical committee (CEN/TC/411) for bio-based products to report the sustainability aspects of bio-based products [33]. These standards are responsible for the determination, declaration and reporting of environmental impact assessment (e.g., EN16751: Bio-based products: sustainability criteria). The scope of EN16751 in particular, despite providing guidance on undertaking impact assessment and reporting on bio-based products, covers the stages from feedstock acquisition up to the feedstock "pre-processing" phase. Lack of guidance on assessment and reporting of environmental burden resulting from "manufacturing" to "end-of-life" phases, and lack of assessment methodologies and thresholds are some of the major gaps and limitations in these standards.

Country-based feedstock preference: Consistency in raw material supply and chain-productivity is essential for the successful uptake of bio-based products and their associated value chains. The guarantee of a promising flow of feedstock to the facilities can only be ensured through the choice of "locally sourced" feedstock. "Locally-sourced" feedstocks generally have established logistics and reporting procedures, which can communicate their point of origin to the economic operator. Moreover, utilisation of such "locally generated feedstock" can be associated with positive social impacts from employing decades of skilled cultivation related knowledge from the local rural community and its established infrastructure, catalysing its development with an innovative biorefinery, subsequently reducing the overall cost of value-chain establishment [34,35].

Multi-sector application: The ability of a bio-based product and its value chain to cover a range of applications (in different industrial sectors) was identified as an important criterion for value chain selection. Undertaking this task for value chains with products that serve a rather smaller demand/specialised demand could make this study highly specific, deviating from the aim of creating a harmonised sustainability framework for horizontal sector application. Therefore, focus is placed on value chains and bio-based products that have the potential to be applied in a variety of sectors (e.g., bio-based mulch film catering to agricultural/horticulture industries and other industries like the landscaping industry; versatile fine chemicals that find application as solvents in paints and coating, adhesives and binders, fuel additives and agrochemicals).

Application of These Value Chain Selection Factors with Multi-Criteria Decision Analysis

First round assessment and selection: A dedicated matrix composed of a combination of the preliminary value chains which are to be assessed from the viewpoints of each of these selection criteria was developed as a part of this first-tier analysis. Since this study is a part of a project with a wider vision, the expertise of the consortium members in bioeconomy was invited. Their recommendations in terms of the performance of the set of preliminary value chains, in combination with the set of selection criteria, presented in Figure 5, was obtained and reviewed. Scores, in the form of rating (scale of 1–10), were allocated to the recommendations (yes/maybe/no) with justification provided by the consortium members.

Secondly, each of the selection criteria were allotted weighting factors based on their relevance and significance to the principles of material circularity and bioeconomy transition. The weighting factors are presented in Table 2. The criteria that have been allocated a weighting of 0.2 are those that directly contribute to innovative or under-represented but resource efficient value chains which subsequently have the potential to encourage the establishment of a circular economy. Criteria with a weighting of 0.1 (preference within EU member states) has been covered further with an elaborate evaluation under second round assessment. Relevant characteristics of each of the bio-based value chains were assessed and reviewed against the weighted selection criteria to finally assign ranks from the first round. The outcomes of the first round of assessment and further discussion on this approach to the identification of promising bio-based value chains have been presented in Section 3.1.

Table 2. Distribution of weighting to the "value-chain selection" criteria.

Selection Criteria	Weighting
Feedstock variability	0.2
Gaps in certification/sustainability schemes	0.2
Multi-sector application	0.15
Variety in End-of-life options	0.2
Multi-regional supply chain	0.15
Preference within EU member states	0.1

Second round assessment and selection: The second round of assessment was initiated with the collation of information and analysis of national policies, bioeconomy initiatives and growth plans established by individual EU member states. This review provides an insight into the bioeconomy strategy adopted by individual states based on their strengths such as natural bio-resources, preference for bioeconomy development, access and development of technological innovations and maturity level. A summarised list of initiatives and action plans associated with each of the EU member states is presented as a part of the supplementary information.

Upon collation, analysis and categorisation of these initiatives, the preference of these member states over the choice of feedstock, bio-refining technology, current and desired products/sector development and techno-economic or social optimisation route were identified and ranked. This information was used to calculate weighted scores called "preference scores" to specific (feedstock-conversion route-bio-product combinations, based on the preference demonstrated by the EU-collective bioeconomy strategies. These scores, (as a % of total number of strategies), have been presented in Table 3. Information on most of the EU-relevant bioeconomy strategies and initiatives collated and analysed as a part of the second round of assessment, is presented in the supplementary section in Table S1. The outcomes of this second round of assessment have also been discussed further under Section 3.1.

Table 3. EU value chain preference scores as a function of strategy type and nature (as a % of total number of EU bioeconomy strategies).

Value Chains Targeted by the Strategies	Strategy Type	EU Chain Preference Scores
Bio energy and fuel production	Renewable energy	0.74
Food and beverage production	Primary food production	0.6
Crop based primary production	Using waste and residue	0.37
Animal based primary products	Using waste and residue	0.32
Forest based primary production	Using waste and residue	0.26
Bio-based material and plastics	Products/Technology and research	0.26
Marine based primary production	Primary food production	0.2
Bio-based chemicals	Products/Technology and research	0.21
Bio-based construction and furniture	Common conversion	0.2
Biorefinery	Products/Technology and research	0.2
Cosmetics and health	Biomass conversion	0.17

2.2. Value Chain Mapping

Within the H2020 programme, an industrial value chain is defined as stages of value creation by enterprises and other organisations as part of the process of designing and delivering goods and services for their users [22]. Nevertheless, new and innovative value chains are not required to be novel value chains but can be seen as new combinations across value chains, an innovative technology/product brought from one sector or context into another resulting in a disruptive effect.

Value chain maps are a valuable, flexible and convenient tool to develop and analyse the scope and performance potential of a bio-based business model by breaking down the various process dynamics into logistics, sectors of application and embedded stakeholders. The strengths, weaknesses, costs and competition from other value chains in the production of specific commodities can be visualised via value chain maps. The next step in identifying such promising value chains is to understand the chain complexities via such a "mapping exercise".

In this study, the initial "cradle-to-grave" value chain mapping provides a generalised yet visual schematic of the dynamics including the resource flow and actors integrated within bio-based value-chains that have been chosen via the assessments above. For the selected and finalised list of bio-based value chains, the following chain characteristics are crucial and relevant to visualising their significance to a circular economy. The characteristics are as follows:

- Material/energy inputs and outputs, including potential products, co-products, waste and emissions;
- Sector-level contributions;
- Technology/conversion routes;
- Chain-actors or stakeholders linkages
- End-of life (variable) characteristics emphasising the fate of the outputs from each of the life cycle stages.

3. Results and Discussions

3.1. Value Chain Selection

A two-tier multi-criteria decision analysis was undertaken to identify and select the most promising value chains and the outcomes of this assessment have been presented in Table 4.

Bio-plastics, bio-based solvents, bio-lubricants, fabrics and fine chemicals followed by bio-based insulation material were chosen to progress to a second round of assessment. Within the second round of assessment, they were subjected to a similar weighted scoring, set against a background of EU-wide bioeconomy initiative. This "bioeconomy preference score" is primarily based on the target-feedstock and technology preferences of the bioeconomy initiatives and other relevant sustainability schemes established/planned with an active interest to transform from a linear economy to circular bio-based economy.

Table 4. Selection of bio-based value chains from first-round "multi-criteria" assessment.

Sector	Value Chain	Score	Rank	Status
Chemical	Cellulose to bio-based solvents	7.44	1	Selected
Food Packaging	Starch to bio-plastics	7.25	2	Selected
Agriculture	Starch to bio-based mulch films	6.62	3	Selected
Fabrication	Starch to bioplastic framing material	6.09	4	Selected
Multiple sectors	Vegetable fats/plant lipids to bio-based lubricants	5.50	5	Selected
Textile	Cellulose to fabric	5.50	6	Selected
Chemical	Solid biomass to fine chemicals	5.20	7	Selected
Construction	Waste agri. biomass to insulation material	4.78	8	Selected
Food packaging	Wood/cellulose to plastic paper cups	4.37	9	-
Food packaging	Straw to food packaging	4.31	10	-
Construction	Solid biomass to wood-plastic composite	4.00	11	-
Agriculture	Algal polysaccharides to phytoprotectives	3.91	12	-

In terms of feedstock preference, EU member states seemed to possess a clear strategy on utilising feedstock generated locally or nationally with minimal logistics and not demanding an additional stream (land-conversion)/infrastructure for feedstock generation (in other words, use excess and residual biomass). As a result, a majority of the initiatives highlight a feedstock preference in the following order: agricultural (63%), forestry (35%), waste stream (organic waste from domestic and commercial waste) (25%). In terms of initiatives, there are those that either focus on pursuing innovative technology routes or prefer a combined approach to utilising biomass with innovative biomass transformation technologies. Preference for bio-based value chains based on the nature and goal of the initiatives assessed as a part of this study was identified and ranked in Table 5.

Table 5. Selection of value chains from a second round of "initiatives-based preference" assessment.

Sector	Value Chain	EU Chain Preference Scores	Final Score	Rank	Status
Food Packaging	Starch to bio-plastics	0.63	4.57	1	Selected
Agriculture	Starch to bio mulch films	0.63	4.17	2	Selected
Fabrication	Starch to frame material	0.63	3.84	3	Selected
Chemicals	Cellulose to bio-based solvents	0.47	3.50	4	Selected
Multiple sectors	Vegetable fats/plant lipids to bio-based lubricants	0.58	3.19	5	Selected
Chemical	Solid biomass to fine chemicals	0.58	3.02	6	-
Construction	Waste agri. biomass to insulation material	0.57	2.72	7	-
Textile	Cellulose to fabric	0.31	1.71	8	-

It is evident that there is a greater emphasis on development and exploitation of bio-based chemicals and bio-plastics in the conceived bioeconomy agenda and the existing bio-based infrastructure. There is a huge array of bioplastics, classified under roughly 10 categories, that are prevalent worldwide [36]. Currently, there is a focus on bioplastics that can be synthesised from one important feedstock, starch. This is due to the current availability of a mature and commercial conversion route that provides a feasible solution to existing "plastic waste management issues". In addition to this, the multi-functionality of bioplastic under study (PLA synthesised from starch), which (in combination with other polymers) may be utilised to create mulch films, disposable cutleries, framing materials and fibres defines its suitability to address the current challenges facing our transition to a fully-functional bioeconomy. Similarly, bio-based chemicals such as solvents, lubricants, dyes and pigments offer a greener and relatively low-environmental impact alternatives to their conventional counterparts facing restrictions of use from regulatory bodies such as REACH (Registration, Evaluation, Authorisation and Restriction of Chemicals) and ECHA (European Chemicals Agency) [37]. In addition to their lessened impact in terms of eco and human toxicity, their promising potential to create opportunities for socio-economic growth and reduce dependence on non-renewable resources creates a sustainable pathway for development. The construction sector is one of the biggest contributors of landfill waste in the EU owing to the design flaws in the product. Construction materials (particularly contemporary insulation material) are seldom created for any kind of responsible end-of-life management [38].

In view of the commercial, environmental and socio-economic potential identified from the value chains selected from the second round of assessment, they were adopted for an elaborate value-chain mapping in the Section 3.2.

3.2. Value Chain Mapping—Case Studies

The five bio-based value chains chosen from the second round of assessment have been adopted to be mapped for full (general) coverage of the resource flows, technology/conversion routes employed, the various stakeholders and the fate of the products or the other waste streams that may result from the value chain. These schematics have been presented in Figures 6–9.

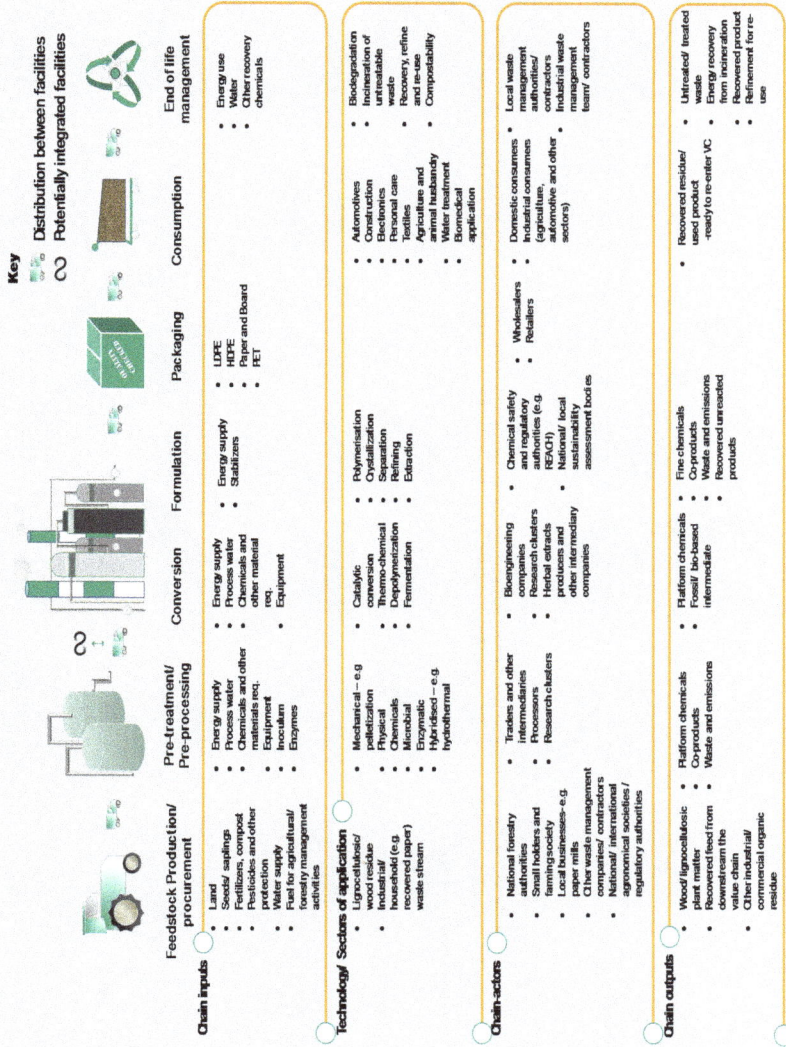

Figure 6. Value chains for solid biomass to bio-based chemicals, mapped for material flow, technology routes and stakeholders.

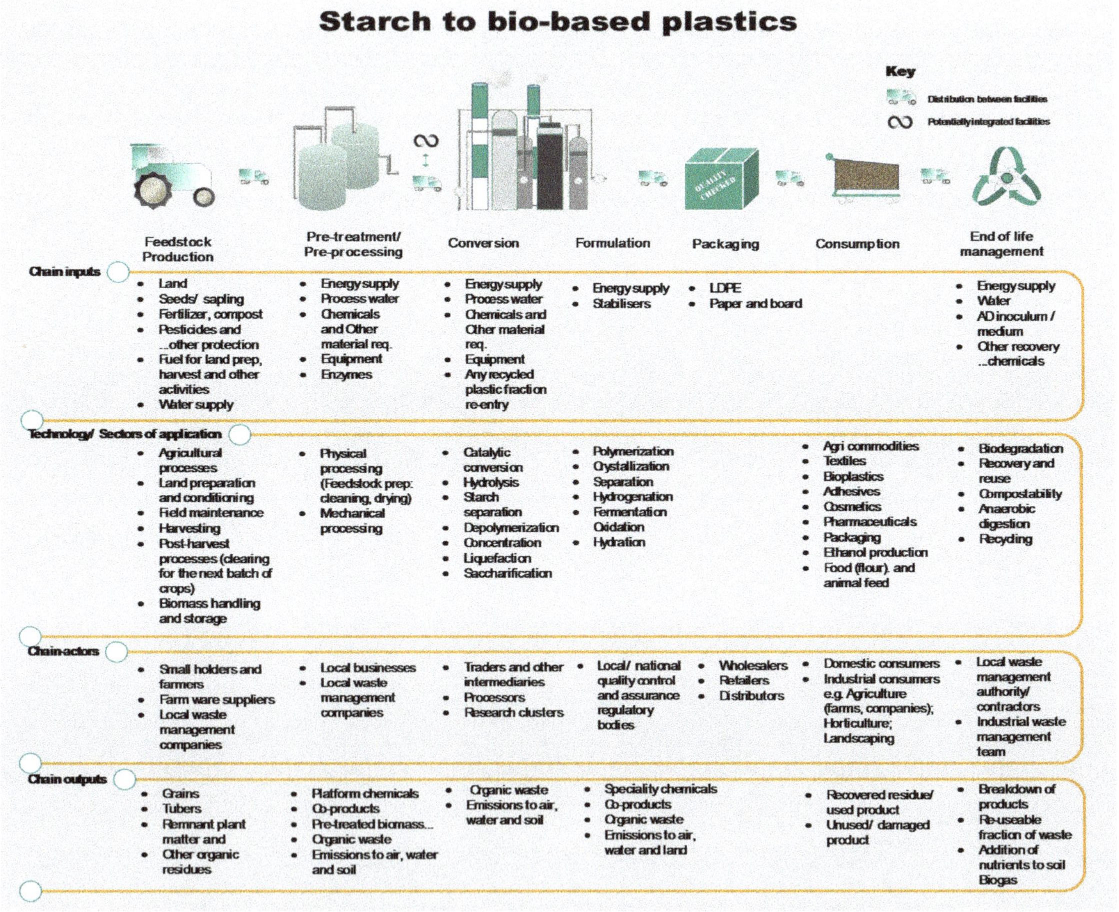

Figure 7. Value chains for starch to bio-based plastics, mapped for material flow, technology routes and stakeholders.

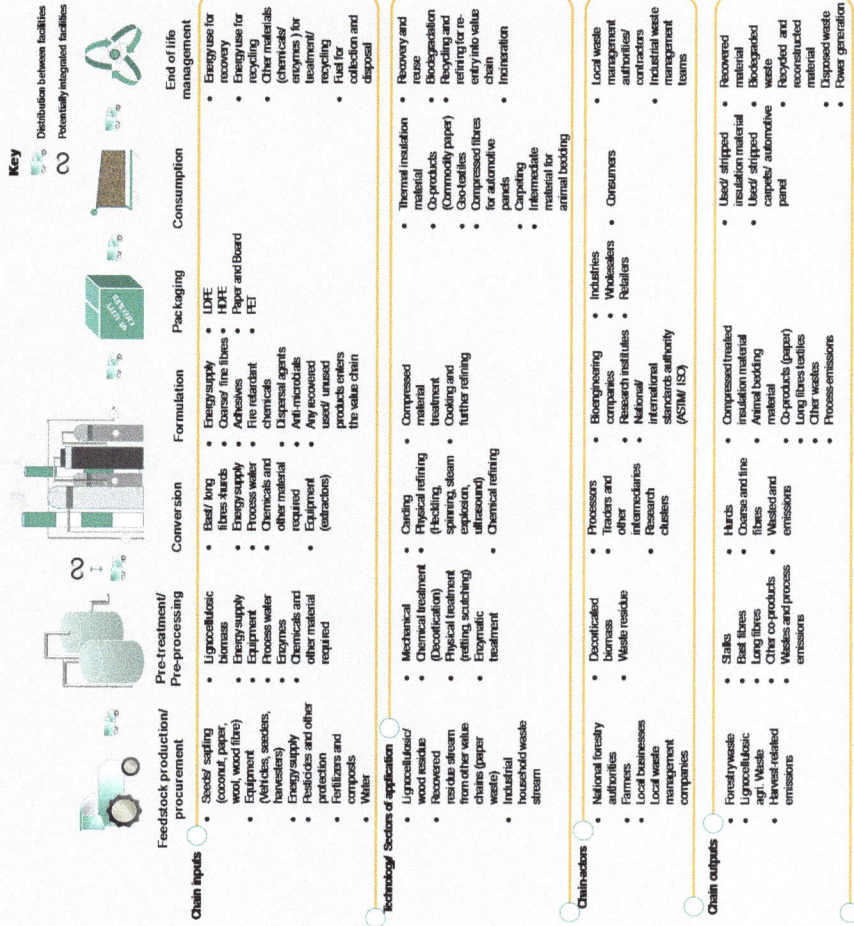

Figure 8. Value chains for solid biomass to insulation material, mapped for material flow, technology routes and stakeholders.

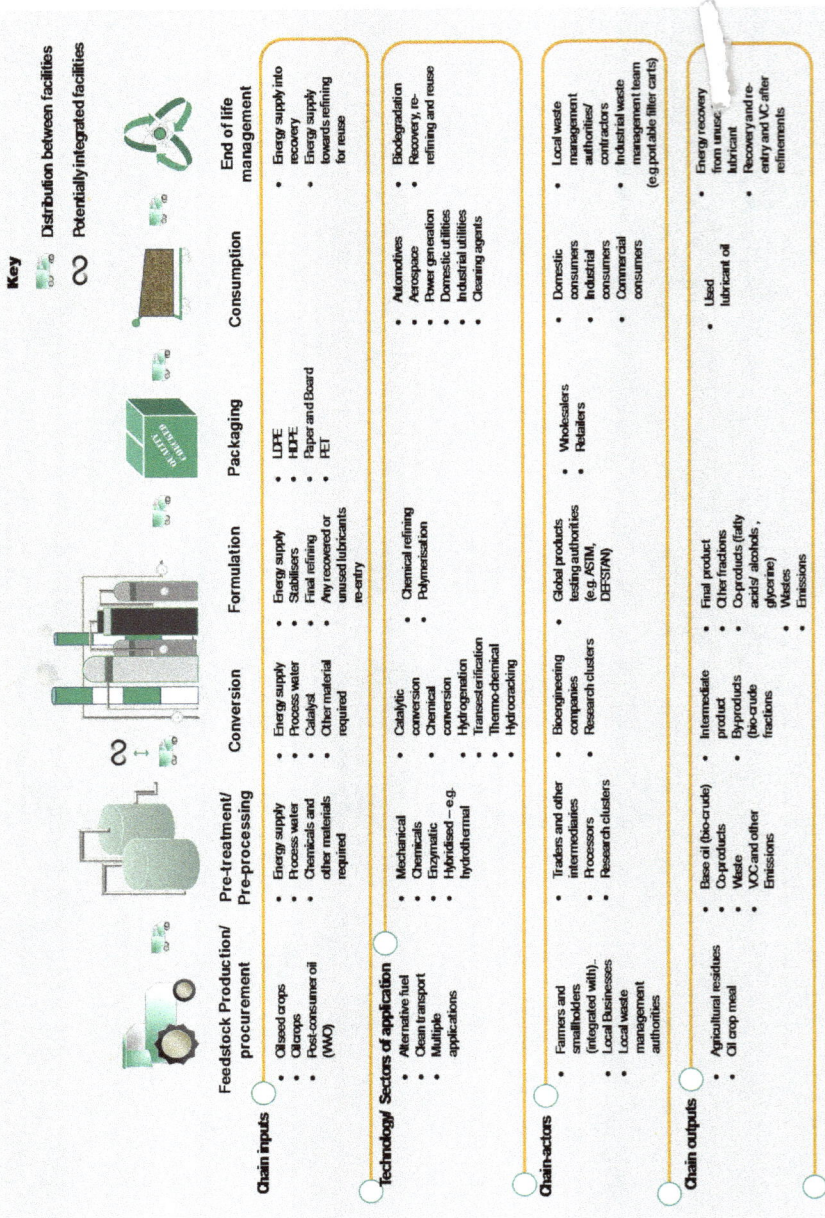

Figure 9. Value chains for solid biomass to bio-based lubricants, mapped for material flow, technology routes and stakeholders.

3.2.1. Bio-Based Chemicals

The market for bio-based chemicals in general is worth $6 billion and at a projected annual growth rate of 16.16%, the market is expected to reach $27 billion by 2025 [39]. Bio-based chemicals include a broad spectrum of products, which may be classified as commodity chemicals, intermediate chemicals and specialty chemicals, based on their application. Commodity chemicals refer to the "high volume-low value" products, sourced from biomass (but not restricted to), such as fatty acids, methyl esters and alcohols. Intermediate products refer to the refined sugar complexes, basic polymers, pigments/dyes, plant oils and other types of starches. Specialty chemicals, synthesised either independently from plant or prepared from intermediate chemicals includes bio-based chemicals such as advanced polymer solvents and other preparations for final formulation in personal care products, pharmaceuticals, paint coatings, additives, domestic/industrial detergents and other applications.

In particular, bio-based solvents are broadly classified into plant-based alcohols, diols, organic acids, glycols and many more. From an economic perspective, according to the above mentioned report [39] the global bio-based solvent market was worth roughly 6 billion USD in December 2016 and it is currently projected to grow at a CAGR of 7.8%, reaching 9 billion by 2024. The versatility of bio-based chemicals, particularly bio-based solvents (for example, in pharmaceutical, cosmetics, agriculture, cleaning, printing inks and adhesive applications), and demand/room for innovation and product development, coupled with stringent regulations on hazardous pollutants released from the use of conventional chemicals have fostered increased research interest and financial investment via national programmes and government support. Moreover, the feedstock variety that can be used to generate a myriad of bio-based chemicals makes these value chains innovative and techno-economically viable, in addition to their improved environmental performance.

3.2.2. Bioplastics

The EU preference to develop a manageable and multifunctional bioplastic category for the commercial market stems from the rapid and unsustainable consumption of conventional plastics for a variety of purposes, at a global level [40]. In addition, the discovery of alarming levels of micro plastics in our food sourced from soil, water and sea, has led to the awareness of the interactions between plastic degradation and the environment (bioaccumulation) [41]. This is evident in a number of initiatives, listed in the supplementary information, that all target withdrawal from fossil-based resources over the next decade with particular focus on energy and plastic consumption. Unlike a decade ago, modern bioplastics are catching up with bio-based solvents in terms of multi-sectoral application (including packaging, agriculture, cosmetics, electronics, construction and automotive) [36]. Evidence of encouragement of bio-based product development and growth can be seen from a plenary meeting of the European parliament that voted in favour of "biodegradable mulch films" during the revision of EU Fertiliser Regulation [42] and a recent increase in "big brands" adopting bio-plastics to appeal to their prominent (high spending power coupled with relatively high environmental awareness) consumer base [43].

3.2.3. Other Bio-Based Products

Bio-lubricants, predominantly synthesised from oil crops, find application in the domestic, industrial, automotive and aviation industry. Among the 220 EU-28 biorefineries assessed as a part of the study undertaken by the Bio-based Industries Consortium and Nova Institute (Figure 3), 20% are dedicated to the manufacture of oleochemical from plant-derived fats. Environmental concerns and strict standards for management of leakage, maintenance and disposal of unused fossil-derived lubricants provide evidence for the growth and development of this sector. A EU-H2020 funded project entitled FIRST2RUN [44] is dedicated to the identification and development of integrated bio refineries that utilise low-input, under-utilised oil crops, grown in marginal lands to synthesise bio-lubricants and bioplastics from vegetable oil. Besides valorisation of marginal lands and low-input biomass, this project envisages the capability of such a value chain to create a skilled labour pool, generate other bio-based products and energy (composting unused parts of the plants), thereby revitalising the local economy.

Agricultural waste transformed into green, low-environmental impact insulation material was the conventional technology until the discovery of fossil resources, which gave rise to relatively inexpensive polymers and materials (e.g., polyurethane, mineral wool). However, some environmental and human health concerns are associated with these insulation materials (from long-term release of aerosols and vapour) such as respiratory issues and eye and skin irritation, particularly in the case of foam insulations. Natural fibre insulation such as cotton wool and wood fibre boards have been identified to perform similarly to their petro-derived counterparts and are particularly advantageous with regards to complying with any environmental building certification schemes [45]. Bio-based binders and other additives, such as polylactic acids (PLA) and polyhydroxyalkanoates (PHA) generated from other starch-based value chains, may be utilised in the preparation of these insulation materials. Dry lignocellulosic biomass can also be processed into compressed fibres for dashboard panels, geotextiles and animal bedding [46].

Limitations: MCDA can be a valuable tool, however, there are a few concerns when it is applied to immensely complex systems. When a process system involves social interactions such as producers, regulatory authorities and consumers, it becomes challenging to call one perspective as more important than the other. MCDA is also capable of overlooking or under-representing some key factors within a complex value chain. An essential and independently functional element within a bio-based value chain, eco-system services, is one such example. It is essential to be able to apply MCDA for smaller scale analysis to be able to predict all possible determinants of a particular chain-stage prior to its broader application. However, this can be time-consuming. Input/output (IO) analyses, which is a methodology that involves monitoring the sectoral trade data to quantify the complex interactions between the different nodes of a given value chain may also be utilised to study the dynamics of a value chains in real-time. However, its data needs depend on statistical information drawn from datasets published by government and international authorities (UN-FAOSTAT, EUROSTAT) [47] and the data may not always be available. Besides being data intensive, it may not always be possible to derive data for bio-based value chains based in rural communities, with IO methodology.

Value chain maps can be laborious and time-consuming to develop, depending on the complexity of the value chain under analysis. The map is only an informative tool for the visualisation of bio-based business models, identification of market opportunities and the scope of the value chain. It may not be able to highlight any changes in the dynamics associated with the factors (chain actors, inputs/outputs and technology routes) presented in the chain.

For this study, the mapping has been carried out to highlight, in general, probable material, wastes/emissions, conversion/refining routes associated with a given feedstock and end bio-product synthesised from it. These maps do not provide explicit information on coverage of these value chains by specific sustainability schemes/certification programmes as there are diverse products and co-products that could be produced as a part of the value chain. To establish this level of detail, the goal, scope and the product of analysis would have to be established beforehand.

3.3. Gaps and Challenges That Can Be Addressed

From having undertaken the multi-criteria decision analysis, a number of key aspects associated with the bio-based business models were brought into consideration, out of which, only a few general criteria were chosen as the selection criteria. Some of the key gaps and challenges that are addressed in this section were drawn based on our analysis and the mapping exercise on the overall. These are some key hurdles faced by existing and new bio-based business models in the current context of our transition to a bio-based economy. These points have been highlighted to encourage holistic systems thinking, starting from the feedstock procurement stage, through product design, which influences the final product cost up to the need for a dedicated end-of life management infrastructure which has an influence on public perception of bio-based products. Such systems thinking will encourage streamlining the innovation, processes and delivery of an incredibly complex bio-based value chains. Application of systems thinking can be customised to the practices in other non-bio-based sectors thereby facilitating our transition to a fully-functional circular economy.

"Food vs. non-food bio products" conflict: We live in an era of a global lack of food security, the supply chain of which alone is highly complex and fraught with unforeseeable risks such as crop failure from climate change or geo-political instability. To add to this, use of agro-food based biomass as the starting feedstock not only puts the bio-based business model at risk but also invites "food vs. bio-based products" conflict, undermining the sustainability characteristics of a bio-based value chain. However, substituting primarily food-based feedstock with waste-based feedstock or by-products would incorporate material circularity e.g., starch rich feedstock can be drawn from food retailers and food processing industries by using vegetable peels instead of starch rich crops. This strategy not only encourages waste valorisation but also reduces both the risks involved and the dependence of bio-based industries on the primary food supply chains.

Feedstock costs: Biomass may seem economically unfeasible compared to cheap petroleum feedstocks and their intermediates. The cost of feedstock which in turn is influenced by its supply/demand ratio in the commercial market makes it even more difficult for bio-based business models to increase profit margins and sometimes to breakeven. However, encouragement and development of innovative multi-functional bio refineries that are capable of utilising low-value, low-cost and waste biomass can prove to be an ideal alternative; such biorefineries must also be able to transform such waste feedstock into product designs that can be disassembled during their end-of-life with the product components being re-circularised back into the manufacturing loop. Re-incorporation of material can have long-term economic benefits to such business models, in addition maintaining the value of all the first-chain process inputs.

Supply risks: The seasonal nature of the biomass supply stemming from the cultivation time-span of biomass is an issue. However, further challenges posed to cross-border biomass supply from climate change, diminishing ecosystem services due to human intervention (e.g., intensive farming), variable food/biomass demand, geopolitical instabilities and social ethical concerns also undermine the "sustainable" characteristics of a bio-based value chains. It is therefore essential to encourage biorefinery and business models that are built on pre-established climate change resilient but low-demand biomass (e.g., millet and sorghum in Africa). From a cost perspective, their low demand reduces biomass costs, from a socio-economic perspective, accessing the knowledge pool of the local rural communities can be mutually beneficial via opportunities for rural development and encourages their interest in maintaining the local ecosystem services.

Interconnectedness of value chains: Uneven distribution or complete lack of technological readiness for a multi-regional/local value chain to be established is a key hurdle in transformation to a bio-based economy. This requires creation of awareness and a strong social connection between inter-chain stakeholders to encourage technological and operational coherence. The thought-process among the different chain actors (particularly among those involved between the feedstock procurement to packaging stage) has to be parallel with each other. Family businesses embed such wider thoughtful thinking and are mostly successful in staying established on a long-term, despite various external shocks, particularly from low-demand, inflation, etc. The key aspect to note is the trust developed with fellow stakeholders and customers with the consistency and quality of product. Occasional lack of co-operation due to process-level disparities between the embedded stakeholders of the value chains could be overcome via the establishment of national standards, covering stakeholders embedded in a value chain, with standardised templates for recording, reporting and communicating information.

Penetration of non-bio-based value chains: A recent shift in consumer behaviour, challenges to penetrating non-bio-based value chains (existing fossil-based supply chains) owing to consumer perception of "brand-value" and relatively cheaper products need to be considered. However, the trend is changing as there is a growing demand among consumers in small businesses owing to the "trust" factor that they are mainly passion driven rather than profit-driven [48].

Public perception of bio-based products: According to a review of various reports by Spatial Foresight, (2017), the general public's perception of bio-based products are highly variable [3]. As much as there is keenness to switch to bio-based products owing to their overall environmental benefits, the sporadic market supply, the expensive nature of such products and sometimes, limited functionality and durability of the bio-based products seem to hinder their acceptance. For members of the public who choose to opt for a more sustainable life, the existing product end-of-life management infrastructure is partially to fully unsuitable to pursue a circular waste management practice, leading to an overall skepticism. Therefore, a holistic approach is needed in our transformation to not only a bio-based economy but also to a circular economy.

"Top-down" and "bottom-up" initiatives: Stronger policy-level improvisation via "top-down" approaches drive chain-actors to encourage good practice such as producer responsibility (e.g., packaging industry) and deploy "re-direct used material" strategies. Similarly, "bottom-up" approaches, such as altering consumer behaviour via responsible and innovative retail design and practices, must also be devised. This may also include incentives/loyalty schemes for consumers who opt for bio-based products that upon consumption can be recovered and re-introduced into another product.

4. Conclusions

The purpose of this paper is to provide a methodology to assess existing or novel bio-based value chains from key angles that are of significance to our journey to attaining a fully functional bio-based circular economy. Due to the limited time and resources, it is essential to distinguish the most promising bio-based business model from the rest, and that is precisely what this paper, with its methodology suggests. EU-based bioeconomy and bio-based value chains are diverse in nature and are not restricted to those value chains that have been considered in this study. The preliminary list of 12 value chains was selected based on their relevance and significance to the bioeconomy, their current activity level/contribution and coverage by various sustainability and certification schemes. These bio-based value chains have been selected to ensure the representation of EU's diverse bio-based value chains in addition to their potential to address the key environmental, techno-economic and socio-economic threats and challenges faced globally. The value chains were selected in two-steps via multi-criteria decision analysis (MCDA), wherein the first step, the preliminary value chains were ranked by placing them against a back drop of five key selection criteria in the current context: feedstock variability; EU feedstock preference; variety of end-of-life options; multisector application and multi-regional supply chains. This step led to the identification of eight bio-based value chains which were subjected to a second round of assessment where five out of the eight value chains showed a promising interest/inclination for bioeconomic development. For the selected five bio-based value chains which included starch to bio-plastics, starch to bio mulch films, starch to frame material, cellulose to bio-based solvents, vegetable fats/plant lipids to bio-based lubricants, elaborate value chain maps were developed to demonstrate the highly informative nature of this tool and its crucial role in understanding the complex interactions among the various process stages, associated processes and stakeholders within a more complex value chain.

Value chain maps provide valuable insight into the integrated activities, actors and technology, in addition to the material flow, and have provided a foresight of the scope and qualitative performance potential (in socio-economic and environmental terms) within each of the life cycle stages. Inability to acquire real-time information from any changes to the process dynamics may be a limitation. However, for a preliminary assessment, value-chain mapping provides an overall breakdown of the various elements presented above. The methodology prescribed in this paper not only helps understand the embedded complexities and overcome them within bio-based value chains but are also equally applicable to complex non-bio-based supply chains that foresee a transformation by closing their loop to encourage product and process circularity. Overall, the suggested method could be used by policy makers, investors, business groups and economists to understand the interdependencies among the various embedded chain-actors, dependence of the chain on material consumption and the various technology routes available for innovative and sustainable production of bio-based products that can potentially replace high-impact fossil-derived products.

Supplementary Materials: The following are available online at http://www.mdpi.com/2071-1050/10/6/1695/s1, Table S1: Bioeconomy initiatives and strategies analysed for the second round of the value chain assessment and selection.

Author Contributions: The concept of analysis, methodology, data acquisition and interpretation and development of the manuscript were all jointly undertaken by K.L., L.L. and L.S.

Funding: The contents of the paper are a part of the findings of the project STAR-ProBio. STAR-ProBio has received funding from the European Union's Horizon 2020 program and innovation programme under grant agreement No. 727740, Work Programme BB-01-2016: Sustainability schemes for the bio-based economy.

Acknowledgments: We would like to convey a special thanks to our partners from the Agricultural University of Athens, Greece, Deutsches Biomasseforschungszentrum, Germany, SQ Consult, Spain and the University of Bologna, Italy who contributed to the research undertaken in this paper.

References

1. European Commission. Towards a Circular Economy. Available online: https://ec.europa.eu/commission/priorities/jobs-growth-and-investment/towards-circular-economy_en (accessed on 28 October 2017).
2. Jurgilevich, A.; Birge, T.; Kentala-Lehtonen, J.; Korhonen-Kurki, K.; Pietikäinen, J.; Saikku, L.; Schösler, H. Transition towards Circular Economy in the Food System. *Sustainability* **2016**, *8*, 69. [CrossRef]
3. Spatial Foresight; SWECO; ÖIR; Nordregio; Berman Group; Infyde. *Bioeconomy Development in EU Regions. Mapping of EU Member States'/regions' Research and Innovation Plans & Strategies for Smart Specialisation (RIS3) on Bioeconomy for 2014–2020*; DG Research & Innovation, European Commission: Brussels, Belgium, 2017.
4. D'Amato, D.; Droste, N.; Allen, B.; Kettunen, M.; Lähtinen, K.; Korhonen, J.; Leskinen, P.; Matthies, B.D.; Toppinen, A. Green, circular, bio economy: A comparative analysis of sustainability avenues. *J. Clean. Prod.* **2017**, *168*, 716–734. [CrossRef]
5. Lieder, M.; Rashid, A. Towards circular economy implementation: A comprehensive review in context of manufacturing industry. *J. Clean. Prod.* **2016**, *115*, 36–51. [CrossRef]
6. Frank, D. The Bioeconomy is Much More than a Circular Economy. Available online: https://www.brain-biotech.de/en/blickwinkel/circular/the-bioeconomy-is-much-more-than-a-circular-economy/ (accessed on 8 May 2018).
7. Geels, D.I.F.W. The dynamics of transitions in socio-technical systems: A multi-level analysis of the transition pathway from horse-drawn carriages to automobiles (1860–1930). *Technol. Anal. Strateg. Manag.* **2005**, *17*, 445–476. [CrossRef]
8. Shove, E. Putting practice into policy: Reconfiguring questions of consumption and climate change. *Contemp. Soc. Sci.* **2014**, *9*, 415–429. [CrossRef]
9. Honegger, M.; Reiner, D. The political economy of negative emissions technologies: Consequences for international policy design. *Clim. Policy* **2018**, *18*, 306–321. [CrossRef]
10. Tukker, A. Product services for a resource-efficient and circular economy—A review. *J. Clean. Prod.* **2015**, *97*, 76–91. [CrossRef]
11. Pan, S.-Y.; Du, M.A.; Huang, I.-T.; Liu, I.-H.; Chang, E.-E.; Chiang, P.-C. Strategies on implementation of waste-to-energy (WTE) supply chain for circular economy system: A review. *J. Clean. Prod.* **2015**, *108*, 409–421. [CrossRef]

12. Greene, J.P. Introduction to Sustainability. In *Sustainable Plastics*; John Wiley & Sons, Inc.: Hoboken, NJ, USA, 2014; pp. 1–14. ISBN 978-1-118-89959-5.

13. Brosowski, A.; Thrän, D.; Mantau, U.; Mahro, B.; Erdmann, G.; Adler, P.; Stinner, W.; Reinhold, G.; Hering, T.; Blanke, C. A review of biomass potential and current utilisation—Status quo for 93 biogenic wastes and residues in Germany. *Biomass Bioenergy* **2016**, *95*, 257–272. [CrossRef]

14. Pick, D.; Dieterich, M.; Heintschel, S. Biogas Production Potential from Economically Usable Green Waste. *Sustainability* **2012**, *4*, 682–702. [CrossRef]

15. Gurría, P.; Ronzon, T.; Tamosiunas, S.; López, R.; García Condado, S.; Guillén, J.; Cazzaniga, N.; Jonsson, R.; Banja, M.; Fiore, G.; et al. *Biomass Flows in the European Union*; European Commission Joint Research Centre: Seville, Spain, 2017.

16. Parisi, C.; Ronzon, T. *A Global View of Bio-Based Industries: Benchmarking and Monitoring Their Economic Importance and Future Developments*; JRC Technical Reports; European Commission: Brussels, Belgium, 2016.

17. Bio-Based Industries Consortium Mapping European Biorefineries I Bio-Based Industries Consortium. Available online: http://biconsortium.eu/news/mapping-european-biorefineries (accessed on 1 December 2017).

18. Fehrenback, H.; Köppen, S.; Kauertz, B.; Detzel, A.; Wellenreuther, F.; Brietmayer, E.; Essel, R.; Carus, M.; Kay, S.; Wern, B.; et al. *Biomass Cascades: Increasing Resource Efficiency by Cascading Use of Biomass—From Theory to Practice*; German Environmental Agency: Heidelberg, Germany, 2017; p. 29.

19. Myllyviita, T.; Holma, A.; Antikainen, R.; Lähtinen, K.; Leskinen, P. Assessing environmental impacts of biomass production chains—Application of life cycle assessment (LCA) and multi-criteria decision analysis (MCDA). *J. Clean. Prod.* **2012**, *29–30*, 238–245. [CrossRef]

20. Zanghelini, G.M.; Cherubini, E.; Soares, S.R. How Multi-Criteria Decision Analysis (MCDA) is aiding Life Cycle Assessment (LCA) in results interpretation. *J. Clean. Prod.* **2018**, *172*, 609–622. [CrossRef]

21. European Parliament. *Circular Economy Package: Four Legislative Proposals on Waste*; European Parliamentary Research Service: Brussels, Belgium, 2016.

22. European Commission. *Bioeconomy Strategy: Innovating for Sustainable Growth: A Bioeconomy for Europe*; European Commission: Brussels, Belgium, 2012.

23. European Commission. 2030 Energy Strategy. Available online: https://ec.europa.eu/energy/en/topics/energy-strategy-and-energy-union/2030-energy-strateg (accessed on 5 January 2018).

24. Meyer-Kohlstock, D.; Schmitz, T.; Kraft, E. Organic Waste for Compost and Biochar in the EU: Mobilizing the Potential. *Resources* **2015**, *4*, 457–475. [CrossRef]

25. STAR-ProBio. *Report on Identified Environmental, Social and Economic Criteria/Indicators/Requirements and Related "Gap Analysis*; STAR-ProBio: York, UK, 2017.

26. Hennig, C.; Brosowski, A.; Majer, S. Sustainable feedstock potential—A limitation for the bio-based economy? *J. Clean. Prod.* **2016**, *123*, 200–202. [CrossRef]

27. Budzinski, M.; Bezama, A.; Thrän, D. Monitoring the progress towards bioeconomy using multi-regional input-output analysis: The example of wood use in Germany. *J. Clean. Prod.* **2017**, *161*, 1–11. [CrossRef]

28. Pagotto, M.; Halog, A. Towards a Circular Economy in Australian Agri-food Industry: An Application of Input-Output Oriented Approaches for Analyzing Resource Efficiency and Competitiveness Potential. *J. Ind. Ecol.* **2016**, *20*, 1176–1186. [CrossRef]

29. CEN European Committee for Standardization. CEN/TS 16295:2012: Bio-Based Products. Requirements for Business-to-Consumer Communication and Claims. 2017. Available online: https://shop.bsigroup.com/ProductDetail/?pid=000000000030330832 (accessed on 15 August 2017).

30. CEN European Committee for Standardization. BS EN 16751:2016: Bio-Based Product. Sustainable Criteria. 2016. Available online: https://standards.cen.eu/dyn/www/f?p=204:22:0::::FSP_ORG_ID,FSP_LANG_ID:874780,25&cs=1D63BAA7EABE56EB230DDAA05D6F2CE70 (accessed on 13 June 2017).

31. CEN European Committee for Standardization. PD CEN/TS 16398:2012: Plastics. Template for Reporting and Communication of Biobased Carbon Content and Recovery Options of Biopolymers and Bioplastics. Data Sheet. 2012. Available online: https://shop.bsigroup.com/ProductDetail/?pid=000000000030258216 (accessed on 28 June 2017).

32. CEN European Committee for Standardization. *Plastics. Determination of Bio-Based Carbon Content*; CEN European Committee of Standardisation: Brussels, Belgium, 2011.

33. CEN European Committee for Standardization. CEN/TC 411. Available online: https://standards.cen.eu/dyn/www/f?p=204:7:0::::FSP_ORG_ID:874780&cs=112703B035FC937E906D8EFA5DA87FAB8 (accessed on 11 January 2018).

34. Keegan, D.; Kretschmer, B.; Elbersen, B.; Panoutsou, C. Cascading use: A systematic approach to biomass beyond the energy sector. *Biofuels Bioprod. Biorefin.* **2013**, *7*, 193–206. [CrossRef]

35. Blennow, K.; Persson, E.; Lindner, M.; Faias, S.P.; Hanewinkel, M. Forest owner motivations and attitudes towards supplying biomass for energy in Europe. *Biomass Bioenergy* **2014**, *67*, 223–230. [CrossRef]

36. Greene, J.P. Sustainable Plastic Products. In *Sustainable Plastics*; John Wiley & Sons, Inc.: Hoboken, NJ, USA, 2014; pp. 145–186, ISBN 978-1-118-89959-5.

37. Dubois, J.-L. Requirements for the development of a bioeconomy for chemicals. *Curr. Opin. Environ. Sustain.* **2011**, *3*, 11–14. [CrossRef]

38. Akanbi, L.A.; Oyedele, L.O.; Akinade, O.O.; Ajayi, A.O.; Davila Delgado, M.; Bilal, M.; Bello, S.A. Salvaging building materials in a circular economy: A BIM-based whole-life performance estimator. *Resour. Conserv. Recycl.* **2018**, *129*, 175–186. [CrossRef]

39. Zion Market Research Global Bio-Solvents Market Will Reach USD 9.43 Billion by 2022: Zion Market Research. Available online: http://globenewswire.com/news-release/2017/09/22/1131467/0/en/Global-Bio-Solvents-Market-Will-Reach-USD-9-43-Billion-by-2022-Zion-Market-Research.html (accessed on 6 November 2017).

40. Greene, J.P. Biobased and Biodegradable Polymers. In *Sustainable Plastics*; John Wiley & Sons, Inc.: Hoboken, NJ, USA, 2014; pp. 71–106, ISBN 978-1-118-89959-5.

41. Carrington, D. Plastic fibres found in tap water around the world, study reveals. *The Guardian*, 6 September 2017.

42. Schwede, K. European Parliament Supports Use of Biodegradable Mulch Films. Available online: http://www.european-bioplastics.org/european-parliament-supports-use-of-biodegradable-mulch-films/ (accessed on 6 November 2017).

43. European Bioplastics. *Bioplastics: Facts and Figures*; European Bioplastics: Berlin, Germany, 2016.

44. Bio-based Industries Consortium FIRST2RUN | Bio-Based Industries Consortium. Available online: http://biconsortium.eu/library/case-studies/first2run (accessed on 14 January 2018).

45. Carus, M.; Eder, A.; Dammer, L.; Korte, K.; Scholz, L.; Essel, R.; Breitmayer, E.; Barth, M. *WPC/NFC Market Study 2014; Wood-Plastic Composites (WPC) and Natural Fibre Composites (NFC)*; Nova-Institut GmbH: Hürth, Germany, 2014.

46. Alkhagen, M.; Samuelsson, A.; Aldaeus, F.; Gimaker, M.; Ostmark, E.; Swerin, A. *Roadmap 2015–2025: Textile Material from Cellulose*; The RISE Research Institute of Sweden: Stockholm, Sweden, 2015.

47. You, F.; Tao, L.; Graziano, D.J.; Snyder, S.W. Optimal design of sustainable cellulosic biofuel supply chains: Multiobjective optimization coupled with life cycle assessment and input–output analysis. *AIChE J.* **2012**, *58*, 1157–1180. [CrossRef]

48. Daneshkhu, S. Consumer Goods: Big Brands Battle with the 'Little Guys'. Available online: https://www.ft.com/content/4aa58b22-1a81-11e8-aaca-4574d7dabfb6 (accessed on 3 May 2018).

sustainability

MDPI

Article

Social Life Cycle Approach as a Tool for Promoting the Market Uptake of Bio-Based Products from a Consumer Perspective

Pasquale Marcello Falcone [1,*] and Enrica Imbert [2]

[1]   Bioeconomy in Transition Research Group, IdEA, Unitelma Sapienza—University of Rome, Viale Regina Elena, 291, 00161 Roma, Italy
[2]   Department of Law and Economics, Unitelma-Sapienza University of Rome, Viale Regina Elena, 295, 00161 Roma, Italy; enrica.imbert@unitelma.it
*   Correspondence: pm.falcone@bioeconomy-in-transition.eu

Received: 23 February 2018; Accepted: 27 March 2018; Published: 30 March 2018

Abstract: The sustainability of bio-based products, especially when compared with fossil based products, must be assured. The life cycle approach has proven to be a promising way to analyze the social, economic and environmental impacts of bio-based products along the whole value chain. Until now, however, the social aspects have been under-investigated in comparison to environmental and economic aspects. In this context, the present paper aims to identify the main social impact categories and indicators that should be included in a social sustainability assessment of bio-based products, with a focus on the consumers' category. To identify which social categories and indicators are most relevant, we carry out a literature review on existing social life cycle studies; this is followed by a focus group with industrial experts and academics. Afterwards, we conduct semi-structured interviews with some consumer representatives to understand which social indicators pertaining to consumers are perceived as relevant. Our findings highlight the necessity for the development and dissemination of improved frameworks capable of exploiting the consumers' role in the ongoing process of market uptake of bio-based products. More specifically, this need regards the effective inclusion of some social indicators (i.e., *end users' health and safety, feedback mechanisms, transparency,* and *end-of-life responsibility*) in the social life cycle assessment scheme for bio-based products. This would allow consumers, where properly communicated, to make more informed and aware purchasing choices, therefore having a flywheel effect on the market diffusion of a bio-based product.

Keywords: social life cycle assessment; bio-based products; social indicators; consumers; sustainability

---

1. Introduction

Social sustainability is an essential component of sustainable development, even though it has been largely under-investigated when compared to economic and environmental components [1]. This is particularly true when restricting the spectrum of the analysis to the bio-based economy [2]. The bio-based economy includes the production of renewable biological resources and the conversion of these resources, residues, by-products and side streams into value added products such as food, feed, bio-based products, services and bioenergy [3]. According to the European Standard (EN 16575:2014), bio-based products are wholly or partly derived from materials of biological origin, excluding materials embedded in geological formations and/or fossilized. These might include chemicals, lubricants, surfactants, enzymes, pharmaceuticals, cosmetics, food additives, etc.

Promoting the use of bio-based products can support the transition from a linear towards a circular economy, creating jobs and enhancing a more sustainable growth [4]. However, while some policy documents (e.g., [5,6]) have supported the production of renewable biological resources and their conversion into value added products and bio-energy, there are also concerns about sustainability [7]. With the aim of ensuring an effective transition towards sustainability, it is of paramount importance that apposite assessment methods exist and are employed to discern pros and cons of different sustainability options [8].

In this respect, life cycle sustainability assessment (LSCA) represents a valuable framework whose transdisciplinary nature clearly demonstrates the importance of integrating not only with economic models but also with ecological and social theories [9]. However, as emphasized by [10], unlike reputable methods for assessing environmental and economic performance with environmental life cycle assessment (ELCA) and life cycle costing (LCC), social life cycle assessment (SLCA) is still in the development phase, and therefore misses the necessary empirical experience [11]. This is due in part to the lack of standardized social indicators for social performance measurements. The development of general SLCA indicators could provide organizations with relevant information to better understand those social factors that might influence their development over time [12]. This, in turn, would support empirical experience and accordingly contribute to the development of standardized LCSA constructs [10]. Moreover, the use of social indicators can assist decision makers in providing a fit-for-purpose social sustainability scheme, including standards, labels and certifications, based on the product-related impacts on the wellbeing of different stakeholders' categories [13].

Apart from deeply analyzing the theoretical foundations of SLCA, some academics have asked for scientific developments in providing improved methods and case studies with regard to the choice of impact categories and related indicators (e.g., [14,15]). Nonetheless, there is still a restricted number of contributions aimed at reviewing which social sustainability aspects are most relevant to consider throughout the SLCA. However, Kühnen and Hahn [10] recently reviewed trends, coherences, inconsistencies, and gaps in research on SLCA indicators across industry sectors. They found only a few sectors receiving adequate empirical attention to draw cautious conclusions that often neglect relevant social issues; this is because they focus mainly on worker- and health-related indicators. Lastly, Martin et al. [8], by means of a systematic review of scientific life cycle studies on bio-based products and an open space workshop with experts from academia and industry, underlined a discrepancy between those indicators found to be relevant, and the indicators that are recurrently included in the studies.

In this article, we complement the recent interest in understanding what key sustainability impacts and related indicators should be considered for the development, production, and market uptake of bio-based products [8]. We do this by focusing on a specific stakeholder category, namely consumers. This could help fill the gap concerning the overall lack of attention of SLCA research to stakeholder categories other than workers (see [10]). In this vein, SLCA is also a valuable tool for positioning a product in the market and to guide consumer purchases (see [16]). This becomes even more relevant when considering public measures for bio-based products, such as public procurement policies, and where mechanisms for establishing a level playing field with fossil-based products have not yet been implemented [4]. Furthermore, even when demand-side measures have been undertaken, such as in Italy (see [17]), the negative consequences originating from a general lack of social acceptance among consumers represent an important warning to all bioeconomy stakeholders (See http://news.bio-based.eu/the-fight-on-plastics-heats-up-in-the-eu/). Accordingly, focusing on consumers is of paramount importance for deepening our knowledge of the main social aspects that may influence future demand and thus the market uptake of bio-based products, which is currently still limited.

The scope of our analysis ranges from scientific publications and official published documents (e.g., conference proceedings and books), including some from fields other than SLCA, to the so-called "grey literature" (e.g., dissertations and reports). These mostly focus on studies concerning the current situation of the bio-based sector in Europe from 2010 to early 2018. Our intended audience includes academics, practitioners and consumer organizations, as well as decision makers who are looking for reliable evaluation tools to enhance the understanding of which social aspects are worth considering, together with the key phases of design, production, marketing and consumption of bio-based products from the perspective of circular bioeconomy.

The rest of the paper is organized as follows: Section 2 frames the social sustainability of bio-based products and presents the research questions. Section 3 describes the methodology. Section 4, starting from SLCA studies applied to bio-based products, proposes a comprehensive list of social impact categories and indicators tailored to bio-based products; the section shows results relevant to the consumers' category. Section 5 contains a discussion of the results. Finally, Section 6 concludes and suggests further developments of the study.

2. The Context of Analysis and Research Questions

In recent years, the socio-economic sustainability of products and processes has gained greater global attention. With reference to the bioeconomy, the European Union (EU) highlighted that bio-based products have a strong socio-economic dimension that needs to be taken into consideration [3,5]. Along this line of reasoning, Fritsche and Iriarte [18] observed that, in the early phases of the development of the bioeconomy, there was a focus on environmental criteria. At present, however, the further expansion of bio-based products makes the inclusion of social and socio-economic criteria a key issue.

Various studies have identified a wide range of social and socio-economic impacts related to bio-based products at different levels (company, local, national and international). In particular, it has been clearly pointed out that for the assessment of social aspects of bio-based products, upstream processes in the agricultural sector have a high social risk potential [19]. Indeed, the production of biomass affects access to land and land use [18] and the price of feedstocks, with direct and indirect effects on food production and security (see [20,21]). Moreover, given that a great percentage of raw materials are produced in countries with lower human rights standards, working conditions in this phase must be carefully monitored (see [20]). Another crucial issue relates to the impact of bio-based products on health and safety (see [22,23]). In this respect, Álvarez-Chávez et al. [24] have focused on the health and safety impacts of bioplastics throughout their life cycle.

Among the varying impacts, the effects of bio-based products on employment and the creation of new jobs (for example, temporary in nature or not), in both rural and industrial areas, also gained particular attention (e.g., [25–28]). Moreover, the literature also calls to other types of impact. Examples include those related to gender issues (see [2,29]).

Overall, the transition towards a bio-based economy is expected to deliver social and socio-economic benefits in a broad spectrum of areas, spanning from health and safety to working conditions, employment and prosperity, access to material and non-material resources, food and energy security, and gender issues (see [2,29]). These areas have been intertwined with Europe 2020 objectives and UN sustainable development goals (SDGs) (see [30,31]). Therefore, measuring these potential social and socio-economic improvements is of utmost importance for ensuring the sustainability of bio-based products while at the same time promoting their market uptake. Accordingly, this paper will seek to answer the following research questions:

RQ1: *Which are the main impact categories and social indicators that should be included in a social sustainability assessment of bio-based products that take into consideration the whole value chain, from a social life cycle perspective?*

RQ2: *Which of the impact categories and social indicators, identified as pertaining to the consumers' category, are most relevant and could therefore, if properly communicated, encourage greater market penetration of bio-based products?*

3. Methodology

To follow our research aims, we carried out a two-step investigation by means of:

(i) a literature review on existing social life cycle studies on bio-based products, accompanied by a focus group to identify and validate the main social indicators pertaining to the consumers' category; and

(ii) semi-structured interviews with bioeconomy experts to ascertain the most relevant social impact categories and indicators from the consumers' perspective.

In the first step of our methodological approach, we conducted an in-depth peer-reviewed literature review from scientific journals and official published documents (e.g., conference proceedings and books), including some from fields other than SLCA. This review was then complemented by information from the so-called "grey literature" (e.g., dissertations and reports), mostly focusing on studies concerning the current situation of the bio-based sector in Europe.

For the examination of environmental, economic and social aspects, the review was carried out by looking at two main academic databases of peer-reviewed literature, namely Scopus (www.scopus.com) and Web of Science (www.webofscience.com); these databases were used because of their wide-ranging coverage of English-language scientific journals in social sciences. To extend our research, in order to also consider studies and reports not published in academic journals, we employed the Google search engine. We carried out a broad keyword search to detect important documents available online at the beginning of 2018. In this respect, we paired some anchor keywords (i.e., bio*, soci*, and sustainab*) with other search strings (i.e., "life cycle", "supply chain", "indicators", and "impacts"). Additionally, by means of an iterative method of search and discussion between the two authors and other scholars belonging to the same research group, additional search words were used with the aim of focusing the analysis mainly on social aspects in the context of bio-based products: bio-based products, bio-based products life cycle, social assessment of bio-based products, social indicators of bio-based products. This exercise allowed us to select studies dealing with the social dimension of bio-based products.

Our literature review shows the presence of more than 1000 studies on LCAs, focusing at a cradle-to-grave level. With the aim of addressing our research aims, and after having screened article abstracts, we found more than 500 papers that were relevant for a social performance assessment of a product, and more than 100 concerning bio-based products. However, the number of SLCA contributions concerning bio-based products was much smaller and amounted to 18 studies, four of which are case studies applied to bio-based products. See more information in Table 1.

To identify the main social indicators and criteria related to consumers, we also conducted a focus group exercise on the sustainability assessment of bio-based products, under the Horizon 2020 funded project STAR-ProBio. The study consisted of a group of 10 purposely sampled participants from different European countries (i.e., state agencies, public procurement experts, standardization and certification organizations, businesses and business associations, NGOs, and academia). They were intentionally selected and invited by means of gate-keepers (i.e., project partners) who were able to recruit people who, although with different professional backgrounds, share knowledge and general expertise on the bio-based economy; this was done to ensure both homogeneity and heterogeneity in the group creation [32].

The second step of our approach was meant to corroborate the preliminary findings that emerged from the literature review and the focus group, concentrating on the consumer stakeholder category. In particular, building from key issues raised in the first two steps, we administered a follow up semi-structured questionnaire to three experts with long-term involvement and expertise in the context of consumers' behavior and attitudes. These representatives were selected from a range of different organizations: (1) a consumer association; (2) a partner of the EU project BIOWAYS, involved in public awareness of the potential benefits of bio-based products; and (3) a public research center involved with consumers' acceptance drivers related to bio-based products. The questionnaire was administered by telephone and lasted approximately one hour. The interviewees were asked to validate and integrate the proposed list of impact categories and social indicators according to their perspectives and knowledge, and then to appraise them in accordance with a five-option Likert scale, by arguing the weight of each answer. This exercise enabled us to elicit the value tree of the impact categories, social indicators, and possible indicators for the consumers' category.

4. Results

4.1. S-LCA Applied to Bio-Based Products

The literature review identified 18 studies concerning the social life cycle assessment for bio-based products. Eleven articles were scientifically published in peer-reviewed journals related to social, environmental, and sustainability topics, and four were case studies performing SLCA on bio-based products. In addition, we found seven contributions on bio-based products pertaining to the so called "grey literature".

Overall, the discussion over the sustainability of bio-based products throughout their life cycle, especially biofuels, has until now focused primarily on environmental issues [33–35]. In recent years, however, social and socio-economic aspects have gained increasing attention and have progressively been included in all sustainability schemes for biofuels [36]. However, when it comes to other bio-based products, the situation still lags behind [37]. This is probably imputable to the fact that bio-based products involve longer and more complex value chains [38] that make the assessment of social and socio-economic impacts extremely challenging.

Nonetheless, it is worth mentioning that there are some examples making increasing efforts to investigate the social and socio-economic impacts of bio-based products within a life cycle perspective. The Global-Bio-Pact EU FP7-funded project proposed a set of indicators and criteria for assessing the socio-economic impacts of biomass production and numerous conversion chains, with the aim of demonstrating the opportunities and limitations of the inclusion of socio-economic criteria in a European/International certification scheme [39]. The selection process was based on:

- a review of the literature;
- screening of socio-economic criteria and indicators in existing certification and standards; and
- indicators for bioenergy sustainability developed by initiatives such as the Global Bioenergy Partnership [40].

The impacts are related to six major categories: (i) contribution to the local economy; (ii) working conditions and rights; (iii) health and safety; (iv) gender; (v) land rights and conflicts; and (vi) food security. Additionally, the H2020 BioSTEP project, which engaged with the screening of social and socio-economic dimensions of bio-based products and processes, has revealed several interesting insights, with specific reference to bio-based plastics, chemicals and lubricants. Specifically, the following aspects were identified as the most risky: (i) the competition for feedstock and potential contribution to food insecurity; (ii) limited understanding among consumers; (iii) limited public perception in the EU; and (iv) job creation [20].

When looking specifically at the use of SLCA, it is worth noting that there is a rapidly growing literature with a strong focus on biofuels (e.g., [36,41–43]). Furthermore, there are also several studies utilizing SLCA to assess the social sustainability of recycling [44], packaging systems [45], and new technological processes (e.g., [46,47]). Table 1 reports a selected list of exemplifying studies performing SLCA on bio-based products. Impact categories, social indicators and scale of the analysis are identified for each of these studies. As can be clearly seen, there are important indicators common to these studies, such as health and rights of workers and contribution to employment, while others such as community engagement are less frequently addressed. Moreover, these studies have often taken different approaches since, as mentioned above and unlike with ELCA, there is still not a standardized methodology for SLCA.

Table 1. SLCA case studies on bio-based products.

Study	Main Objectives	Followed Approach	Impact Categories	Social Indicators	Scale
[36]	- Comparing potential social and socio-economic impacts of four types of vehicle fuels: two bio-based (biodiesel and bioethanol) and two fossil-fuel (diesel and petrol) utilized in the EU, especially in Northern Europe and Sweden - Identifying potential social hotspots.	Use of the Social Hotspot Database, focusing on mostly risky aspects (screening S-LCA)	A. Human rights; B. Labor; C. Health and safety; D. Community; E. Governance	A → Indigenous rights; high conflicts; gender equity; human health issues B → Child labor; forced labor; excessive working time; wage assessment; poverty; migrant labor; freedom of association, etc. Unemployment; labor laws C → Injuries and fatalities; toxics and hazards D → Hospital beds; drinking water; Sanitation; children out of school; smallholder or conventional farms E → Legal systems; Corruption	Generic level: country and/or sector level data of fuels within the EU, especially in Northern Europe and Sweden
[41]	- Social and socio-economic impacts of palm oil biodiesel in a province of Indonesia	- Impact categories and criteria grounded on UNEP-SETAC (2009), preliminary survey and literature review - Weighting of the criteria through experts' evaluation (by questionnaire) to ensure further applicability to MCDA	A. Human rights; B. Working conditions; C. Cultural heritage; D. Socio-economic repercussions; E. Governance	A → Free from the employment of child labor; free from the employment of forced labor; equal opportunities; free from discrimination. B → Freedom of association and collective bargaining; fair salary; Decent working hours; occupational health and safety; social benefit. C → Land acquisition, delocalization, migration; respect of cultural heritage and local wisdom; respect of customary rights of indigenous people; community engagement; safe and healthy living conditions; access to material resources; access to non-material resources; Transparency of social/environmental issues D → Contribution to local employment; contribution to economic development; food security; horizontal conflict; transfer of technology and knowledge E → Public commitments to sustainability; fair competition; free from corruption	Regional level
[42]	- Pointing out the difference between performances, effects and impacts in conducting an SLCA - investigating social impacts/effects performing a scenario analysis on biodiesel, comparing different raw materials, i.e., palm oil, forest biomass and algae	- Approach based on Weidema (2006), E-LCA, Kim and Hur (2009), Hofstetter and Norris (2003), Norris (2006)	A. Health; B. Well-being	A → *Company level*: Health of the population; health of workers; health in foreign countries *Regional level*: Health of workers in the region; health of the population in the region; *State level*: Health of the national population; occupational health B → *Regional level*: Well-being of the region's population (no tool available); *State level*: Welfare (e.g., changes in poverty) of national population (no tool available); Welfare of foreign populations	General company, region, state level
[44]	- Assessing the social impacts of three Peruvian recycling systems	- Impact categories and subcategories grounded on UNEP-SETAC (2009) and context specific topics;	A. Human rights; B. Working conditions; C. Socioeconomic repercussions	A → Child labor; discrimination; freedom for association and collective bargaining B → Working hours; minimum income; fair income; recognized employment relationships and fulfilment of legal social benefits; physical working conditions; psychological working conditions C → Education	General/regional level

Source: own elaboration.

In this respect, the approach developed by the EU-FP7 funded project "Prospective Sustainability Assessment of Technologies" (PROSUITE) has recently attracted great interest among scholars involved in bio-based product sustainability assessment. It integrates the social assessment within a comprehensive framework that brings together the three dimensions of sustainability. By identifying five main impact categories: (1) human health; (2) social well-being; (3) prosperity; (4) natural environment; and (5) exhaustible resources, PROSUITE defined a set of indicators with the objective of minimizing potential overlapping. With reference to social well-being, and in line with [48], PROSUITE took four core aspects into consideration: (i) autonomy; (ii) safety, security and tranquillity; (iii) equality; and (iv) participation and influence. This work was subsequently taken up by [29] who suggested a general "modified systemic approach for a social sustainability impact assessment", tailored to the bio-based economy. In particular, the authors paid particular attention to the second step of the SLCA, i.e., the inventory analysis. Within this step, the developed approach emphasized the importance of identifying, through the involvement of experts, the impact of (sub-)categories and indicators associated with the stakeholders' categories. Indeed, the identification of criteria for selecting these categories and indicators has been recognized as one of the most critical issues in conducting an SLCA (see, among others, [44]), which is also influenced by different perspectives and local contexts [49]. Taking this into account, the main indicators must also be selected and/or validated by the stakeholders (see [36]). Moreover, with reference to context-specific SLCA, various scholars [50,51] suggest the integration of top-down, universally recognized social sustainability aspects with bottom-up context-specific social aspects (for example, drawing on national and regional sustainability strategies, sector-specific issues and stakeholders' interests). In this vein, Mattila et al. [52] suggested an approach based on the integration of global methods with participatory methods involving local stakeholders.

In recent years, great efforts have been also made by the European Committee for Standardization (CEN) in defining social sustainability of bio-based products, in particular by the CEN technical committee for bio-based products (CEN/TC 411). When setting social criteria for the bio-based part of bio-based products (excluding food, feed and energy), EN 16751:2016 focused on:

(1) labor rights (including indicators on bargaining rights, elimination of forced labor, child labor and discrimination, safe working conditions for employees, knowledge required and training, living conditions, and satisfaction of the basic needs of employees);
(2) land use rights and land use change (including indicators related to respect for land use rights and on food security);
(3) water use rights in areas with water scarcity (including indicators on the identification of potential negative impacts related to water resources and measures to address them); and
(4) local development (description of measures undertaken to address local development).

It is worth noting, however, that access to data on bio-based products represents a major challenge. In this context, the Social Hotspot Database (SHDB), developed by Benoit-Norris et al. [53], represents a reference point for SLCA practitioners. However, as Ekener-Petersen et al. [36] pointed out, this database provides data at the sector level but not on specific sites/plants and products. This limitation makes it possible to identify only potential impacts (i.e., the identification of aspects where there are significant risks of social impact) but not actual impacts. Currently, another database on "Product Social Impact Life-Cycle Assessment" (PSILCA) has been developed by the sustainability consulting and software company GreenDelta. This database covers 88 indicators in total, addressing 25 main indicators (sub-categories). However, as emphasized by Rafiaani et al. [29], there are no specific data for bio-based products.

4.2. List of Social Impact Categories and Indicators Tailored to Bio-Based Products

At the end of this extensive review of social sustainability with a specific focus on bio-based products, a list of social impact categories and associated indicators can now be proposed (see Appendix A). This list is built on a set of frameworks that have already been applied by the literature. These are the BioSTEP project [20], the UNEP-SETAC guidelines and methodological sheets [54,55], the PROSUITE approach [46], the Global-Bio-Pact project [39], and the Global Bioenergy Partnership (GBEP) [40].

The list is composed of eight impact categories (see Figure 1), identified as relevant to bio-based products, i.e., health and safety, social acceptability, food security, employment, income, human rights and working conditions, gender issues and discrimination, and access to material resources and land use change.

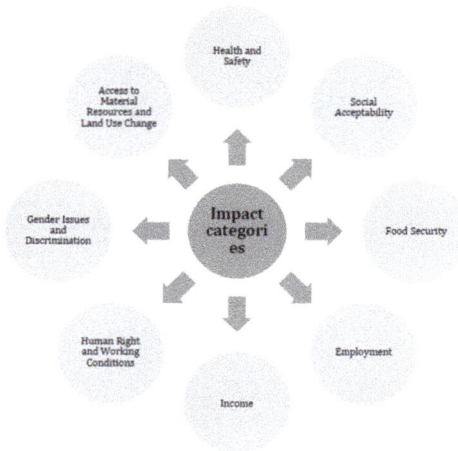

Figure 1. Impact categories tailored to bio-based products. Source: own elaboration.

Each impact category is related to different potentially affected categories of stakeholders [54], i.e., workers, consumers, local community, value chain actors and society. Consequently, each stakeholder category connected to different impact categories can, in turn, be associated to a wide range of indicators (Figure 2).

Figure 2. General framework for S-LCA tailored for bio-based products. Source: adapted from [54].

4.3. Focus Group Exercise

During the focus group, the experts were asked to identify the key sustainability criteria to be included in a sustainability assessment scheme for bio-based products. With reference to social categories and indicators to be included in an overall assessment, several issues were brought into discussion. First, it was outlined that, although environmental criteria are more evident for consumers, socio-economic criteria should also be considered. More specifically, it was pointed out that, even though the obligatory inclusion of social criteria for industry might be perceived as an obstacle to the creation of a level playing field with fossil-based products, it has also been stressed that properly communicating social impacts might be a key factor for increasing consumers' demand for bio-based products.

Another interesting point that came from the discussion relates to environmental criteria, i.e., the bio-based content of products. This content is strongly interlinked with one of the impact categories identified in the previous section, i.e., consumers' social acceptability. Specifically, it was stressed that a product that is not 100% bio-based may destabilize the consumer, not only for its potential detrimental effect on the environment but also for social reasons, in particular for possible negative health impacts. However, this is an issue that remains open since many products are not 100% bio-based; it has been emphasized that this represents a critical issue for consumers' social acceptability. The relevance of both social acceptability and health and safety for consumers, is in line with some preliminary results of a survey carried out in the framework of an H2020 project in which one of the focus group participants was involved.

The discussion then moved to consumers' perception towards bio-based products. Particularly, as clearly emerged, there is still no clear understanding of the bio-based idea, whose meaning might often be confused with "organic", "biodegradable" and "compostable" ideas. However, it has been stressed that consumers are usually aware that they would pay a higher price for bio-based products, and, therefore, their willingness to pay for such products must be supported by adequate information, including social criteria.

4.4. Validation of Social Impact Categories and Indicators Related to Consumers

To elicit consumers' perspectives about the most significant social impact categories and indicators to be included in the SLCA of bio-based products, we performed semi-structured interviews with experts. The interviews were carried out by telephone in February 2018, following an ice-breaking approach. More specifically, after clarifying our research goals, the respondents were asked to express their personal views about the role of consumers in defining the most relevant factors to enhance the market uptake of bio-based products. This was the opening question and enabled us to involve the respondents in the topic under investigation. We then illustrated our selected impact categories and social indicators (see Figure 3) in order to check their relevance (i.e., validation) for the consumers and, thus, to ascertain whether our experts believed they must be included in the social assessment of bio-based products related to this specific stakeholder category. Subsequently, they were asked to explore the possible integration of such social aspects.

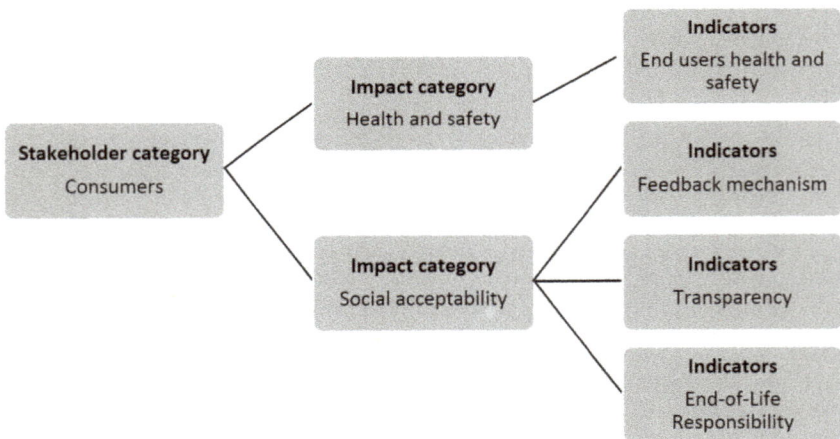

Figure 3. Value tree of the social impact categories and indicators for consumers' category. Source: own elaboration.

Overall, the respondents found the identified impact categories and social indicators relevant in the assessment of social sustainability of bio-based products, since they adequately describe the social aspects that characterize the investigated context. In particular, each question was assessed according to a five-option Likert scale (from -2 = not important to 2 = very important) to measure their relevance and to allow any possible neutral answers on an odd-numbered scale (see Figure 4).

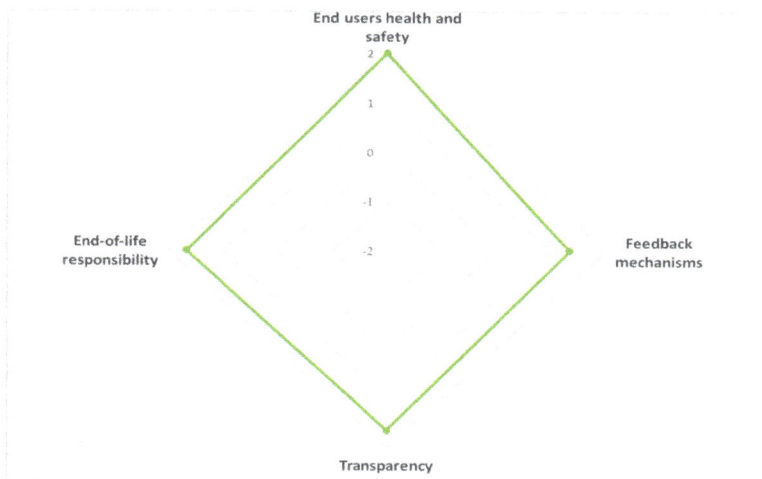

Figure 4. Five-option Likert scale for social indicators. Source: own elaboration.

All respondents recognized the *health and safety* impact category as "very important" since it is considered to be a key determinant for consumers. In particular, they emphasized that consumers' willingness to pay is strongly affected by this category. In line with the findings that emerged from the focus group, this has been outlined as a crucial factor for justifying the additional price that consumers are willing to pay for a product just for the fact that it is commonly understood as "eco-friendly" or, as for our case, "bio-based" (i.e., green premium). (In the framework of the European project BIOFOREVER (www.bioforever.org), nova-Institute is conducting a number of surveys on GreenPremium prices for bio-based products.) Moreover, one respondent stressed the importance of considering health and safety as separate impact categories since they explicitly reflect two different social themes. The social item found to be relevant for the *Health and Safety* impact category from the literature review, i.e., *end-users health and safety*, was ranked as "very important" by all respondents.

The second impact category, i.e., *social acceptability*, was found to be relevant by all respondents. In particular, two out of three recognized it as "very important" for the social assessment of bio-based products. Consumer social acceptability is driven by some specific social indicators that are worth taking into account, namely, feedback mechanisms, transparency and end-of-life responsibility. The presence of *feedback mechanisms* was found to be "important" for two respondents, and "very important" for the other. This highlights the significant role of feedback for consumers since they represent paths to communicate and signal their (dis)satisfaction to the organization in the use of bio-products. Moreover, as discovered by the interviews, those consumers oriented towards bio-based products often share a specific pro-environmental and social attitude that reflects their sensibility towards current concerns (i.e., climate change, depletion of resources, working conditions, quality of life, etc.); they wish to share these concerns with companies to gain more information on the products.

With reference to *transparency*, two respondents acknowledged this as "very important", while the third defined it as "important". In this vein, the presence of clear sustainability reports, labels and certification highlighting the (over)compliance with existing regulations, enable an informed choice for the consumer without intent to mislead or conceal. Moreover, according to respondents' opinions, the presence of strong *transparency* is another necessary condition to balance the higher prices of bio-based products compared to conventional ones with the same technical performance, as also already emphasized in the focus group exercise.

Finally, the *end-of-life responsibility* namely, the disposal, re-use or recycling of bio-based products, was rated by two respondents as "very important", highlighting the increasing diffusion among consumers of the circular economy concept. Conversely, according to an interviewee, the *end-of-life responsibility* social indicators, although recognized as "important", are a mere dimension on which consumers want to be informed for conscious recycling and disposal of the products.

5. Discussion

Following a general tendency of the literature, most of the studies on bio-based products entailing a life cycle perspective have focused on some fundamental themes recognized as crucial to be protected, i.e., human health, natural environment, natural resources, and man-made environment [56]. These have been defined as areas of protection (AoPs) and are addressed by the ELCA. However, as outlined by Reitinger et al. ([15], p. 381), "*we are faced with the paradoxical situation of avoiding harm to environment and human health while ignoring other aspects of human life and thus the aims of sustainability*". It was therefore suggested (i.e., [48,51]) that SLCA should complement the existing AoPs analyzed in ELCA by focusing on social well-being, which has been considered the primary AoP of LSCA [57]. Accordingly, the social impact assessment would involve impacts on the stakeholders' well-being that are associated with a wide range of indicators measuring the quality of life of people on both an individual and a collective level [46].

In this vein, the development of general SLCA indicators could provide organizations with relevant information to better understand those important social factors for positioning a product in the market and to guide consumer purchases (see [16]). It is worth noting that the use of biomass does not make bio-based products automatically sustainable or even, by definition, more sustainable than fossil-based products [58]. To this end, an SLCA therefore becomes crucial to compare "*options, especially when complex supply chains are involved*" [59]. In fact, for a market uptake of bio-based products, certain necessary conditions must be achieved, not only in terms of the implementation of appropriate policies (see, e.g., [17,60]) but also on the conditions that relate directly to consumer acceptability (see [61]).

As became apparent from the case studies considered in our review, and according to several scholars [8,19], SLCA studies on bio-based products have focused mainly on social indicators relating to the worker stakeholder category while overlooking other stakeholders. Indeed, focusing on consumer perspectives is very important to deepen our knowledge on the main social sustainability indicators able to guarantee consumer well-being and therefore enhance the market development of bio-based products. In this regard, although it was stressed during the focus group exercise and expert interviews that consumers still have very little understanding of "bio-based" as a concept [62], they are at the same time aware of the higher pricing of bio-based products in comparison with more traditional products. In line with the literature (see [61,63]), we found that the willingness to pay represents a key factor. More specifically, the health and safety of consumers appears to be the factor that most influences the consumer's willingness to pay. In fact, moving towards the choice of bio-based products is also viewed as a change of consumer propensities. Far from dismissing the relevance of the health and safety impact category for both bio-based and traditional products, we observed that respondents emphasized that consumers expect bio-based products to perform their intended functions better, and to not pose risks to their health and safety. Therefore, as reported in Appendix A, both the presence of labels or standards (e.g., GRI 416) certifying an organization's systematic efforts to address health and safety along the life cycle of a product, and the organization's adherence to customer health and safety regulations, could have a flywheel effect on the market diffusion of a bio-based product. Furthermore, the analysis suggests that willingness to pay is also strongly related to social acceptability, whose lack represents a dramatic barrier (see [64]). This is in line with recent literature that proves that consumers, when confronted with eco-friendly purchase alternatives, respond not only rationally but also emotionally [65]; this could also be reflected in consumer acceptance of bio-based products [63]. Accordingly, the willingness to pay must be supported by adequate information, through certification, labels and development of standards for LCSA studies that are the core transparency (see Appendix A). Furthermore, the dynamic and active involvement of companies in providing user-friendly feedback mechanisms, in particular by means of social networks media, has been highlighted as a further catalyst for the market development of the bio-based products.

Finally, the end-of-life responsibility indicators also emerge as an important factor since, in several European countries, separate waste collection is mandatory or advocated. In this context, designing products in a smarter way, extending their useful lives, and providing complete and clear information for consumers regarding sustainable end-of-life options represent necessary changes for going well beyond the traditional resource efficiency and recycling of waste.

The discussion above clearly indicates the need for the development and dissemination of improved methods capable of exploiting the consumers' role in the ongoing process of market uptake of bio-based products. This need regards the effective inclusion of some social indicators (i.e., health and safety, and social acceptability, feedback mechanisms, strong transparency, and end-of-life responsibility) in the assessment scheme for bio-based products. This allows consumers, where properly communicated, to make more informed and aware purchasing choices.

6. Conclusions and Recommendations for Further Analysis

Together with environmental sustainability, demonstrating that bio-based products are also sustainable from a social and socio-economic perspective is critical to augment public acceptance and increase demand (see [66]). Overall, the proposed study provides an in-depth understanding of the impact categories and social indicators that should be included in an SLCA for bio-based products.

Our investigation allows the identification of a preliminary list of impact categories and social indicators grounded on an extensive literature review. Additionally, the focus group conducted under the Horizon 2020 funded project STAR-ProBio, and the interviews with experts, give us the opportunity to better explore and validate the impact categories and associated social indicators for consumers.

Several findings emerged from our investigation and can be synthetized as follows:

1. Eight impact categories have been identified as relevant for SLCA tailored to bio-based products. Moreover, a wide range of social indicators have been associated with different impact categories for potentially affected stakeholder categories.
2. Although the inclusion of social criteria in the assessment scheme for bio-based products might be perceived by the industry as an obstacle towards the creation of a level playing field with fossil-based products, if properly communicated, it might be a key factor for increasing consumer demand for bio-based products. Furthermore, as consumers are willing to pay a higher price for bio-based products, that willingness to pay should be supported by adequate information, including social life cycle impacts of the product.
3. The elicited experts' perspectives about the most significant consumer impact categories and social indicators to be included in the SLCA of bio-based products seem to endorse the findings that emerged from the literature review. In fact, both *health and safety* and *social acceptability* of consumers are perceived as very important impact categories on which to focus to achieve a comprehensive social assessment from the demand side. Moreover, the presence of adequate feedback mechanisms, strong transparency and end-of-life responsibility might allow consumers to make more informed and aware purchasing choices.

Our findings support evidence for going beyond the traditional life cycle assessment of products. From this perspective, SLCA provides the opportunity to accomplish a comprehensive social sustainability assessment in which all the involved parties (from industry to policy makers) should make an effort to tackle the proposed social indicators along the life cycle of the product. Its adherence to customer acceptance and health and safety could have a flywheel effect on the market diffusion of bio-based products.

The main limitation of this study rests on the restricted sample size of experts interviewed. However, it represents the first attempt in the literature to assess the relevance of social impact categories and social indicators from the perspective of consumers. Future empirical studies could apply our two-step methodology to other stakeholder categories, or other sectors that are relevant for transition towards sustainability.

Acknowledgments: The authors are very grateful to the STAR-ProBio project (*Sustainability Transition Assessment and Research of Bio-based Products*) for their financial support. The project is funded by the European Union's Horizon 2020 Research and Innovation Programme under Grant Agreement No. 727740, Work Programme BB-01-2016: Sustainability schemes for the bio-based economy.

Author Contributions: Both authors conceived and designed the study methodology and contributed to the primary share of the article writing. Enrica Imbert wrote the context of analysis and discussions sections. Pasquale Marcello Falcone wrote the introduction, methodology and conclusions. Both authors contributed to write results section.

Conflicts of Interest: The authors declare no conflict of interest.

Appendix A

Table A1. List of impact categories and social indicators suggested in the framework of SLCA within the bio-based economy.

Impact Categories	Reference	Stakeholder Categories	Social Indicators
Health and Safety	BioSTEP (2016)	Workers	*Health:* Exposure to agrochemicals; Numbers of multi-resistant organisms; Toxicity of "green" vs. "grey" industrial products *Health and safety:* Generic analysis (Hotspots): occupational accident rate by country: number/percentage of injuries or fatal accidents in the organization by job qualification inside the company Number of injuries per level of employees. Presence of a formal policy concerning health and safety. Adequate general occupational safety measures are taken. Preventative measures and emergency protocols exist regarding accidents and injuries. Preventative measures and emergency protocols exist regarding pesticide and chemical exposure. Appropriate protective gear required in all applicable situations; number of serious/non-serious Occupational Safety and Health Administration (OSHA) violations reported within the past three years and status of violations; education, training, counseling, prevention and risk control programs in place to assist workforce members, their or community members regarding serious diseases
	UNEP-SETAC (2009; 2013)	Consumers	*End-users health and safety:* Generic analysis (Hotspots): Quality of or number of information/signs on product health and safety; presence of consumer complaints (at national, sectorial, organizational level); total number of incidents of non-compliance with regulations and voluntary codes concerning health and safety impacts of products and services and type of outcomes (GRI PR2) Specific analysis: Presence of labels on health and safety; number of consumer complaints; GRI 416
		Local community	*Safe and healthy living conditions:* Generic analysis (Hotspots): Burden of Disease by Country; Pollution Levels by Country; Presence/Strength of Laws on Construction Safety Regulations by Country Specific Analysis: Management oversight of structural integrity; Organization's efforts to strengthen community health (e.g., through shared community access to organization health resources); Management effort to minimize use of hazardous substances
	PROSUITE (2013)	Occupational health	Number of: non-fatal accidents at work; fatal accidents at work; occupational diseases
		Environmental Human health	Climate change; ozone depletion; human toxicity; respiratory inorganics; ionizing radiation
	Global-Bio-Pact (2012)	Workers	Work related accidents and diseases: Number of work related accidents per person days of employment per year, number of work related diseases/person days of employment per year. Personal protective equipment: Percentage of workers that use appropriate personal protective equipment. OSH training: Percentage of employees that have received OSH (Occupational Safety and Health) training
	GBEP (2011)	Not proposed	Change in mortality and burden of disease attributable to indoor smoke. Incidence of occupational injury, illness and fatalities

Table A1. *Cont.*

Impact Categories	Reference	Stakeholder Categories	Social Indicators
	UNEP-SETAC (2009; 2013)	Consumers	*Feedback Mechanism:* Presence of a mechanism for customers to provide feedback. Management measures to improve feedback mechanisms. Practices related to customer satisfaction, including results of surveys measuring customer satisfaction. *Transparency:* Compliance with regulations regarding transparency; publication of a sustainability report; divulgence of results on ELCA and SLCA; Number of certifications and labels *End-of-Life Responsibility:* Presence of clear information provided to consumers on end-of-life; number of incidents of non-compliance with regulatory labeling requirements
Social Acceptability		Value chain actors	*Promoting Social Responsibility:* Presence of explicit code of conduct that protects human rights of workers among suppliers. Membership of an initiative that promotes social responsibility along the supply chain
		Society	*Public Commitment to Sustainability Issues:* Presence of publicly available documents as promises or agreements on sustainability issues. Formalized commitment of the organization to prevent corruption, referring to recognized standards.
		Workers	Involvement of smallholders or small suppliers. Percentage of feedstock that originates from associates, smallholders, and out-growers.
	Global-Bio-Pact (2012)	Local community	*Contribution to local economy:* Amount invested in community investment projects (e.g., CSR) (percent of annual revenue) and qualitative description of investments including any projects specific for women.
	GBEP (2011)	Workers	Training and requalification of workforce (i.e., share of trained workers in the bio-energy sector out of total bio-energy workforce, and share of re-qualified workers out of the total number of jobs lost in the bio-energy sector)
	BioSTEP (2016)	Not proposed	Use of agrochemicals (including fertilizers) and GMO crops; change in food prices (and its volatility); malnutrition, risk of hunger; macronutrient intake/availability
		Workers and Local community	Availability of food: Perceived change in availability of food after the beginning of bio-energy operations Time spent in subsistence agriculture: Change in time spent in subsistence agriculture in the household
Food Security	Global-Bio-Pact (2012)	Processing company or plantation	Land that has been converted from staple crops (ha) Edible feedstock diverted from food chain to bio-energy: amount of edible raw material diverted into bio-energy production (t)
		Government and NGOs	Land that has been converted from staple crops (ha)

Table A1. *Cont.*

Impact Categories	Reference	Stakeholder Categories	Social Indicators
	GBEP (2011)	Not proposed	Price and supply of a national food basket, allocation and tenure of land for new bio-energy production (percentage of land used for new bio-energy production). Change in income (wages paid for employment into bio-energy sector in relation to comparable sectors; net income from the sale, barter and/or own consumption of bio-energy products, including feedstocks, by self-employed households/individuals) Bio-energy used to expand access to modern energy services (total amount and percentage of increased access to modern energy services gained through modern bio-energy, measured in terms of energy and numbers of households and businesses. Jobs in the bio-energy sector as a result of bio-energy production and use (total number of jobs in the bio-energy sector and percentage adhering to nationally recognized labor standards consistent with the principles enumerated in the ILO Declaration on Fundamental Principles and Rights at Work, in relation to comparable sectors).
	BioSTEP (2016)	Not proposed	Change in employment rate; full time equivalent jobs; job quality; need for/lack of highly specialized work force
		Local community	*Local employment:* Generic analysis (Hotspots): Unemployment and poverty statistics by region Specific analysis: Percentage of workforce hired locally; Strength of policies on local hiring preferences; percentage of spending on locally based suppliers
	UNEP-SETAC (2009, 2013)	Workers	*Hours of works:* Generic analysis (Hotspots): excessive hours of work Specific analysis: Number of hours effectively worked by employees (at each level of employment); number of holidays effectively used by employees; clear communication of working hours and overtime arrangements; respect of contractual agreements concerning overtime
Employment		Society	*Contribution to economic development:* Economic situation of the country/region (GDP, economic growth, unemployment, wage level, etc.); Relevance of the considered sector for the (local) economy (share of GDP, number of employees in relation to size of working population, wage level, etc.)
	PROSUITE (2013)	Social wellbeing	Safety, security and tranquility (knowledge-intensive jobs, total employment)
	Global-Bio-Pact (2012)	Workers and processing company	*Contribution to local economy:* Total number of employees and person days of employment per year. Number of workers that have received training (for skills development, education, etc.) each year, number of working days spent in training provided by the operation each year, type of training.
	Global-Bio-Pact (2012)	Government and Local community	*Contribution to local economy:* Ratio of employment from local area/outside local area per category of employment. Percentage of workers that have a fixed contract employment per category of employment
	BioSTEP (2016)	Households income	Income of employees in bio-economy sector (total); distribution of income
	PROSUITE (2013)	Social wellbeing	Global Income Inequalities between GDP levels around the world.
Income	Global-Bio-Pact (2012)	Workers and processing company	*Working conditions and rights:* Average income of employees by category of employment (EUR)
	Global-Bio-Pact (2012)	Workers and local community	*Working conditions and rights:* Income spent on basic needs (percentage of worker's disposable income spent on fulfilling basic needs (food, accommodation and transport)

Table A1. *Cont.*

Impact Categories	Reference	Stakeholder Categories	Social Indicators
Human Rights and Working Conditions	UNEP-SETAC (2009; 2013)	Workers	*Freedom of associations and collective bargaining:* Generic analysis (Hotspots): Evidence of restriction to freedom of association and collective bargaining. Specific analysis: Employment is not conditioned by any restrictions on the right to collective bargaining; presence of unions within the organization is adequately supported; workers are free to join unions of their choosing: *Child labor:* Generic analysis (Hotspots): Percentage of children working by country and sector Specific analysis: Absence of working children under the legal age or 15 years old; children are not performing work during the night. *Fair salary:* Generic analysis (Hotspots): Living Wages in the US by state, county, community Specific analysis: Lowest paid worker, compared to the minimum wage; the lowest paid workers consider that their wages meet their needs. Presence of suspicious deductions on wages: *Hours of works:* Generic analysis (Hotspots): Excessive Hours of work Specific analysis: Number of hours effectively worked by employees; Number of holidays effectively used by employees; Clear communication of working hours and overtime arrangements *Forced labor:* Generic analysis (Hotspots): Risk of forced labor used for production of commodity Specific analysis: Workers voluntarily agree on employment terms; Workers are free to terminate their employment within the prevailing limits, etc.
		Local community	*Delocalization and migration:* Strength of organizational procedures for integrating migrant workers into the community *Community engagement:* Freedom of Peaceful Assembly and Association; Diversity of community stakeholder groups that engage with the organization. *Respect of Indigenous Rights:* Human rights issues faced by indigenous peoples. Indigenous land rights conflicts/land claims. Strength of policies in place to protect the rights of indigenous community members.
		Society	*Prevention and Mitigation of Conflicts:* Is the organization doing business in a region with on-going conflicts? Organization's role in the development of conflicts; Disputed products.
		Value chain actors	*Promoting social responsibility:* Presence of explicit code of conduct that protect human rights of workers among suppliers. Industry code of conduct in the sector. *Respect of Intellectual Property Rights:* General intellectual property rights and related issues associated with the economic sector
	PROSUITE (2013)	Social wellbeing	*Autonomy:* (child labor and forced labor) number of children under legal age who perform hazardous work with companies. Amount of forced labor under the menace of any penalty and not undertaken voluntarily by the person.
	Global-Bio-Pact (2012)	Workers, local community and processing companies	*Working conditions and rights:* Freedom of association (existence of labor unions); Employments benefits (e.g., housing, health care, holidays) provided by operations.

Table A1. Cont.

Impact Categories	Reference	Stakeholder Categories	Social Indicators
Gender Issues and Discrimination	BioSTEP (2016)	Not proposed	Property rights; access to land; quality of life (equality of genders)
	UNEP-SETAC (2009; 2013)	Workers	*Equal opportunities/Discrimination:* Generic analysis (Hotspots): Women in the labor force participation rate by country; country gender index ranking Specific analysis: Presence of formal policies on equal opportunities; total numbers of incidents of discrimination and actions taken; ratio of basic salary of men to women by employee category
		Value chain actors	*Promoting Social Responsibility:* Integration of ethical, social, environmental and regarding gender equality in purchasing policy, distribution policy and contract signatures
	Global-Bio-Pact (2012)	Workers and processing companies	*Gender:* Benefits created for woman (i.e., maternity leave, and others)
		Community and processing companies	*Contribution to local economy:* Investments in projects (percent of annual revenue) including any programs specific for women
Access To Material Resources and Land Use Change	BioSTEP (2016)	Land access	Land prices, Land tenure, Property rights, Access to land.
	UNEP-SETAC (2009; 2013)	Local community	*Access to Material Resources:* Generic analysis (Hotspots): Changes in land ownership. Levels of industrial water use. Extraction of material resources. Percentage of population (urban, rural, total) with access to improved sanitation facilities. Specific Analysis: Has the organization developed a project related infrastructure with mutual community access and benefit? Strength of organizational risk assessment with regard to potential for material resource conflict. Does the organization have a certified environmental management system?
	PROSUITE (2013)	Not proposed	Water use, terrestrial; Land use
	Global-Bio-Pact (2012)	Processing companies	*Land rights and conflicts:* Legal title of land right—has a legal title/concession for the land that is not challenged? Communal/ public land and land conflicts—has the operation had any land use conflicts, if so, what caused them, how were they resolved?
	Government and NGOs		*Land rights and conflicts:* Legal title of land right—operation has a legal title/concession for the land that is not challenged. Area of land currently under dispute, land conflict. Has the operation had any land use conflicts, if so, what caused them, how were they resolved?
	GBEP (2011)	Not proposed	Access to land, water and other natural resources: Allocation and tenure of land for new bio-energy production

Source: own elaboration.

164

References

1. Hutchins, M.J.; Sutherland, J.W. An exploration of measures of social sustainability and their application to supply chain decisions. *J. Clean. Prod.* **2008**, *16*, 1688–1698. [CrossRef]
2. Sillanpää, M.; Ncibi, C. *A Sustainable Bioeconomy: The Green Industrial Revolution*; Springer: Cham, Switzerland, 2017.
3. European Commission. Bio-Based Economy in Europe: State of Play and Future Potential-Part 2 Summary of Position Papers Received in Response to the European Commission's public On-line Consultation, 2011. Available online: https://ec.europa.eu/research/consultations/bioeconomy/bio-based-economy-for-europe-part2.pdf (accessed on 9 January 2018).
4. EuropaBio. Building a Bio-Based Economy for Europe for 2020, EuropaBio Policy Guide. 2011. Available online: http://www.europabio.org/industrial-biotech/publications/building-bio-based-economy-europe-2020 (accessed on 9 January 2018).
5. European Commission. *Innovating for Sustainable Growth—A Bioeconomy for Europe*; European Commission, Directorate Research and Innovation: Brussels, Belgium, 2012.
6. EuropaBio, 2011, Bioeconomy from a Vision to a Realty. Available online: https://www.europabio.org/sites/default/files/bieconomy_-_from_vision_to_reality.pdf (accessed on 9 January 2018).
7. Falcone, P.M.; Lopolito, A.; Sica, E. Policy mixes towards sustainability transition in the Italian biofuel sector: Dealing with alternative crisis scenarios. *Energy Res. Soc. Sci.* **2017**, *33*, 105–114. [CrossRef]
8. Martin, M.; Røyne, F.; Ekvall, T. Moberg, Å. Life Cycle Sustainability Evaluations of Bio-based Value Chains: Reviewing the indicators from a Swedish Perspective. *Sustainability* **2018**, *10*, 547. [CrossRef]
9. Zamagni, A. Life cycle sustainability assessment. *Int. J. Life Cycle Assess.* **2012**, *17*, 373–376. [CrossRef]
10. Kühnen, M.; Hahn, R. Indicators in Social Life Cycle Assessment—A Review of Frameworks, Theories, and Empirical Experience. *J. Ind. Ecol.* **2017**, *21*, 1547–1565. [CrossRef]
11. Baumann, H.; Arvidsson, R.; Tong, H.; Wang, Y. Does the production of an airbag injure more people than the airbag saves in traffic? *J. Ind. Ecol.* **2013**, *17*, 517–527. [CrossRef]
12. Lamberton, G. Sustainability accounting: A brief history and conceptual framework. *Account. Forum* **2005**, *29*, 7–26. [CrossRef]
13. Jørgensen, A.; Dreyer, L.C.; Wangel, A. Addressing the effect of social life cycle assessment. *Int. J. Life Cycle Assess.* **2012**, *17*, 828–839. [CrossRef]
14. Mathe, S. Integrating participatory approaches into social life cycle assessment: The SLCA participatory approach. *Int. J. Life Cycle Assess.* **2014**, *19*, 1506–1514. [CrossRef]
15. Reitinger, C.; Dumke, M.; Barosevcic, M.; Hillerbrand, R. A conceptual framework for impact assessment within SLCA. *Int. J. Life Cycle Assess.* **2011**, *16*, 380–388. [CrossRef]
16. Parent, J.; Cucuzzella, C.; Revéret, J.P. Revisiting the role of LCA and SLCA in the transition towards sustainable production and consumption. *Int. J. Life Cycle Assess.* **2013**, *18*, 1642–1652. [CrossRef]
17. Imbert, E.; Ladu, L.; Morone, P.; Quitzow, R. Comparing policy strategies for a transition to a bioeconomy in Europe: The case of Italy and Germany. *Energy Res. Soc. Sci.* **2017**, *33*, 70–81. [CrossRef]
18. Fritsche, U.R.; Iriarte, L. Sustainability criteria and indicators for the bio-based economy in Europe: State of discussion and way forward. *Energies* **2014**, *7*, 6825–6836. [CrossRef]
19. Spierling, S.; Knüpffer, E.; Behnsen, H.; Mudersbach, M.; Krieg, H.; Springer, S.; Albrecht, S.; Herrmann, C.; Endres, H.J. Bio-based plastics—A review of environmental, social and economic impact assessments. *J. Clean. Prod.* **2018**. [CrossRef]
20. Hasenheit, M.; Gerdes, H.; Kiresiewa, Z.; Beekman, V. Summary Report on the Social, Economic and Environmental Impacts of the Bioeconomy, BioSTEP. 2016. Available online: http://www.bio-step.eu/results.html (accessed on 8 September 2017).
21. Kline, K.L.; Msangi, S.; Dale, V.H.; Woods, J.; Souza, G.M.; Osseweijer, P.; Clancy, J.S.; Hilbert, J.A.; Johnson, F.X.; McDonnell, P.C.; et al. Reconciling food security and bioenergy: Priorities for action. *GCB Bioenergy* **2017**, *9*, 557–576. [CrossRef]
22. BIOCHEM Project D2.3 Report on the Assessment of the Bio-Based Products Market Potential for Innovation. Available online: http://www.biochem-project.eu/download/toolbox/innovation/06/Bio-based%20product%20market%20potential.pdf (accessed on 6 September 2017).
23. Massawe, E.; Geiser, K.; Ellenbecker, M.; Marshall, J. Health, safety, and ecological implications of using biobased floor-stripping products. *J. Environ. Health* **2007**, *69*, 45. [PubMed]

24. Álvarez-Chávez, C.R.; Edwards, S.; Moure-Eraso, R.; Geiser, K. Sustainability of bio-based plastics: General comparative analysis and recommendations for improvement. *J. Clean. Prod.* **2012**, *23*, 47–56. [CrossRef]

25. OECD. *Bio-Based Chemicals and Bioplastics: Finding the Right Policy Balance*; OECD Science, Technology and Industry Policy Papers, No. 17; OECD Publishing: Paris, France, 2014.

26. Ronzon, T.; Santini, F.; M'Barek, R. *The Bioeconomy in the European Union in Numbers. Facts and Figures on Biomass, Turnover and Employment*; European Commission, Joint Research Centre, Institute for Prospective Technological Studies: Seville, Spain, 2015; p. 4.

27. Reinshagen, P. *Bioeconomy: Much More Employment in Bio-Based Chemicals Than in Biofuels*; Bio Based Press: Amsterdam, The Netherlands, 2015; Available online: http://www.bio-basedpress.eu/2015/06/bioeconomy-much-moreemployment-in-bio-based-chemicals-than-in-biofuels/ (accessed on 10 January 2018).

28. Piotrowski, S.; Carus, M.; Carrez, D. European Bioeconomy in Figures. 2016. Available online: http://biconsortium.eu/sites/biconsortium.eu/files/downloads/20160302_Bioeconomy_in_figures.pdf (accessed on 18 September 2017).

29. Rafiaani, P.; Kuppens, T.; Van Dael, M.; Azadi, H.; Lebailly, P.; Van Passel, S. Social sustainability assessments in the bio-based economy: Towards a systemic approach. *Renew. Sustain. Energy Rev.* **2018**, *82*, 1839–1853. [CrossRef]

30. Anand, M. Innovation and Sustainable Development: A Bioeconomic Perspective; Brief for Global Sustainable Development Report 2016. Available online: https://sustainabledevelopment.un.org/content/documents/982044_Anand_Innovation%20and%20Sustainable%20Development_A%20Bioeconomic%20Perspective.pdf (accessed on 10 January 2018).

31. German Bioeconomy Counsil, Nature commentary: Bioeconomy important for SDGs, 2016. Available online: http://biooekonomierat.de/en/news/the-bioeconomy-is-central-to-the-achievement-of-climate-protection-and-sdgs/ (accessed on 10 January 2018).

32. Morgan, D.L. *Focus Groups as Qualitative Research*, 2nd ed.; SAGE Publications: Thousand Oaks, CA, USA, 1997.

33. Pawelzik, P.; Carus, M.; Hotchkiss, J.; Narayan, R.; Selke, S.; Wellisch, M.; Weiss, M.; Wicke, B.; Patel, M.K. Critical aspects in the life cycle assessment (LCA) of bio-based materials–Reviewing methodologies and deriving recommendations. *Res. Conserv. Recycl.* **2013**, *73*, 211–228. [CrossRef]

34. Hottle, T.A.; Bilec, M.M.; Landis, A.E. Sustainability assessments of bio-based polymers. *Polym. Degrad. Stab.* **2013**, *98*, 1898–1907. [CrossRef]

35. Tsiropoulos, I.; Faaij, A.P.; Lundquist, L.; Schenker, U.; Briois, J.F.; Patel, M.K. Life cycle impact assessment of bio-based plastics from sugarcane ethanol. *J. Clean. Prod.* **2015**, *90*, 114–127. [CrossRef]

36. Ekener-Petersen, E.; Höglund, J.; Finnveden, G. Screening potential social impacts of fossil fuels and biofuels for vehicles. *Energy Policy* **2014**, *73*, 416–426. [CrossRef]

37. Rutz, D.; Janssen, R. Summary Report of the Global-Bio-Pact Project, Global Assessment of Biomass and Bioproduct Impacts on Socio-Economics and Sustainability. 2013. Available online: http://www.globalbiopact.eu/ (accessed on 5 September 2017).

38. Bell, G.; Schuck, S.; Jungmeier, G.; Wellisch, M.; Felby, C.; Jørgensen, H.; Spaeth, J. *IEA Bioenergy Task42 Biorefining*; IEA Bioenergy: Wageningen, The Netherlands, 2014.

39. Diaz-Chavez, R.; Rettenmaier, N.; Rutz, D.; Janssen, R. Global-Bio-Pact Set of Selected Socio-Economic Sustainability Criteria and Indicators, Imperial College; Report of the FP7 Global-Bio-Pact Project. 2012. Available online: http://www.globalbiopact.eu/socio-economic-impacts.html (accessed on 4 September 2017).

40. FAO. *The Global Bioenergy Partnership (GBEP) Sustainability Indicators for Bioenergy*, 1st ed.; GBEP Secretariat e FAO: Rome, Italy, 2011; 223p, Available online: http://www.globalbioenergy.org/fileadmin/user_upload/gbep/docs/Indicators/The_GBEP_Sustainability_Indicators_for_Bioenergy_FINAL.pdf (accessed on 4 September 2017).

41. Manik, Y.; Leahy, J.; Halog, A. Social life cycle assessment of palm oil biodiesel: A case study in Jambi Province of Indonesia. *Int. J. Life Cycle Assess.* **2013**, *18*, 1386–1392. [CrossRef]

42. Macombe, C.; Leskinen, P.; Feschet, P.; Antikainen, R. Social life cycle assessment of biodiesel production at three levels: A literature review and development needs. *J. Clean. Prod.* **2013**, *52*, 205–216. [CrossRef]

43. Raman, S.; Mohr, A.; Helliwell, R.; Ribeiro, B.; Shortall, O.; Smith, R.; Millar, K. Integrating social and value dimensions into sustainability assessment of lignocellulosic biofuels. *Biomass Bioenergy* **2015**, *82*, 49–62. [CrossRef] [PubMed]
44. Aparcana, S.; Salhofer, S. Application of a methodology for the social life cycle assessment of recycling systems in low income countries: Three Peruvian case studies. *Int. J. Life Cycle Assess.* **2013**, *18*, 1116–1128. [CrossRef]
45. Albrecht, S.; Brandstetter, P.; Beck, T.; Fullana-i-Palmer, P.; Grönman, K.; Baitz, M.; Deimling, S.; Sandilands, J.; Fischer, M. An extended life cycle analysis of packaging systems for fruit and vegetable transport in Europe. *Int. J. Life Cycle Assess.* **2013**, *18*, 1549–1567. [CrossRef]
46. PROSUITE. Handbook on a Novel Methodology for the Sustainability Impact Assessment of New Technologies. 2013. Available online: www.prosuite.org (accessed on 4 September 2017).
47. Valente, C.; Saur Modahl, I.; Askham, C. Method Development for Life Cycle Sustainability Assessment (LCSA) of New Norwegian Biorefinery. Project Title: Nytt Norsk Bioraffineri, Report No.: OR.39.13. 2013, p. 62. Available online: https://www.ostfoldforskning.no/media/1141/3913.pdf (accessed on 8 January 2018).
48. Weidema, B. The integration of economic and social aspects in life cycle impact assessment. *Int. J. Life Cycle Assess.* **2006**, *11*, 89–96. [CrossRef]
49. De Luca, A.I.; Iofrida, N.; Strano, A.; Falcone, G.; Gulisano, G. Social life cycle assessment and participatory approaches: A methodological proposal applied to citrus farming in Southern Italy. *Integr. Environ. Assess. Manag.* **2015**, *11*, 383–396. [CrossRef] [PubMed]
50. Siebert, A.; Bezama, A.; O'Keeffe, S.; Thrän, D. Social life cycle assessment: In pursuit of a framework for assessing wood-based products from bioeconomy regions in Germany. *Int. J. Life Cycle Assess.* **2018**, *23*, 651–662. [CrossRef]
51. Dreyer, L.; Hauschild, M.; Schierbeck, J. A framework for social life cycle impact assessment. *Int. J. Life Cycle Assess.* **2006**, *11*, 88–97. [CrossRef]
52. Mattila, T.J.; Judl, J.; Macombe, C.; Leskinen, P. Evaluating social sustainability of bioeconomy value chains through integrated use of local and global methods. *Biomass Bioenergy* **2018**, *109*, 276–283. [CrossRef]
53. Benoit-Norris, C.; Cavan, D.A.; Norris, G. Identifying social impacts in product supply chains: Overview and application of the social hotspot database. *Sustainability* **2012**, *4*, 1946–1965. [CrossRef]
54. UNEP-SETAC. *Guidelines for Social Life Cycle Assessment of Products*; United Nations Environment Programme: Paris, France, 2009.
55. UNEP SETAC. The Methodological Sheets for Subcategories in Social Life Cycle Assessment (S-LCA) Pre-Publication Version. 2013. Available online: http://www.lifecycleinitiative.org/wp-content/uploads/2013/11/S-LCA_methodological_sheets_11.11.13.pdf (accessed on 3 July 2017).
56. De Haes, H.U.; Finnveden, G.; Goedkoop, M.; Hauschild, M.; Hertwich, E.G.; Hofstetter, P.; Jolliet, O.; Klöpffer, W.; Krewitt, W.; Lindeijer, E.; et al. *Life-Cycle Impact Assessment: Striving Towards Best Practice*; Society of Environmental Toxicology and Chemistry (SETAC): Pensacola, FL, USA, 2002.
57. Schaubroeck, T.; Rugani, B. A revision of what life cycle sustainability assessment should entail: Towards modeling the Net Impact on Human Well-Being. *J. Ind. Ecol.* **2017**, *21*, 1464–1477. [CrossRef]
58. Sherwood, J.; Clark, J.H.; Farmer, T.J.; Herrero-Davila, L.; Moity, L. Recirculation: A New Concept to Drive Innovation in Sustainable Product Design for Bio-Based Products. *Molecules* **2016**, *22*, 48. [CrossRef] [PubMed]
59. Sala, S.; Vasta, A.; Mancini, L.; Dewulf, J.; Rosenbaum, E. *Social Life Cycle Assessment-State of the Art and Challenges for Supporting Product Policies*; Publications Office of the European Union: Luxemburg, 2015.
60. Falcone, P.M.; Lopolito, A.; Sica, E. The networking dynamics of the Italian biofuel industry in time of crisis: Finding an effective instrument mix for fostering a sustainable energy transition. *Energy Policy* **2018**, *112*, 334–348. [CrossRef]
61. Hagemann, N.; Gawel, E.; Purkus, A.; Pannicke, N.; Hauck, J. Possible futures towards a wood-based bioeconomy: A scenario analysis for Germany. *Sustainability* **2016**, *8*, 98. [CrossRef]
62. Sijtsema, S.J.; Onwezen, M.C.; Reinders, M.J.; Dagevos, H.; Partanen, A.; Meeusen, M. Consumer perception of bio-based products—An exploratory study in 5 European countries. *NJAS-Wagening. J. Life Sci.* **2016**, *77*, 61–69. [CrossRef]

63. Vandermeulen, V.; Van der Steen, M.; Stevens, C.V.; Van Huylenbroeck, G. Industry expectations regarding the transition toward a biobased economy. *Biofuels Bioprod. Biorefin.* **2012**, *6*, 453–464. [CrossRef]
64. McCormick, K.; Kautto, N. The bioeconomy in Europe: An overview. *Sustainability* **2013**, *5*, 2589–2608. [CrossRef]
65. Koenig-Lewis, N.; Palmer, A.; Dermody, J.; Urbye, V. Consumers' evaluations of ecological packaging—Rational and emotional approaches *J. Environ. Psychol.* **2014**, *37*, 94–105. [CrossRef]
66. Elghali, L.; Clift, R.; Sinclair, P.; Panoutsou, C.; Bauen, A. Developing a sustainability framework for the assessment of bioenergy systems. *Energy Policy* **2007**, *35*, 6075–6083. [CrossRef]

sustainability

MDPI

Review

Life Cycle Sustainability Evaluations of Bio-based Value Chains: Reviewing the Indicators from a Swedish Perspective

Michael Martin [1,*], Frida Røyne [2], Tomas Ekvall [1,3] and Åsa Moberg [1]

[1] IVL Swedish Environmental Research Institute, P.O. Box 210 60, 100 31 Stockholm, Sweden; tomas.ekvall@ivl.se (T.E.); asa.moberg@ivl.se (A.M.)
[2] RISE-Research Institutes of Sweden, Eklandagatan 86, 412 61 Gothenburg, Sweden; frida.royne@ri.se
[3] Department of Technology Management and Economics, Division of Environmental Systems, Chalmers University of Technology, Vera Sandbergs Allé 8, 412 96 Gothenburg, Sweden
* Correspondence: michael.martin@ivl.se

Received: 7 February 2018; Accepted: 15 February 2018; Published: 20 February 2018

Abstract: Policymakers worldwide are promoting the use of bio-based products as part of sustainable development. Nonetheless, there are concerns that the bio-based economy may undermine the sustainability of the transition, e.g., from the overexploitation of biomass resources and indirect impacts of land use. Adequate assessment methods with a broad systems perspective are thus required in order to ensure a transition to a sustainable, bio-based economy. We review the scientifically published life cycle studies of bio-based products in order to investigate the extent to which they include important sustainability indicators. To define which indicators are important, we refer to established frameworks for sustainability assessment, and include an Open Space workshop with academics and industrial experts. The results suggest that there is a discrepancy between the indicators that we found to be important, and the indicators that are frequently included in the studies. This indicates a need for the development and dissemination of improved methods in order to model several important environmental impacts, such as: water depletion, indirect land use change, and impacts on ecosystem quality and biological diversity. The small number of published social life cycle assessments (SLCAs) and life cycle sustainability assessments (LCSAs) indicate that these are still immature tools; as such, there is a need for improved methods and more case studies.

Keywords: sustainability; life cycle assessment; SLCA; social; economic; LCC; LCSA; bio-based; bioeconomy

1. Introduction

Policymakers in many countries have developed goals and strategies for the development of bio-economies as a means to reach sustainability goals, secure energy supplies, and develop competitive, innovative products [1–4]. Sweden, in particular, with vast resources of biomass, has created increased optimism on the emergence of a bio-based economy [5]. In recent years, the use of renewable energy has dramatically increased compared with other European member states, and the share of renewables in the Swedish transport sector is dominated by biofuels [6]. In addition, the use of bio-based materials in other sectors has also continually increased to meet demands [5].

Nonetheless, while several policy documents have promoted the bio-based economy in Sweden on many positive premises [3], there are also concerns that the expectations created for the bio-based economy may undermine the sustainability of the transition [7]. Examples include the overexploitation of biomass resources, and indirect impacts of land use [8,9].

In order to ensure a sustainable transition to a bio-based economy, it is important that appropriate assessment methods exist and are applied for assessing the advantages and disadvantages of available options from a sustainability perspective. To reduce the risk of sub-optimization and burden-shifting, methods should have a broad systems perspective. Addressing such concerns, life cycle-based tools have been developed, including tools to review environmental impacts through life cycle assessments (LCA), economic indicators through life cycle costing (LCC), and social indicators throughout the life cycle using social life cycle assessments (SLCA).

The past decade saw the development of different frameworks for life cycle sustainability assessments (LCSA) in order to combine the environmental, economic, and social perspectives; e.g., Klöpffer [10] and Guinée et al. [11]. While the importance of using the life cycle perspective is not contested, the way in which such an assessment can and should be conducted is still debated, as LCSA struggles with applicability challenges [12]. Examples of challenges include identifying the scope of the economic, social and environmental impacts that need to be assessed, understanding and managing the complex relationship between these impacts, selecting indicators for the impacts, and finding input data [13,14]. Therefore, it is important to understand how LCSA is being used, as well as its development and potential for improvement. In this study, we focus on the choice of indicators in life cycle studies relevant to a bio-based Swedish economy.

There is a limited number of reviews specific to the indicators, their selection, etc. in the literature for LCSA. However, Kühnen and Hahn [15] recently presented a systematic review of indicators in the global scientific SLCA literature across all of the sectors, finding that the social aspects most commonly accounted for relate to workers' health and safety. A review of environmental life cycle assessments on biofuels in Sweden found that these studies typically include a small number of environmental indicators; see Lazarevic and Martin [16]. Several previous reviews with a broad scope have criticized such limitations, and recommend including a wider range of indicators [17–19] as the different environmental impacts do not necessarily correlate and trade-offs can occur [20–22]. This argument grows even stronger when the scope of the study is expanded to also include social and economic impacts.

2. Aim and Scope

This study expands on the review by Lazarevic and Martin [16] by expanding from biofuels to bio-based products in general, and from environmental impacts to sustainability indicators. Our aim, then, is to investigate to what extent important sustainability indicators are already used in the life cycle studies of bio-based products. We also discuss how LCSA can be expanded or improved in order to better contribute to the transition to a sustainable bio-based economy.

The scope of our investigation is limited to scientific publications between 2010 and 2015. We focus on assessments that are relevant to decisions made in Sweden, i.e., assessments of products that are produced or used in Sweden. The discussion on how LCSA can be improved is relevant also beyond the Swedish decision-making context.

The target audience of the article is the scientific community, practitioners who seek tools for evaluating the value chains of bio-based products, and decision-makers in industry and governmental institutions with a drive to understand what sustainability aspects are most important to consider in the development, production, and promotion of bio-based products.

3. Methodology

3.1. Identifying Important Sustainability Indicators

To investigate to what extent relevant sustainability indicators are used in life cycle studies, we first need to establish which indicators are important. There is no objective truth regarding what impacts should be included in a sustainability assessment. Instead, the perception regarding the importance of impacts and indicators is subjective. In the context of this paper, we combine stakeholder processes in order to identify prominent indicators. We use the following literature to establish what environmental and social indicators are important in sustainability assessments in general:

- impacts addressed by the planetary boundaries (PB) framework [23],
- the default list of impact categories in the European guide for Product Environmental Footprints [24], and
- impacts covered by the United Nations Environmental Programme and Society of Environmental Toxicology and Chemistry UNEP-SETAC Life Cycle Initiative guide for social life cycle assessment [25].

The planetary boundaries are chosen because of the widespread impact of this framework. We chose to use the guides for Product Environmental Footprint (PEF) and UNEP-SETAC SLCA because each of them is a product of a process involving a broad range of researchers involved in life cycle studies, and in the case of PEF, many stakeholders are involved as well.

We also carried out an 'Open Space' workshop (OSW) with experts in Sweden to identify and validate important indicators in the sustainability assessments of bio-based products. This serves the purpose of capturing aspects that are specific to bio-based products and/or to the Swedish context. Information on how the OSW was carried out in will be provided later in this paper.

Life cycle sustainability assessments include the social, environmental, and economic pillars of sustainability. However, the aforementioned approaches and frameworks do not cover the economic dimension. Thus, in addition to the indicators from these sources, we claim the life cycle cost to be an important sustainability indicator. This is the only economic indicator in the LCSA framework presented by Klöpffer [10]. It can be regarded as relevant because environmentally preferable products often have a higher acquiring cost, but can be cheaper over the span of a life cycle due to, for example, lower energy demand; see, e.g., Klöpffer and Ciroth [26]. Another reason is that the competitiveness of sustainable products on the market depends on their production costs [27]. Bio-based products often compete with cheaper alternatives, and an assessment of the production cost may be important in order to estimate what market penetration is possible. However, it has been argued that the life cycle cost indicator might not be sufficient to assess the economic dimension of sustainability [13], as it does not reflect the extent to which products affect the economic capital available for future generations [28]. Ekvall et al. [29] also demonstrated that economic sustainability can include aspects such as business opportunities and business risks.

3.2. Open Space Workshop

Open Space is a self-organizing technique that aims to generate creativity and informal discussion on a common theme [30]. Open Space workshops begin without a fixed agenda beyond this overall theme; specifying the agenda is instead one of the tasks assigned to the workshop participants. Ekvall et al. [29] previously used this method to identify important indicators and research questions in a LCSA of a 50-km pipeline for residual heat.

Invitations to our Open Space workshop were distributed mainly to researchers and industry practitioners in Sweden. The 19 participants who attended the workshop were primarily researchers in academia, institutes, and industry, with a background in environmental life cycle assessment (LCA) or energy systems analysis.

At the beginning of the workshop, the participants generated ideas for important sustainability aspects and indicators. After an initial allotted time for individual brainstorming, the participants formed five small groups, each of which selected three to five sustainability aspects or indicators that they considered important for assessments of bio-based products. The ideas were presented for the rest of the workshop participants and posted on a wall, and the overlaps were eliminated. From these aspects and indicators, the participants selected eight for in-depth group discussions with an aim to agree on why the indicator is important, and on what aspects or indicators should be considered and accounted for in a sustainability assessment of bio-based products.

At the end of the workshop, each participant was given six *yes*-votes and three *no*-votes to freely distribute among all of the ideas for sustainability indicators, in addition to all of the aspects identified in the previous group discussions. We interpreted the result of the voting as an indication of what the workshop participants considered important to account for in a sustainability assessment of bio-based products.

3.3. Inventory of Existing Life Cycle Studies

We performed a systematic literature review (cf. [16,31]) to identify what sustainability indicators are included in scientifically published life cycle studies. This included LCSAs, but also LCAs, SLCAs, and LCCs.

We used a Boolean string to find peer-reviewed publications in the Scopus database (www.scopus.com). This string included the following terms (see Supplementary material for the exact Boolean search strings):

- "bio", "biomass", and "bio-based", because they are common words in discussions of bio-based products,
- "forest" and "wood" because forestry is an important industry in Sweden,
- "bioenergy", "biofuel", "biogas", "biodiesel", "ethanol", Hydrogenated vegetable oil "HVO" and Fatty Acid Methyl Esters "FAME", because they denote the most common biofuels in Sweden [6], and
- "district heating", because this is an important sector for the use of solid biofuel.

The abstracts found were reviewed to find papers that applied at least a cradle-to-gate perspective. These included more than 900 LCAs. In order to facilitate the analysis of the papers, we excluded LCA papers that did not have any co-author from a Swedish research institution. We gave priority to research at Swedish institutions because it is often funded by the Swedish government or industry, which are likely to have an interest in the feedstock, processes and products that are relevant for the Swedish market, which limited the number of studies to 63.

The number of LCSAs, SLCAs, and LCCs was much smaller, and we included not only papers with contributions from Swedish institutions, but also from countries that are important for imports from and/or exports to Sweden of bio-based products, including: Norway, the United Kingdom (UK), Germany, Finland, Denmark, the Netherlands, Russia, the United States of America (USA), Poland, Portugal, Latvia, Estonia, Italy, China, Brazil, Australia, Indonesia, Ukraine, and Lithuania; see information on trade statistics; see e.g., Swedish Energy Agency [6] and SLU [32]. See more information in Tables A1–A4 Appendix A.

The selected papers were reviewed, and key information about each paper was compiled, including the goal of the study, indicators, methodological considerations, system boundaries, stakeholders, etc. This information was collected in a separate matrix for LCAs, SLCAs, LCCs, and LCSAs (see Supplementary materials for more details). We used it to determine the extent to which important sustainability indicators are considered in the studies. Furthermore, we merged some indicators. For example, we included assessments of changes in soil organic carbon in the broader category "direct land use change". All of the indicators related to toxicity (e.g., human and freshwater toxicity) were grouped as "toxicity potential".

4. Results

4.1. Important Indicators

For the context of this paper, we identified the following sustainability indicators and aspects to be important, besides the life cycle cost:

- from the PB framework [23]: climate change, stratospheric ozone depletion, chemical pollution, atmospheric aerosol concentration, nitrogen and phosphorus emissions, acidification of oceans, freshwater consumption, land-system change, and biodiversity loss;
- from the PEF guide [24]: climate change, ozone depletion, freshwater eco-toxicity, human toxicity (cancer and non-cancer impacts), emissions of particulate matter, human health impacts of radiation, photochemical ozone formation, acidification, eutrophication (terrestrial and aquatic), water depletion, depletion of fossil and mineral resources, and land transformation;
- from the UNEP-SETAC guide for SLCA [25]:
  - impacts on workers: freedom of association, child labor, fair salary, working hours, forced labor, discrimination, health and safety, and social benefits and security;
  - impacts on local community: access to material and immaterial resources, delocalization and migration, cultural heritage, safe and healthy living conditions, indigenous rights, community engagement, local employment, and secure living conditions;
  - impacts on consumers: health and safety, feedback mechanisms, consumer privacy, transparency, and end-of-life responsibility;
  - impacts on other value-chain actors: fair competition, social responsibility, supplier relationships, and intellectual property rights; and
  - impacts on society overall: public commitments to sustainability issues, contribution to economic development, the prevention and mitigation of armed conflicts, technology development, and corruption.
- from the Open Space workshop [33] (see Table 1): climate change, biodiversity, working conditions, water use, ecosystem functions, and the use of various resources.

There is a large overlap between the PB framework and the impact categories in the PEF guide, because both focus on environmental sustainability and its impacts. The overlap is clear in the impact categories of climate change and ozone depletion. Ocean acidification is closely linked to climate change, since carbon dioxide emissions drive both. Chemical pollution and aerosols in the PB framework can affect the climate, but also dominate the toxicity impacts in the PEF guide. Eutrophication is dominated by emissions of nitrogen and phosphorus.

The PB framework and the PEF guide also differ on some points. The PB indicator for water use relates to the global consumptive use of freshwater, while the PEF indicator takes regional water scarcity into account. The PB indicator for particles includes solid and liquid particles that reside in the atmosphere, while the PEF indicator focuses on solid particles that are emitted to the atmosphere. The PEF indicator for land transformation focuses on changes in organic soil content, while the corresponding indicator in the PB framework has a broader scope related to the function, quality, and spatial distribution of the land cover. This has implications for biodiversity, which also is an explicit indicator in the PB framework, but is not included in the PEF impact categories. The PEF guide, on the other hand, includes more detailed indicators that are related to toxicity impacts and also several impacts that are not in the PB framework, such as: ionizing radiation, photochemical ozone creation, and the depletion of resources other than water.

The results from the Open Space workshop, with its focus on impacts related to bio-based products and the Swedish context, also strongly overlaps the PB framework and PEF guide, because the indicators identified as important are dominated by environmental aspects. However, note that the workshop participants identified biodiversity as one of the top environmental indicators for sustainability assessments of bio-based products, indicating that the impact categories listed in the PEF guide might not be enough for this purpose.

The indicator "working conditions" was the non-environmental aspect that was given top priority in the workshop. This indicator really includes a broad range of aspects and impacts. The UNEP-SETAC guide for SLCA, for example, divides work-related social aspects into the sub-categories Freedom of Association and Collective Bargaining, Child Labor, Fair Salary,

Working Hours, Forced Labor, Equal opportunities/Discrimination, Health and Safety, and Social Benefits/Social Security.

Notably, the only clearly economic indicator that was identified as even potentially important is not life cycle cost, but regional value creation. This indicator also received a few votes in the end.

Table 1. Sustainability aspects identified by the participants in the Open Space workshop as potentially important to include in a sustainability assessment of bio-based products, and the result of the voting indicating which indicators were considered important [33].

Indicator	Selected for Group Discussion	Yes-Votes	No-Votes
Climate impact	yes	12	0
Biodiversity	yes	10	0
Working conditions	yes	10	0
Water use	yes	9	0
Ecosystem functions	no	9	1
Resource use	yes	7	0
Emissions of particulates	yes	2	0
Odor	no	4	1
Human health	no	3	0
Corruption/Human rights	yes, as a joint topic	3	0
Regional value creation	no	2	0
Resource availability	no	0	0
Eutrophication	no	0	0
Intragenerational and intergenerational human well-being	yes	2	3

4.2. Life Cycle Studies Identified

Our literature search resulted in more than 100 scientifically published life cycle studies of bio-based products: 63 LCAs, 30 LCCs, seven SLCAs, and six LCSAs. Most of these were assessments of products produced from wood or food crops (such as cereals and vegetable oils), but the literature also covered the assessment of products made from several other biological feedstock materials (Table 2). In many cases, the biological material was part of an assessment that also included other materials. In some cases, the exact feedstock material was not clear in the published paper. In several cases, the articles identified in the literature review applied more than one of the life cycle methods, i.e., two or three of the LCA, LCC, and SLCA methods (see e.g., [34–36]). In this case, the feedstock type and product type in Tables 2 and 3 were applied to only one column. For those reviewing all three tools, these were applied in the LCSA column, and for those reviewing environmental and economic tools, these were applied in the LCA column.

Table 2. Type of feedstocks in the life cycle studies found in the literature review. Figures shown in parenthesis, e.g., (+10), refer to studies that include the biological feedstock as one of two or more feedstock types. LCA: life cycle assessment; LCC: life cycle costing; LCSA: life cycle sustainability assessments; SLCA: social life cycle assessments.

Feedstock Types	LCA	LCC	SLCA	LCSA	Total
Wood	21 (+ 10)	7 (+ 3)	2	2	32 (+ 13)
Food crop	15 (+ 9)	6 (+ 2)	4	1	26 (+ 11)
Non-food crop	5 (+ 8)	3 (+ 2)	0	1	9 (+ 10)
Algae	1	1	0	0	2
Animal-based	1	0	0	0	1
Waste	3 (+ 4)	2 (+ 2)	0	1	6 (+ 6)
Manure	0 (+ 3)	1 (+ 2)	0	0	1 (+ 5)
Wool	0	1	0	0	1
Not specified	1	5	0	1	7

A majority of the studies were LCAs or LCCs of energy products, but a broad range of other products were also assessed (Table 3). Energy products included various biofuels (e.g., ethanol, biogas, and Fischer–Tropsch diesel) and heating systems (e.g., district heating, pellets, and briquettes). Construction products included individual products, such as coatings, as well as whole building systems. Commodities included products such as bioplastics, cups, and fertilizers.

Table 3. Type of products assessed in the different methods from the literature review.

Product Type	LCA	LCC[1]	SLCA	LCSA	Total
Energy	37	21	7	2	67
Construction	5	3	0	0	8
Commodity	19	5	0	3	27
Mixed	2	2	0	1	5

The total number of products for the LCC studies included 31, while only 30 studies were found. This is because one study (Zhang et al., 2013) assessed several products.

The products assessed are produced in various regions of the world. Approximately a quarter of the LCA studies and the majority of the LCC studies were assessments of products produced outside Europe. The SLCAs, for example, include products produced in Brazil [37,38], Indonesia [39], Australia [40,41], China [34,41], and the UK [42], as well as France, the USA, and Lithuania [37]. The LCSAs were primarily of European origin, but examples from China [34] and Mexico [35] were also present. More details about the distribution of the literature can be found in the Supplementary Materials.

4.3. Indicators Present in Life Cycle Assessments

Climate change is accounted for in all 63 LCA papers in our review (see Table 4). Energy use is also included in most LCAs. Acidification and eutrophication are both included in about half of the studies. Many studies also include direct land use and/or land use change. The former is primarily calculated as land occupation in square meters, while the latter usually focus on changes in the organic carbon content of the soil, as proposed in the PEF guide (and ultimately related to greenhouse gas emissions). Overall, though, the LCAs of bio-based products typically include few impact categories.

Compared to the environmental impacts and indicators that we identified as important, nearly all of the important indicators are covered in at least one case study. However, there is a divergence between what impacts are important, and how often they are used. Climate change and water use or depletion was explicitly listed as a significant impact in all three sources: the PB framework, the PEF guide, and the Open Space workshop. Climate change was included in all of the LCAs, but only two case studies accounted for water depletion. Acidification, which neither the PB framework or the Open Space workshop listed as very important, was included in more than half of the LCAs. Ecosystem quality, resource use, and indirect land use change are important according to two or all three sources, but were only included in very few scientifically published LCAs. Impacts on biodiversity were explicitly mentioned as important in the PB framework and were one of the top-priority indicators according to the Open Space workshop; still, it was not accounted for in any of the LCAs in our literature review.

Only eight of the 63 LCA articles included in our literature review provide reasons for the choice of impact categories. Five studies specified the product type (e.g., biofuel or fertilizer) as the cause of the selection, while one study also specified the biomass type (i.e., Jatropha) as the cause of selection.

None of the studies justified their choices through referring to the geographical context of the assessment. Which indicators are important can otherwise depend on where in the world the biomass is extracted and converted into products. For example, Lazarevic and Martin [16] indicated that environmental challenges in regions outside of Sweden can be different to those in regions that are more commonly assessed with LCA (Europe and USA), and that the choice of impact categories normally coincides with European environmental problems. The geographical context can significantly affect the technique and technology used, and how the bio-based product is treated at the end-of-life. It can also affect the impacts of a specific environmental intervention (emission or resource extraction).

Table 4. Environmental impact categories included in the 63 LCA publications and their explicit (exp.) or implicit (imp.) importance according to the planetary boundaries (PB) framework [23], the European Union (EU) guide to Product Environmental Footprints (PEF) [24], and our Open Space workshop (Table 1).

Impact Category	No. of Studies	% of Studies	PB Framework	PEF Guide	Open Space
Climate change	63	100%	exp.	exp.	exp.
Energy use	44	70%		imp.	imp.
Acidification	35	56%	imp.	exp.	
Eutrophication	34	54%	imp.	exp.	
Photochemical oxidant formation	21	33%		exp.	
Direct land use change	18	29%	exp.	exp.	imp.
Direct land use	13	21%	imp.		imp.
Toxicity impacts	8	13%	imp.	exp.	
Abiotic depletion	7	11%		exp.	imp.
Ozone depletion	6	10%		exp.	
Particles	4	6%	imp.	exp.	
Human health	3	5%	imp.	imp.	
Ecosystem quality	3	5%	imp.	imp.	exp.
Resources	3	5%		exp.	exp.
Indirect land use change	2	3%	exp.	exp.	
Water depletion	2	3%	exp.	exp.	exp.
Resource use	1	2%		exp.	exp.
Biodegradability	1	2%			
Gross calorific values	1	2%			
Human health damage by particles and ozone	1	2%	imp.	imp.	
Biodiversity	0	0%	exp.		exp.

4.4. Indicators Present in LCCs

It is clear from the number of LCCs found that life cycle costs are included in many LCA studies. Of the 30 identified LCCs, 17 were included in studies that combined it with LCA; only 13 solely reviewed the LCC. Economic indicators included the life cycle cost, which were shown as different monetary values per functional unit, and other economic indicators to show the economic viability and sustainability of the different systems. While most studies reviewed the life cycle cost from a producer's (company) perspective, several studies also outlined the societal costs and benefits when comparing different technologies. Similar to the indicator "regional value creation" identified in the OSW, these studies took a larger perspective in order to understand the implications of regional production and consumption.

4.5. Indicators Present in Social LCAs

Several of the seven identified SLCAs (c.f. [34,37–42]) used the UNEP-SETAC guide [25] as the basis for their choice of impact categories. Furthermore, Ekener-Petersen et al. [37] used the Social Hotspots Database, where the categories are also based on the UNEP-SETAC guidelines, although they were somewhat adjusted. The other studies defined their own impacts categories. While the number of aspects or impact categories accounted for differences between the studies, working conditions and socio-economic repercussions, such as local employment, food security, and energy security, are reoccurring. Other aspects such as human rights (e.g., indigenous rights, child labor, etc.), governance (e.g., corruption, public commitments to sustainability, etc.) and cultural heritage (e.g., land acquisition, community engagement, etc.) are also addressed in several studies. The social aspects assessed in the articles reviewed cover both positive and negative impacts. Positive impacts relate, for example, to the increase in numbers of jobs (e.g., [38,42]) or public commitment to the sustainability of businesses [39].

All of the articles covered impacts on the stakeholder groups, or "workers", which was indicated as important both in the UNEP-SETAC guidelines and in the results from our Open Space workshop (Table 1). All of the papers also included impacts on the "local community". Impacts on the society at large and value chain actors are covered in only a selected few of the studies; c.f. [34,37,39]. The stakeholder categories were not explicitly selected and justified in most of the case studies reviewed. It is also important to note that the reviewed case studies primarily covered stakeholders who were involved in the production phase of the life cycle.

4.6. Indicators Present in Life Cycle Sustainability Assessments

The six LCSA articles found in the literature review [34,35,43–46] each applied in total between eight and 29 indicators in the assessment of environmental, economic, and social sustainability. A total of eight environmental impact categories were included in the studies reviewed: climate change, energy use, acidification, eutrophication, abiotic depletion, photochemical oxidant creation, toxicity, and particulate emissions. All of these were also included in multiple LCA studies (see Table 4). All of the articles covered climate change. Other commonly used environmental impact categories included acidification and energy use. Eutrophication, abiotic depletion, photochemical oxidant creation, toxicity, and particle emissions were less frequently used. All of the environmental impact categories found in the LCSA studies were recommended in the PEF guide. Most of them are also part of the PB framework, although the latter does not address abiotic depletion. Similar to the results from the review of LCA studies, none of the LCSA studies accounted for biodiversity, although both the OSW and PB frameworks identified these as important.

The LCSAs included, in total, four economic indicators. Life cycle cost was the most prevalent of these. Other economic indicators included investment cost and net present value. The social sustainability was assessed using a range of quantitative and qualitative indicators. The most common indicators in the LCSA studies reflected different aspects of accidents/safety risks, economic development, and education. Keller et al. [44] also accounted for impacts on a higher systems level through the indicator "socio-economic repercussions". Furthermore, Santoyo-Castelazo and Azapagic [35] used qualitative indicators such as public acceptability and diversity of supply. They modeled health impacts through the use of the indicator "human health potential", and suggested that global warming potential and abiotic resource depletion be used to account for social impacts on future generations. This creates a link between LCA and SLCA in the LCSA.

Many of the studies justified their choice of impact categories and indicators. The majority of the studies used stakeholder input in order to identify relevant indicators and aspects for the LCSA. Several of the studies discussed the challenges of combining all three pillars. These challenges were primarily related to developing a methodology for several of the impacts and indicators, combining quantitative and qualitative indicators, and adapting the assessments to the objects under review.

5. Discussion

5.1. Why Important Indicators are Missing

Our literature review suggests that the current practices of impact category selection for LCAs of bio-based products, and for the environmental dimension of LCSAs, focus on a limited, reoccurring set of indicators with limited justifications provided. These results are coherent with the recently published review of LCAs on biofuels in Sweden [16].

There is a discrepancy between how often an impact category is used in the LCAs, and how important it is according to the sources we have used. In particular, water depletion, ecosystem quality, indirect land use change, and biodiversity are important according to two or all three of our sources, and also implied by, for example, Lewandowski [8] and O'Brien et al. [9]. Still, they are excluded from most or all of the published LCAs. Impact categories that are less clearly important for bio-based products, e.g., acidification impacts, are included in a much larger number of studies. This suggests that the choice of impact categories in LCAs of bio-based products is not primarily decided by the importance of the environmental impact.

A simple explanation for the discrepancy between the importance and frequency of the indicators would be that the indicators we identified as important in our project are, in fact, not really that important. This explanation is contradicted by, for example, the United Nations (UN) Sustainable Development Goals, which state that water-use efficiency should substantially increase, and water scarcity should be addressed in order to substantially reduce the number of people suffering from water scarcity [47]. They also call for urgent and significant actions in order to reduce the degradation of natural habitats and halt the loss of biodiversity, including the integration of ecosystem and biodiversity values into national and local planning [48]. From the Swedish perspective, water scarcity is not yet an urgent matter in most parts of the country; however, the national environmental objective related to groundwater mentions the increased demand for, and hence pressure, on groundwater reservoirs [49]. One of the 16 national environmental objectives focuses entirely on biodiversity [50], and another four focus on land use and its impacts on ecosystems [51–54].

Another explanation for why important impact categories and indicators were excluded would be that no methods (or data) exist that allow for including all of them in an LCA. This is contradicted by all of the important indicators, except biodiversity, being included in at least one study. Methods also exist for taking impacts on biodiversity into account [55,56]. Hence, rudimentary methods, at least, exist to account for all of the important environmental impacts in LCAs. However, data availability in order to allow for the review of certain systems will need to be improved; see e.g., discussions by Martin and Brandão [57].

A more plausible explanation for why several important environmental indicators are often missing in the LCAs is that the methods that exist generate results that are regarded as poor indications of the actual impacts. For example, several of the studies recognized the importance of biodiversity, but referred to methodological immaturity for the exclusion of the indicator; see also [55,56].

The existing methods can also be difficult to apply. Indirect land use change can, for example be quantified through the use of the general equilibrium model from the Global Trade Analysis Project (GTAP); see Kløverpris et al. [58]. However, learning to run such a model might require more time and resources than are available for a specific LCA.

All of the SLCAs include impacts on workers, which is the stakeholder group that was given the highest priority in the Open Space workshop. This is consistent with the finding of Kühnen and Hahn [15] that indicators related to workers, in particular their health and safety, are the most common type of indicators in the global SLCA literature. The small total number of published SLCAs and LCSAs means that no social indicator is included in a large number of life cycle studies of bio-based products. This might be because the expertise needed to carry through SLCAs is still scarce, and/or because of a lack of mature methods. The latter reason is supported by three of the seven SLCAs having the explicit aim of developing the SLCA methodology.

5.2. The Need for Improved Indicator Modeling Methods

The discussion above indicates a need for the continued development and dissemination of operational methods in order to model specific, important impacts. This need concerns several environmental impacts and indicators, in particular water depletion, ecosystem quality, indirect land use change, and biodiversity.

As stated in Section 4.3, the impacts of a specific emission or resource extraction can significantly vary, depending on where it occurs. This may be accounted for through spatially explicit impact assessment methods. Lazarevic and Martin recommended taking regional differences into account when reviewing impacts in a Swedish context [16] through using, e.g., the Swedish Environmental Objectives [51]; however, the use of such methods is still not common practice in LCA databases [16,22].

The need for improved life cycle impact assessment methods includes social impacts in general, and impacts on workers in particular. Social impacts also depend on the geographical context. Our study takes a Swedish perspective, which includes a well-developed social welfare system; as a result, negative social impacts might not be as important to address as in other regions. However, the biomass that is employed to produce products consumed in Sweden is often sourced from other parts of the world, thus, it has implications that extend outside of Sweden [59,60]. An accurate assessment of the social sustainability needs to take this into account.

There is also a need for development of methods in order to assess the economic impacts, and associated indicators. The relevance of the life cycle cost as an indicator was that the competitiveness of a product depends on its production costs [27], and that environmentally preferable products often have a higher acquiring cost, but can be cheaper to use because of lower energy demand [26]. This indicates the need for two different economic cost indicators: a cradle-to-consumer calculation of the costs of production and distribution, and a life cycle calculation of the costs for the consumer. In addition, as indicated in Section 3.1, operational methods should perhaps be developed and disseminated for estimating the business risks and opportunities [29], and the impact of the product on the economic capital of future generations [28].

Supplementary Materials: The following are available online at www.mdpi.com/2071-1050/10/2/547/s1. Excel File: The excel file includes details from the matrices produced to document the articles reviewed in the literature review for the LCA, LCC, SLCA and LCSA articles.

Acknowledgments: The research in this project has been funded through the Swedish Research Council for Sustainable Development (FORMAS Grant Number 2015-14057). The final analysis and writing of the paper was also co-funded through the EU ERA-Net Sumforest project BenchValue (Formas Grant Number 2016-02113). The authors would also like to thank participants in the open space workshop for their participation, and the valuable input provided by Miguel Brandão to further develop the article. Finally, we would like to acknowledge the support provided by Anja Karlsson and Albin Pettersson of IVL in the literature review conducted for this research project.

Author Contributions: Michael Martin, Frida Røyne and Tomas Ekvall conceived and designed the study methodology and contributed to the primary share of the article writing. Åsa Moberg contributed in the SLCA sections, revisions and discussions. All authors were involved in the data collection, revisions and study design.

Conflicts of Interest: The authors declare no conflict of interest. The funding sponsors had no role in the design of the study; in the collection, analyses, or interpretation of data; in the writing of the manuscript, nor in the decision to publish the results.

Appendix A. Literature Search Details

The following subsections provide details on the literature search, terms used for the searches used to identify the articles in Scopus and details about input into the matrices.

Appendix A.1. Life Cycle Assessment (LCA)

Table A1. LCA literature search specifications.

Document Search Settings	Specification
Boolean string (article title, abstract, keywords)	*"LCA" OR "life cycle assessment" OR "life cycle analysis"* AND *"bio" OR "biomass" OR "biobased" OR "forest" OR "wood" OR "biofuel" OR "biodiesel" OR "biogas" OR "ethanol" OR "HVO" OR "FAME" OR "bioenergy" OR "district heating"*
Date range (inclusive) published	2000 to 2015
Country/territory limited to	Sweden

Appendix A.2. Life Cycle Sustainability Assessment (LCSA)

Table A2. LCSA literature search specifications.

Document Search Settings	Specification
Boolean string (article title, abstract, keywords)	*"Life Cycle Sustainability Assessment"* OR *"life cycle sustainability analysis"* OR *"life cycle"* OR *"sustainability assessment"* OR LCSA OR *"sustainability analysis"* OR *"Economic"* OR *"social"* OR *"environmental"* AND *"bio"* OR *"biomass"* OR *"biobased"* OR *"forest"* OR *"wood"* OR *"biofuel"* OR *"biodiesel"* OR *"biogas"* OR *"ethanol"* OR *"HVO"* OR *"FAME"* OR *"bioenergy"* OR *"district heating"*
Date range (inclusive) published	2000 to 2015
Country/territory limited to	Sweden, Norway, UK, Germany, Finland, Denmark, The Netherlands, Russia, USA, Poland, Portugal, Latvia, Estonia, Italy, China, Brazil, Australia, Indonesia, Ukraine, Lithuania.

Appendix A.3. Life Cycle Costing (LCC)

Table A3. LCC literature search specifications.

Document Search Settings	Specification
Boolean string (article title, abstract, keywords)	*"Life Cycle"* OR *"LCC"* OR *"life cycle cost"* OR *"life cycle costing"* OR *"Life Cycle Cost Analysis"* OR LCCA OR *"Life cycle cost assessment"* OR *"Life Cycle Economic Analysis"* AND *"bio"* OR *"biomass"* OR *"biobased"* OR *"forest"* OR *"wood"* OR *"biofuel"* OR *"biodiesel"* OR *"biogas"* OR *"ethanol"* OR *"HVO"* OR *"FAME"* OR *"bioenergy"* OR *"district heating"*
Date range (inclusive) published	2000 to 2015
Country/territory limited to	Sweden, Norway, UK, Germany, Finland, Denmark, The Netherlands, Russia, USA, Poland, Portugal, Latvia, Estonia, Italy, China, Brazil, Australia, Indonesia, Ukraine, Lithuania.

Appendix A.4. Social Life Cycle Assessment (SLCA)

Table A4. SLCA literature search specifications.

Document Search Settings	Specification
Boolean string (article title, abstract, keywords)	*"Life Cycle"* OR *"SLCA"* OR *"social life cycle"* OR *"life cycle assessment"* OR *"Socio-economic"* OR S-LCA OR *"life cycle analysis"* OR *"LCA"* OR *"social impacts"* OR *"social sustainability"* AND *"bio"* OR *"biomass"* OR *"biobased"* OR *"forest"* OR *"wood"* OR *"biofuel"* OR *"biodiesel"* OR *"biogas"* OR *"ethanol"* OR *"HVO"* OR *"FAME"* OR *"bioenergy"* OR *"district heating"*
Date range (inclusive) published	2000 to 2015
Country/territory limited to	Sweden, Norway, UK, Germany, Finland, Denmark, The Netherlands, Russia, USA, Poland, Portugal, Latvia, Estonia, Italy, China, Brazil, Australia, Indonesia, Ukraine, Lithuania.

Appendix A.5. Limitations for the Literature Search

The search was also limited to a set of specified feedstock materials and biomass value chains. In addition to the general keywords "bio", "biomass" and "biobased", the specification is based on three Swedish bio-based markets:

Appendix A.5.1. The Forest Product Market (Forest Products are Here Defined as Products Derived from Forest Biomass)

Sweden is covered by almost 70% forest [61], and forestry is an important industry in Sweden. Sweden has both export and import of forest products (round wood, chips, pellets, wooden products (e.g., particle boards, paper, and pulp)) with several countries (more than 5000 M SEK per year: Norway, Germany, Finland, Poland, Denmark, Latvia, Estonia, The Netherlands, China, Russia, UK, Portugal, USA, and Italy) [62]. The keywords "forest" and "wood" were therefore added to the Boolean string.

Appendix A.5.2. The Biofuel Market

The biofuel market in Sweden mainly consists of three markets: a local (Swedish); biogas, a regional (European); biodiesel, and a global; ethanol [63]. The keywords "biofuel*", "biogas", "biodiesel" and "ethanol" were therefore added to the Boolean string. Figure A1 shows the biofuels with the highest consumption volumes in Sweden. Based on the figure, the keywords "HVO" and "FAME" were also added to the Boolean string.

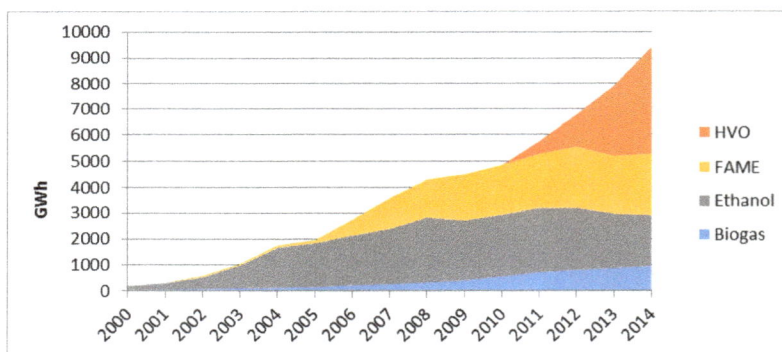

Figure A1. Biofuel consumption in Sweden from 2000 to 2014 (shown in GWh annually). Figure from [60].

Appendix A.5.3. The Bioenergy Market

Main energy sources for district heating are residues from the forest industry, forest residues, recovered waste wood, refined wood fuels, municipal and industrial biogenic waste, bio-oils, and peat. Bioenergy also plays an important role in industry and electricity production [64]. Sweden also imports waste, mainly from Norway and the UK, of which 85% goes to energy recovery [65]. The keywords "bioenergy" and "district heating" were therefore added to the Boolean string.

Appendix A.5.4. Date Range, Document Type, and Subject Areas

The project focuses on documents published 2000–2015.

Appendix A.5.5. The Swedish Context

The project focuses on research which is relevant for Swedish conditions. We therefore limited the literature review to articles written by one or more authors from a Swedish institution. Our rationale is that Sweden based authors (often) are financed by the Swedish state or industry. These have an interest in feedstock, processes and products relevant for the Swedish market. For LCC and SLCA, a limitation to Sweden resulted in too few studies. If this is the case, the scope can be broadened to include the countries that are most important in a Swedish import/export context of forestry products [62], biofuels [60,66,67] and waste [65]. These included Norway, UK, Germany, Finland, Denmark, The Netherlands, Russia, USA, Poland, Portugal, Latvia, Estonia, Italy, China, Brazil, Australia, Indonesia, Ukraine and Lithuania.

Appendix A.5.6. Criteria for Selection

From the studies identified from the literature searches above, only those studies meeting the following guidelines were chosen for the review. These included:

(1) Following the LCA methodology (excluding LCI studies),
(2) focusing on biomass value chains (for example, the waste for district heating should be (mainly) bio-derived),
(3) involving case studies (e.g., no review studies, discussions) and
(4) following a peer-review process.

Thereafter, each article abstract was reviewed to ensure that it was relevant for the study. As identified in the text, the articles were then reviewed and relevant information and details on the articles, and key information such as the goal of the study, methodological considerations, system boundaries, stakeholders, impact categories and specific aspects covered were compiled. The information was used to determine the extent and details of important sustainability aspects considered in LCA, LCC, SLCA and LCSAs applied to biomass value chains of relevance for Swedish conditions, see the Supplementary materials for a copy of the matrices for the respective life cycle based methods.

References

1. EuropaBio. *Building A Bio-Based Economy for Europe in 2020, EuropaBio Policy Guide*; European Association for Bioindustries (EuropaBio): Brussels, Belgium, 2011.
2. European Commission. *Innovating for Sustainable Growth: A Bioeconomy for Europe*; European Association for Bioindustries (EuropaBio): Brussels, Belgium, 2012.
3. McCormick, K.; Kautto, N. The Bioeconomy in Europe: An Overview. *Sustainability* **2013**, *5*, 2589–2608. [CrossRef]
4. Staffas, L.; Gustavsson, M.; McCormick, K. Strategies and Policies for the Bioeconomy and Bio-Based Economy: An Analysis of Official National Approaches. *Sustainability* **2013**, *5*, 2751–2769. [CrossRef]
5. Formas. Swedish Research and Innovation Strategy for a Bio-based Economy Report: R3:2012. Available online: http://www.formas.se/PageFiles/5074/Strategy_Biobased_Ekonomy_hela.pdf (accessed on 20 February 2018).
6. Swedish Energy Agency. *Drivmedel Och BiobräNslen 2015—MäNgder, Komponenter Och Ursprung Rapporterade I Enlighet Med Drivmedelslagen Och HåLlbarhetslagen*; ER 2016:12; Swedish Energy Agency: Stockholm, Sweden, 2016.
7. Hedlund-de wit, A. An integral perspective on the (un)sustainability of the emerging bio-economy: Using the integrative worldview framework for illuminating a polarized societal debate. In Proceedings of the Integral Theory Conference, San Francisco, CA, USA, 18–21 July 2013.
8. Lewandowski, I. Securing a sustainable biomass supply in a growing bioeconomy. *Glob. Food Secur.* **2015**, *6*, 34–42. [CrossRef]
9. O'Brien, M.; Schütz, H.; Bringezu, S. The land footprint of the EU bioeconomy: Monitoring tools, gaps and needs. *Land Use Policy* **2015**, *47*, 235–246. [CrossRef]
10. Kloepffer, W. Life cycle sustainability assessment of products. *Int. J. Life Cycle Assess.* **2008**, *13*, 89–94. [CrossRef]
11. Guinée, J.; Heijungs, R.; Huppes, G.; Zamagni, A.; Masoni, P.; Buonamici, R.; Ekvall, T.; Rydberg, T. Life cycle assessment: Past, present, and future. *Environ. Sci. Technol.* **2011**, *45*, 90–96. [CrossRef] [PubMed]

12. Finkbeiner, M.; Schau, E.M.; Lehmann, A.; Traverso, M. Towards life cycle sustainability assessment. *Sustainability* **2010**, *2*, 3309–3322. [CrossRef]
13. Zamagni, A.; Pesonen, H.-L.; Swarr, T. From LCA to Life Cycle Sustainability Assessment: Concept, practice and future directions. *Int. J. Life Cycle Assess.* **2013**, *18*, 1637–1641. [CrossRef]
14. Schaubroeck, T.; Rugani, B. A revision of what life cycle sustainability assessment should entail: Towards modeling the Net Impact on Human Well-Being. *J. Ind. Ecol.* **2017**, *21*, 1464–1477. [CrossRef]
15. Kühnen, M.; Hahn, R. Indicators in Social Life Cycle Assessment—A Review of Frameworks, Theories, and Empirical Experience. *J. Ind. Ecol.* **2017**, *21*, 1547–1565. [CrossRef]
16. Lazarevic, D.; Martin, M. Life cycle assessments, carbon footprints and carbon visions: Analysing environmental systems analyses of transportation biofuels in Sweden. *J. Clean. Prod.* **2016**, *137*, 249–257. [CrossRef]
17. Van der Voet, E.; Lifset, R.J.; Luo, L. Life-cycle assessment of biofuels, convergence and divergence. *Biofuels* **2010**, *1*, 435–449. [CrossRef]
18. Weiss, M.; Haufe, J.; Carus, M.; Brandão, M.; Bringezu, S.; Hermann, B.; Patel, M.K. A Review of the Environmental Impacts of Biobased Materials. *J. Ind. Ecol.* **2012**, *16*, S169–S181. [CrossRef]
19. Von Blottnitz, H.; Curran, M.A. A review of assessments conducted on bio-ethanol as a transportation fuel from a net energy, greenhouse gas, and environmental life cycle perspective. *J. Clean. Prod.* **2016**, *15*, 607–619.
20. Janssen, M.; Xiros, C.; Tillman, A.M. Life cycle impacts of ethanol production from spruce wood chips under high-gravity conditions. *Biotechnol. Biofuels* **2016**, *9*, 1–19. [CrossRef] [PubMed]
21. Pascual-González, J.; Guillén-Gosálbez, G.; Mateo-Sanz, J.M.; Jiménez-Esteller, L. Statistical analysis of the ecoinvent database to uncover relationships between life cycle impact assessment metrics. *J. Clean. Prod.* **2016**, *112*, 359–368. [CrossRef]
22. Steinmann, Z.J.N.; Schipper, A.M.; Hauck, M.; Huijbregts, M.A.J. How Many Environmental Impact Indicators Are Needed in the Evaluation of Product Life Cycles? *Environ. Sci. Technol.* **2016**, *50*, 3913–3919. [CrossRef] [PubMed]
23. Steffen, W.; Richardson, K.; Rockström, J.; Cornell, S.E.; Fetzer, I.; Bennett, E.M.; Biggs, R.; Carpenter, S.R.; de Vries, W.; de Wit, C.A.; et al. Planetary boundaries: Guiding human development on a changing planet. *Science* **2015**, *347*, 1259855. [CrossRef] [PubMed]
24. EC. Commission Recommendation of 9 April 2013 on the use of common methods to measure and communicate the life cycle environmental performance of products and organisations. European Commission (EC). *Off. J. Eur. Union* **2013**, *54*, 124.
25. Benoît, C.; Mazijn, B. (Eds.) *Guidelines for Social Life Cycle Assessment of Products—A Social and Socio-Economic LCA Code of Practice Complementing Environmental LCA and Life Cycle Costing, Contributing to the Full Assessment of Goods and Services Within the Context of Sustainable Development*; United Nations Environment Programme: Paris, France, 2009.
26. Klöpffer, W.; Ciroth, A. Is LCC relevant in a sustainability assessment? *Int. J. Life Cycle Assess.* **2011**, *16*, 99–101. [CrossRef]
27. Hannouf, M.; Assefa, G. Comments on the relevance of life cycle costing in sustainability assessment of product systems. *Int. J. Life Cycle Assess.* **2016**, *21*, 1059–1062. [CrossRef]
28. Jørgensen, A.; Herrmann, I.T.; Bjørn, A. Analysis of the link between a definition of sustainability and the life cycle methodologies. *Int. J. Life Cycle Assess.* **2013**, *18*, 1440–1449. [CrossRef]
29. Ekvall, T.; Ljungkvist, H.; Ahlgren, E.; Sandvall, A. *Participatory Life Cycle Sustainability Analysis*; Report B2268; IVL; Swedish Environmental Research Institute: Stockholm, Sweden, 2016.
30. Owen, H. *Open Space Technology: A User's Guide*, 3rd ed.; Berrett-Koehler Publishers: San Fransisco, CA, USA, 2008.
31. Zumsteg, J.M.; Cooper, J.S.; Noon, M.S. Systematic Review Checklist: A Standardized Technique for Assessing and Reporting Reviews of Life Cycle Assessment Data. *J. Ind. Ecol.* **2012**, *16*, S12–S21. [CrossRef] [PubMed]
32. SLU. *Forest Statistics 2015*; Official Statistics of Sweden, Swedish University of Agricultural Sciences (SLU): Genevå, Switzerland, 2014.
33. Ekvall, T. *Open Space Workshop on Sustainability Indicators for Bio-Based Products*; Report B238; IVL Swedish Environmental Research Institute: Stockholm, Sweden, 2017.

34. Ren, J.; Manzardo, A.; Mazzi, A.; Zuliani, F.; Scipioni, A. Prioritization of bioethanol production pathways in China based on life cycle sustainability assessment and multicriteria decision-making. *Int. J. Life Cycle Assess.* **2015**, *20*, 842–853. [CrossRef]

35. Santoyo-Castelazo, E.; Azapagic, A. Sustainability assessment of energy systems: Integrating environmental, economic and social aspects. *J. Clean. Prod.* **2014**, *80*, 119–138. [CrossRef]

36. La Rosa, A.D.; Cozzo, G.; Latteri, A.; Recca, A.; Björklund, A.; Parrinello, E.; Cicala, G. Life cycle assessment of a novel hybrid glass-hemp/thermoset composite. *J. Clean. Prod.* **2013**, *44*, 69–76. [CrossRef]

37. Ekener-Petersen, E.; Höglund, J.; Finnveden, G. Screening potential social impacts of fossil fuels and biofuels for vehicles. *Energy Policy* **2014**, *73*, 416–426. [CrossRef]

38. Souza, A.; Watanabe, M.D.B.; Cavalett, O.; Ugaya, C.M.L.; Bonomi, A. Social life cycle assessment of first and second-generation ethanol production technologies in Brazil. *Int. J. Life Cycle Assess.* **2016**, *23*, 1–12. [CrossRef]

39. Manik, Y.; Leahy, J.; Halog, A. Social life cycle assessment of palm oil biodiesel: A case study in Jambi Province of Indonesia. *Int. J. Life Cycle Assess.* **2013**, *18*, 1386–1392. [CrossRef]

40. Weldegiorgis, F.; Franks, D. Social dimensions of energy supply alternatives in steelmaking: Comparison of biomass and coal production scenarios in Australia. *J. Clean. Prod.* **2014**, *84*, 281–288. [CrossRef]

41. Hu, J.; Lei, T.; Wang, Z.; Wang, Y.X.; Shi, X.; Zaifeng, L.; Xiaofeng, H.; Zhang, Q. Economic, environmental and social assessment of briquette fuel from agricultural residues in China: A study on flat die briquetting using corn stalk. *Energy* **2014**, *64*, 557–566. [CrossRef]

42. Stamford, L.; Azapagic, A. Life cycle sustainability assessment of UK electricity scenarios to 2070. *Energ. Sustain. Dev.* **2014**, *2*, 194–211. [CrossRef]

43. Albrecht, S.; Brandstetter, P.; Beck, T.; Fullana-i-Palmer, P.; Grönman, K.; Baitz, M.; Deimling, S.; Sandilands, J.; Fischer, M. An extended life cycle analysis of packaging systems for fruit and vegetable transport in Europe. *Int. J. Life Cycle Assess.* **2013**, *18*, 1549–1567. [CrossRef]

44. Keller, H.; Rettenmaier, N.; Reinhardt, G.A. Integrated life cycle sustainability assessment—A practical approach applied to biorefineries. *Appl. Energ.* **2015**, *154*, 1072–1081. [CrossRef]

45. Colodel, C.M.; Kupfer, T.; Barthel, L.P.; Albrecht, S. R&D decision support by parallel assessment of economic, ecological and social impact—Adipic acid from renewable resources versus adipic acid from crude oil. *Ecol. Econ.* **2009**, *68*, 1599–1604.

46. Vinyes, E.; Oliver-Solà, J.; Ugaya, C.; Rieradevall, J.; Gasol, C.M. Application of LCSA to used cooking oil waste management. *Int. J. Life Cycle Assess.* **2013**, *18*, 445–455. [CrossRef]

47. UN a. Goal 6: Ensure Access to Water and Sanitation for All. United Nations, 2017. Available online: http://www.un.org/sustainabledevelopment/water-and-sanitation/ (accessed on 14 November 2017).

48. UN b. Goal 15: Sustainably Manage Forests, Combat Desertification, Halt and Reverse Land Degradation, Halt Biodiversity Loss. United Nations, 2015. Available online: http://www.un.org/sustainabledevelopment/biodiversity/ (accessed on 14 November 2017).

49. SEPA. *Swedish Environmental Objectives. Swedish Environmental Objectives*; Swedish Environmental Protection Agency (SEPA): Stockholm, Sweden, 2015.

50. SEPA a. Objective 9. Good-Quality Groundwater. Swedish Environmental Protection Agency. 2017. Available online: http://www.miljomal.se/Environmental-Objectives-Portal/Undre-meny/About-the-Environmental-Objectives/9-Good-Quality-Groundwater/ (accessed on 14 November 2017).

51. SEPA b. Objective 16. A Rich Diversity of Plant and Animal Life. Swedish Environmental Protection Agency. 2017. Available online: http://www.miljomal.se/Environmental-Objectives-Portal/Undre-meny/About-the-Environmental-Objectives/16-A-Rich-Diversity-of-Plant-and-Animal-Life/ (accessed on 14 November 2017).

52. SEPA c. Objective 11. Thriving Wetlands. Swedish Environmental Protection Agency. 2017. Available online: http://www.miljomal.se/Environmental-Objectives-Portal/Undre-meny/About-the-Environmental-Objectives/11-Thriving-Wetlands/. (accessed on 14 November 2017).

53. SEPA d. Objective 12. Sustainable Forests. Swedish Environmental Protection Agency. 2017. Available online: http://www.miljomal.se/Environmental-Objectives-Portal/Undre-meny/About-the-Environmental-Objectives/12-Sustainable-Forests/ (accessed on 14 November 2017).

54. SEPA e. Objective 13. A Varied Agricultural Landscape. Swedish Environmental Protection Agency. 2017. Available online: http://www.miljomal.se/Environmental-Objectives-Portal/Undre-meny/About-the-Environmental-Objectives/13-A-Varied-Agricultural-Landscape/ (accessed on 14 November 2017).

55. De Baan, L.; Curran, M.A.; Rondinini, C.; Visconti, P.; Hellweg, S.; Koellner, T. High-resolution assessment of land use impacts on biodiversity in life cycle assessment using species habitat suitability models. *Environ. Sci. Technol.* **2015**, *49*, 2237–2244. [CrossRef] [PubMed]

56. Teixeira, R.F.M.; De Souza, D.M.; Curran, M.P.; Antón, A.; Michelsen, O.; Milá I Canals, L. Towards consensus on land use impacts on biodiversity in LCA: UNEP/SETAC Life Cycle Initiative preliminary recommendations based on expert contributions. *J. Clean. Prod.* **2016**, *112*, 4283–4287. [CrossRef]

57. Martin, M.; Brandão, M. Evaluating the Environmental Consequences of Swedish Food Consumption and Dietary Choices. *Sustainability* **2017**, *9*, 2227. [CrossRef]

58. Kløverpris, J.; Wenzel, H.; Nielsen, P.H. Life cycle inventory modeling of land use induced by crop consumption. Part 1: Conceptual analysis and methodological proposal. *Int. J. Life Cycle Assess.* **2008**, *13*, 13–21.

59. Harnesk, D.; Brogaard, S. Social Dynamics of Renewable Energy—How the European Union's Renewable Energy Directive Triggers Land Pressure in Tanzania. *J. Env. Dev.* **2016**, *26*, 156–185. [CrossRef]

60. Martin, M.; Larsson, M.; Oliveira, F.; Rydberg, T. Reviewing the environmental implications of increased consumption and trade of biofuels for transportation in Sweden. *Biofuels* **2017**, 1–15. [CrossRef]

61. SLU. Official Statistics of Sweden & Swedish University of Agricultural Sciences Umeå (SLU). In *Forest Statistics 2015*; The Swedish University of Agricultural Sciences: Uppsala, Sweden, 2015.

62. Swedish Forest Agency. *Skogsstatistisk årsbok 2014 [Swedish Statistical Yearbook of Forestry]*. Available online: https://www.skogsstyrelsen.se/globalassets/statistik/historisk-statistik/skogsstatistisk-arsbok-2010-2014/skogsstatistisk-arsbok-2014.pdf (accessed on 20 February 2018).

63. Bioenergiportalen. Andelen Biodrivmedel öKar. [The Share of Biofuels Is Increasing]. Available online: http://www.bioenergiportalen.se/?p=1443 (accessed on 21 January 2015).

64. Svebio. The Swedish Experience. How bioenergy became the largest energy source in Sweden. Available online: https://www.svebio.se/sites/default/files/Bioenergy%20in%20Sweden_web.pdf (accessed on 10 March 2016).

65. SEPA. Import Och Export Av Avfall 2004–2013 [Import and Export of Waste 2004–2013. Swedish Environmental Protection Agency (SEPA). Available online: https://www.naturvardsverket.se/Sa-mar-miljon/Statistik-A-O/Avfall-import-och-export-2004-2013/# (accessed on 23 February 2016).

66. Swedish Energy Agency. Analys Av Marknaderna FöR Biodrivmedel. Tema: Fordonsgasmarknaden. [Analysis of the Biofuel Markets. Theme: The Biofuel Market]. Available online: https://www.energimyndigheten.se/globalassets/nyheter/2013/analys-av-marknaderna-for-biodrivmedel.pdf2013 (accessed on 24 February 2016).

67. Swedish Energy Agency. Hållbara Biodrivmedel Och Flytande BiobräNslen 2014. [Sustainable Biofuels and Liquid Bioenergy 2014]. Available online: https://energimyndigheten.se.2014 (accessed on 23 November 2015).

sustainability

MDPI

Article

Gaps and Research Demand for Sustainability Certification and Standardisation in a Sustainable Bio-Based Economy in the EU

Stefan Majer [1,*], Simone Wurster [2], David Moosmann [1], Luana Ladu [2], Beike Sumfleth [1] and Daniela Thrän [1,3]

[1] Deutsches Biomasseforschungszentrum (DBFZ), Leipzig 04347, Germany; david.moosmann@dbfz.de (D.M.); Beike.Sumfleth@dbfz.de (B.S.); daniela.thraen@ufz.de (D.T.)

[2] Technische Universität Berlin (TU Berlin), Berlin 10623, Germany; simone.wurster@tu-berlin.de (S.W.); luana.ladu@tu-berlin.de (L.L.)

[3] Helmholtz Centre for Environmental Research (UFZ), Leipzig 04318, Germany

[*] Correspondence: stefan.majer@dbfz.de; Tel.: +49-341-2434-411

Received: 29 April 2018; Accepted: 10 July 2018; Published: 13 July 2018

Abstract: The concept of the bio-based economy has gained increasing attention and importance in recent years. It is seen as a chance to reduce the dependency on fossil resources while securing a sustainable supply of energy, water, and raw materials, and furthermore preserving soils, climate and the environment. The intended transformation is characterized by economic, environmental and social challenges and opportunities, and it is understood as a social transition process towards a sustainable, bio-based and nature-oriented economy. This process requires general mechanisms to establish and monitor safeguards for a sustainable development of the bio-based economy on a national and EU level. Sustainability certification and standardisation of bio-based products can help to manage biogenic resources and their derived products in a sustainable manner. In this paper, we have analysed the current status of sustainability certification and standardisation in the bio-based economy by conducting comprehensive desktop research, which was complemented by a series of expert interviews. The analysis revealed an impressive amount of existing certification frameworks, criteria, indicators and applicable standards. However, relevant gaps relating to existing criteria sets, the practical implementation of criteria in certification processes, the legislative framework, end-of-life processes, as well as necessary standardisation activities, were identified which require further research and development to improve sustainability certification and standardisation for a growing bio-based economy.

Keywords: bio-based economy; sustainability; certification; standardisation; sustainability criteria; gaps; sustainability assessment

1. Introduction

The bio-based economy (BBE) is seen as a chance to supersede the era of fossil resources and technologies, to foster health and nutrition of a growing world population, and to secure a sustainable supply of energy, water, and raw materials, while preserving soils, climate and the environment [1]. The intended transformation is characterized by economic, environmental and social challenges and opportunities, and it is understood as a social transition process towards a sustainable, bio-based and nature-oriented economy. Currently, more than 40 countries have defined specific strategies for the development of a BBE or the bioeconomy (BE) [2]. Even though there is a slight difference between the two terms BBE and BE, both concepts are often used synonymously in the literature [3]. For the purpose of this study, we refer to the concept of the BBE as defined by the EU, which suggests that a "bio-based economy integrates the full range of natural and renewable biological resources—land and sea resources, biodiversity and biological materials (plant, animal and microbial), through to the processing and the consumption of these bio-resources" [4]. Management and control of the intended transitions needs appropriate measurement, information, and tools to cover not only the BE as a whole, but also the different dimensions of BE development [5]. This requires general mechanisms to establish and monitor safeguards for a sustainable development of the BBE economy on a national and EU level.

Sustainability certification of bio-based products can help to manage biogenic resources and their derived products in a sustainable manner. The development and use of sustainability assessment schemes for bio-based products which contribute to a clear and evidence-based view of the economic, social and environmental impact/benefits of bio-based solutions is, therefore, an important goal in current research activities.

The concept of sustainable resource management was originally developed for the organization of forestry resources [6]. The management concepts tried to ensure that extraction of timber did not surpass regenerative capacity; thus, future generations can benefit from the forests the same way the present generation does [7]. It is not surprising that the beginnings of sustainability certification, which started by the end of last century, are also related to forest management systems. Additionally, as described by [8], early sustainability certification activities are described for agricultural production processes. To fight global deforestation and especially the destruction of native forests in the tropics and subtropics, the Forest Steward Ship Council® (FSC®) was founded in 1993. Starting with the certification of forest management, the certification was expanded stepwise to the whole supply chain of products, including transportation steps and processing of biomass, but also to biomass resources outside the forests. This so-called chain of custody certification provided a basis for sustainability certifications for all kinds of lignocellulosic and non-lignocellulosic biomass-based supply chains. This enables complete traceability of the biomass, because each supply chain element is subject to certification. FSC® has become very successful. As of today, there are more than 200 million ha of forests certified globally and until now 33,829 chain of custody certificates were issued [9]. Although this number seems to be significant, it represents only ca. 5% of the global forest area. Nonetheless, there is an increasing demand for FSC® certified materials and products. FSC® certified products and materials have become a standard in some industries. In any case, sustainability certification has reached a stage at which it is being officially accepted to ensure the implementation of sustainability requirements laid down in laws and regulations [10]. There are, for instance, Green Public Procurement requirements that include forest certification according to internationally recognised schemes (e.g., FSC® and PEFC™) as a proof of fulfilment of sustainability criteria.

The EU biofuel market is a good example of a regulated market with binding sustainability requirements based on EU legislation (i.e., [11]). As a reaction to intense debate about the possible impact of national and international biofuel policies (e.g., the 10% target of the Renewable Energy Directive (RED)) on price increases of agricultural commodities [12] or land use changes ([13], legally binding sustainability criteria for liquid biofuels and bioliquids were developed and were, since 2009, implemented in the RED). The mandatory sustainability criteria included in this directive are mainly limited to environmental criteria. Economic or social criteria are not included, mostly because of the existing EU background regulations and potential conflicts with WTO regulations in case they would be applied outside the EU [14]. The main goal of the implementation of the RED criteria is (1) to prevent that areas with high ecological value or high carbon stocks are converted to agricultural areas dedicated to the cultivation of crops used for biofuel production; and (2) that a certain GHG emission saving from biofuels in comparison to the use of fossil fuel is ensured. An expansion of the sustainability criteria from liquid biofuels to all sectors of bioenergy including heat and power production from solid, liquid and gaseous biomass from 2021 is under discussion [15]. Currently, each market actor along the supply chain of biofuels, for example, biomass producers, traders, processors, biofuel refineries, petroleum companies, etc., must fulfill the mandatory sustainability requirements of the RED. To show compliance with the criteria, market actors have to obtain a certificate. The European Commission has officially recognized a number of so-called voluntary schemes for that purpose. The respective schemes differ according to the applicability concerning the geographic and feedstock scope as well as the parts of the supply chain elements being covered. Furthermore, the criteria and indicators used by sustainability certification schemes in the EU BBE can vary significantly. Mostly, the topics and even more the criteria and indicators covered by the various schemes do not reflect the complete and holistic understanding of the sustainability concept, covering social, economic and environmental impacts. Instead, the currently available certification schemes are often the result of stakeholder processes. Consequently, the criteria and indicators included represent the individual perception of the stakeholders involved regarding the relevance of certain criteria and indicators regarding the sustainability of the products to be certified. Furthermore, effective sustainability certification activities are often embedded in a specific regulatory framework, which can already set standards regarding specific sustainability requirements (e.g., EU labour rights conditions, cross compliance regulations, etc.) [16]. After a few years in operation, especially the more advanced schemes were revised and further developed. In addition, their scope was extended to applications and products beyond liquid biofuels (e.g., solid biofuels, bioplastics, biochemicals, etc.).

While the current sustainability certification activities (e.g., in the context of the RED) are focused mainly on environmental and social aspects, product standardisation supports market implementation by reducing economic hurdles [17]. Both tools (certification and standardisation) can play an important role in enabling the existing regulatory system to adapt and support innovation [18]. Based on the EU mandate M/429 from 2008, the European Committee for Standardization (CEN) established a standardisation programme for bio-based products. Consequently, CEN's Technical Committee (TC) 411 was created in 2011. Its scope comprises horizontal aspects of the BBE, including a common terminology, methods for determining bio-based content in a product, Life Cycle Assessments (LCA), sustainability of biomass and guidance on the use of existing standards for the end-of-life (EOL) options. Based on the European standardisation Mandates M/491 [19] and M/492 [20], TC 411 has been developing standards to help specific sectors move towards higher renewable biomass content.

In addition, other CEN TCs deal with specific bio-based products and applications. For example, TC 249 is responsible for the development of standards for biopolymers and TC 19 is tasked with creating standards for bio-based lubricants [21], while CEN/TC 383 works on European standards establishing sustainability criteria for biofuels. This standardisation work relates to relevant legislation supporting the development of sustainable bioenergy [11,22].

The development of harmonized and vertical standards for the BBE on an EU level will support the future sustainability of the sector despite the many challenges that still exist [17]. Nevertheless, observers have called for more advanced tools for conducting sustainability assessments of bio-based products [17]. The goal is to better demonstrate both bio-based products potential to solve important sustainability challenges and reduce concerns about possible negative implications [17].

The existing certification schemes and standardisation approaches provide a contribution to the establishment of the EU BBE. However, it is difficult to identify, if these approaches cover the different aspects of sustainability that influence consumption decisions. With a growing BBE, these aspects will become more and more important (e.g., [23]). Therefore, this paper aims to provide a systematic identification of potential gaps in terms, for example, of important aspects not covered by existing sustainability certification and standardisation tools within the EU BBE and come up with a set of first recommendations to overcome these gaps. The results presented were developed as part of the EU-funded H2020 research project STAR-ProBio.

2. Materials and Methods

Our analysis of existing sustainability certification and standardisation activities was based on a threefold approach. Firstly, we conducted a comprehensive analysis of existing sustainability assessment activities in different sectors of the BBE. This included the development of a general understanding regarding the frequency and market penetration of the various sustainability criteria used by certification frameworks, as well as the identification of differences between sectors of the EU BBE. Secondly, a series of interviews with experts (policy makers, policy advisers, scientists, certification schemes, certification bodies) were conducted to identify mega-trends regarding current and future gaps in sustainability certification. Thirdly, the comprehensive overview of criteria and indicators which was generated with the first approach was used to verify and generalise the trends and opinions which were formulated during the interviews.

2.1. Analysis of Existing Sustainability Certification Schemes, Labels, and Initiatives in the EU Bio-Bio-Based Economy

This analysis was conducted in three steps. Firstly, sustainability frameworks (the expressions framework and system are used synonymously in this context), with a focus on the EU BBE, were identified by desktop research and by using available databases such as Standards Map and Label Online (a). Secondly, the selected relevant frameworks were analysed to build an inventory of the sustainability criteria and indicators used in current sustainability certification (b). Thirdly, relevant standardisation activities were analysed (c).

During the first step, which aimed to picturise the sustainability frameworks presently available for application within the BBE, a preliminary list was generated. This process followed a three-stage course of action, which is illustrated in Figure 1. During steps one and two, a desktop research using web databases (ITC Standards Map, Label Online) was used to identify frameworks relevant for the BBE. With this approach, we could identify ~100 existing sustainability certification frameworks with relevance for the BBE.

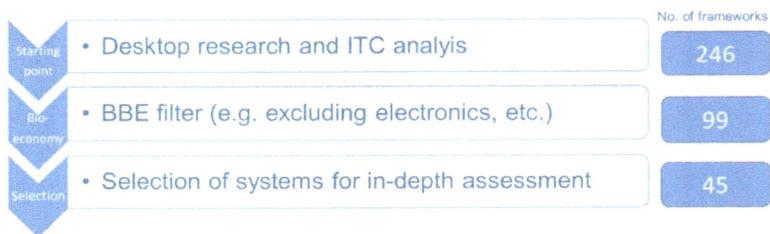

Figure 1. Analysis of relevant frameworks currently available for application in EU BBE–procedure.

The resulting list of frameworks was supplemented by a review of scientific literature. A simple database spreadsheet was produced to allow for a first characterisation (e.g., with regards to parameters such as: regional focus, product focus, sectors of the BBE covered by the framework, etc.) of the identified certification frameworks. In the final step, based on a set of criteria, the sample was reduced by half, in order to make an in-depth analysis of the frameworks feasible. For this purpose, the following criteria were determined among STAR-ProBio project partners: Scope of the certification framework, overall transparency of the system (in terms of accessibility of all relevant system documents and publication of certificate holders), comprehensiveness relating to the three sustainability dimensions (environmental, social, and economic), relevance of the framework (in terms of number of issued certificates).

The described procedure resulted in a list of sustainability frameworks for further elaboration. The frameworks were arranged within a table (see Figure 5) that includes information on the kind of framework (label, initiative, scheme, please see Table 1 for a set of working definitions), the BBE sector addressed, the supply chain coverage (single supply chain elements, full supply chain, etc.), the geographic scope (national, global), the feedstock scope (some frameworks are limited to certain feedstock or feedstock groups), and the sustainability dimension (social, environmental, economic) addressed.

Table 1. Definition of the different types of frameworks included in the analysis.

Framework	Definition
Labels	Labels communicate the guarantee of certain product characteristic to the consumer, which ideally is described in an adequate level of transparency. A certification process can be a precondition for the labelling of a product. However, there are products self-labelled by the producer.
Initiatives	Sustainability initiatives are herein referred to as initiatives compiling sets of sustainability criteria and indicators for a particular purpose. They might be organised as a heterogeneous group of people with different background and with different interests. The goal of this type of initiative is to reach a consensus between the different parties. In the resulting set of criteria, the different interests are reflected. This type of initiative is often referred to as "multi-stakeholder initiative" or "roundtable". The second type of initiative considered is consisting of a group of people belonging to one party. They can have a varying backgrounds and interests. The one objective, quality sustainability initiatives have in common is the outcome/product, which is a set of criteria for further unspecified or specified use. The outcome can be used internally, e.g., for the sustainability strategy of an organisation or may be picked up by others.
Certification schemes	Certification schemes are based on a normative framework. The output of initiatives may be used as the basis for a certification scheme. Sustainability initiatives therefore sometimes turn into a certification scheme holder over time as it happened with different roundtables. The most important characteristic of a certification scheme, as it is understood in this context, is that it includes a third-party verification of the sustainability criteria, stipulated in the system documents. Also, the whole certification process is usually based on accreditation standards (e.g., ISO 19011 or ISO 17065), in which the separation of evaluation and certification is to mention an important feature. As a result of the certification process, a label on a product shows compliance with the respective certification scheme.

During this in-depth analysis, an inventory of the sets of criteria and indicators from the selected certification frameworks was developed (please see [24]). Since criteria and indicator sets are often adjusted and revised frequently by the certification frameworks, the latest versions of the core system documents were obtained for each framework. The inventory database which was developed throughout this process can be considered some kind of meta-standard including all criteria and indicators identified from the analysis. Since criteria and indicators are named and defined quite differently between the various frameworks, it was necessary to harmonise and structure the criteria and indicator sets. For this purpose, firstly the sustainability criteria and (if applicable) respective indicators were isolated from each framework. Gradually, terminology was harmonised and the criteria and respective indicators were organised according to the three sustainability dimensions (i.e., environmental, social and economic).

The assignment of criteria to the three dimensions of sustainability was done using a hierarchic structure with thematic categories and main principles summarising different criteria and their respective indicators (for an example, please see Figure 2). This was a necessary step to reach a point at which a manageable compilation could be formed out of the multitude of criteria of the analysed frameworks. Furthermore, several frameworks express equal criteria in slightly different ways, at different levels of detail or aggregation. In addition, a differentiation between criteria and indicators was not given for every framework analysed, due to the very diverse presentations of criteria sets in the available system documentation. The final database [24] allows tracing back the original, more detailed wording of a certain criteria.

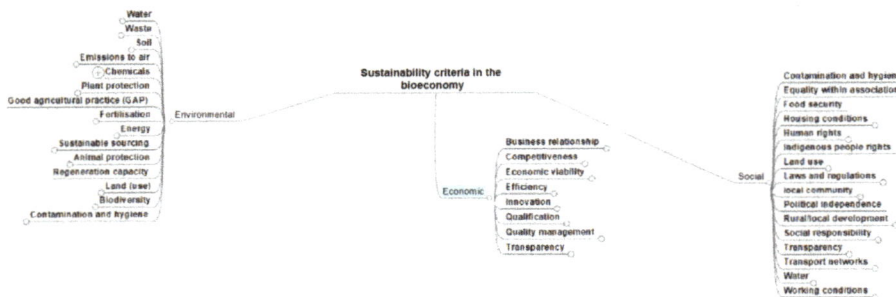

Figure 2. Example of criteria list generated from the analysis of sustainability certification frameworks.

2.2. Expert Interviews

To complement the detailed analysis of the framework documents, we conducted a number of interviews with experts from industry, science, policy making, standardisation and certification bodies, as well as non-governmental organizations. The main objective of these interviews was to develop a better understanding of the current discussion regarding the status of sustainability certification and standardisation in the EU BBE. Furthermore, the individual feedback received during the expert interviews provided valuable insights into the expert's perceptions of the relevance of specific topics. The points mentioned by the experts were discussed and checked against the results from the analysis of existing sustainability certification frameworks and standardisation activities.

The selection of interviewees followed principles of the grounded theory approach: theoretical sampling and theoretical saturation. According to [25], theoretical sampling uses samples that are relevant for a given research question based on individual selections. The sampling must be carried out until theoretical saturation is reached, i.e., until no new or significant information seems to appear in relation to the relevant questions. Following this approach, and under consideration of the available project resources, 25 experts were identified for interviews. In the context of our research, it was important to receive input from experts with a general expertise regarding bio-based products and sustainability certification and standardisation. In most cases, these persons belong to research institutes or universities—for example, professors—or they work as experts at European organisations. Additionally, we wanted to learn more about the specific views of producers and consumers, as well as those of representatives of standardisation and certification bodies. Experts from various European countries were selected. They represent the above-mentioned high-level experts for bio-based products, as well as producers of bio-based products, consumers, and members of standardisation committees.

Following the above-mentioned concept [25], a sampling series is finished if a repetition of the answers is experienced and the extent of new information based on additional samples is low. These repetitions could be experienced, for example, regarding the suggestions to learn from specific other standards and to adopt the good practice of the RED with certain modifications for bio-based products as well. However, specific questions on standardisation issues beyond the scope of these interviews remained, requiring specific information exchanges with representatives of standardisation TCs. In total, 20 interviews were conducted in the first interview series. It is clear that this cannot necessarily be considered a fully representative sample size. Furthermore, it is possible that the focus of the discussion points mentioned during the interviews could differ with a different proportional distribution of the stakeholder groups represented by the experts. More research is recommended to deepen our results and/or to derive new conclusions. Regarding specific standards issues, an additional series of information exchange activities with experts was conducted, based on specific questions, see Section 2.3.

The 20 interviews were conducted using a standardised questionnaire (see Appendix B). Figures 3 and 4 provide impressions regarding the background of the experts, interviewed during the process of gap assessment. It is important to note that the total number differs between the two figures. The reason for this is that a couple of experts gave multiple answers regarding their background and associated stakeholder group.

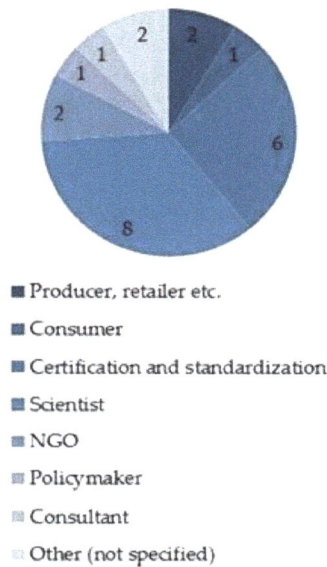

Figure 3. Stakeholder group of the interviewed experts and their mentioned amount (multiple responses allowed).

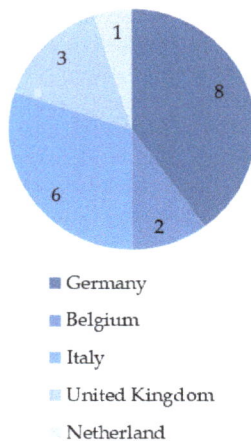

Figure 4. Country of origin of the interviewed experts and their mentioned amount.

2.3. Analysis of Existing Sustainability Standards in the EU Bio-Based Economy

In addition to the analysis in the field of sustainability certification, the focus of the third analytical step was on existing sustainability standards for bio-based products and standards in related areas. The database Perinorm (https://www.perinorm.com) and relevant documents were analysed for this purpose using 22 keywords. Perinorm contains approximately two million records of the European and global key facts on standards, technical regulations and legislation. It includes documents from the European and international standards bodies, the European Committee for Standardization (CEN), the European Committee for Electrotechnical Standardization (CENELEC), the European Telecommunications Standards Institute (ETSI), the International Organization for Standardization (ISO), the International Electrotechnical Commission (IEC) and the International Telecommunication Union (ITU), as well as those from standard bodies in Japan, China, USA, Jordan, South Africa, Canada, and Brazil. Based on the results obtained, more specific information can be obtained regarding the classification (type of standard, industry), document maturity (draft standards, standards etc.), valid records, country of origin and relation to European and international standards. Table 2 shows keywords used for the Perinorm search. Search results were further specified. For example, entries in the following fields were deleted: sustainable tourism, tractors, and machinery for agriculture and forestry, aluminium structures and electronic signatures. Several standards on environmental management, such as JS 14040 and JIS Q 14040, were also deleted, because both refer to the ISO standard with the same name.

Table 2. Keywords of the Perinorm search.

Keyword	Total	Keyword	Total	Keyword	Total
algae products	75	biofuel	94	life cycle assessment	1046
animal-based	3	biofuels	1.527	plant-based	30
bio-based	60	biomass	1.420	starch-based	1
bio-based	300	bioplastics	27	sustainability	3361
biochemicals	290	cellulose-based	95	sustainable	487
biodegradable	445	end-of-life	410	value chain	65
biodiesel	283	footprint	126		
bioenergy	16	forest products	88		

Considering the importance of the issues of direct and, in particular, indirect land use change (iLUC) related to the sustainability of bio-based materials and bioenergy (see, e.g., [26]), three additional terms were analysed afterwards. For the term "indirect land use change" and its abbreviation "ILUC", no results were obtained. On the other hand, the term "land use change" led to 17 hits, and to the identification of a relevant international standard; the latter was analysed in greater detail: ISO 13065 [27]. In addition to this, four national standards that deal with this topic [28–31] were identified. Due to the low number of results, the suitability of an additional search term "land use" was analysed. On the European and international level, the search for valid standards led to 153 hits. However, most of them were focused on other areas, such as agricultural machinery, tractors, etc., and were not relevant for the purpose of this assessment. Exemptions are EN 16214-4 and ISO 14055-1. The Perinorm results were further analysed, screened and clustered to get deeper insight in the current standards landscape of bio-based products (see Tables A13 and A14 in Appendix D).

Our paper aims at addressing the need for more advanced tools for conducting sustainability assessments by integrating its work appropriately in this existing standardisation landscape. For that reason, not only experts for bio-based products in general, but also experts from the field of standardisation were interviewed. Thirteen European and international TCs and PCs (Project Committees) were identified as relevant for this work. In particular, attention was drawn to three technical committees: CEN/TC 411–Bio-based products, ISO/TC 207–Environmental management and ISO/PC 248–Sustainability Criteria for Bioenergy.

3. Results

3.1. Overview on Existing Certification Frameworks

Existing activities regarding sustainability certification in the EU BBE can differ significantly with regard to their operability, stakeholder involvement, scope, etc. To allow for a differentiation, we have distinguished three types of sustainability frameworks (for our working definitions please see Table 1): (a) labels, (b) sustainability initiatives, and (c) certification schemes.

Analysing the currently existing sustainability certification activities in the EU BBE, we found a broad range of existing certification frameworks addressing different sectors, and scopes regarding feedstocks, the completeness of the supply chain as well as coverage of geographic areas. Figure A1 gives a wide overview of available sustainability certification frameworks in the EU BBE. Figure 5 provides a summary of the frameworks selected for an in-depth assessment during our analysis. Additional information can be found in Appendix A.

The criteria and indicators included in the analysed certification frameworks were assessed and structured into a database [24] which is available on the Homepage of the STAR-ProBio project. This collection of information allowed for a comparison between the statements made by experts (e.g., regarding criteria and indicator gaps in current sustainability certification) and the currently available certification frameworks.

3.2. Gaps in Current Sustainability Certification Activities

The assessment approach described under Section 3.1 revealed a number of areas with needs and demands for further research regarding assessment tools and general frameworks, as well as criteria and indicators for certification activities in the BBE.

During the interviews conducted (see Section 2.2 and Appendix C), the experts mentioned several areas of potential gaps in sustainability certification. The statements made by the experts were compared to the criteria and the inventory database which was prepared during the analysis of the existing certification frameworks (see [24]). Based on this process, we summarised the identified discussion points into seven main topics with demand for future research and development. A comprehensive overview on the expert statements received during the process and the allocation of these statements to the seven main topics is included in Appendix C, Tables A6–A12.

The main topics identified can be summarised as follows:

- Gaps and weaknesses in criteria and indicator sets
- Harmonisation in criteria assessment and operationalisation
- Legislation and consensus for minimum criteria in all BBE sectors
- Leakage effects from EU BBE policies
- New innovative, inter-sectoral products
- EOL
- Traceability of sustainability and certificates along the value chain

In the following paragraphs of this Section, we will describe the findings for each of the topics identified.

Sector		Label	Initiative	Certification scheme	Name	Feedstock	Geographic	Supply Chain	Social	Environment	Economic
Bioenergy	Liquid biofuels	x		x	International Sustainability & Carbon Certification (ISCC)	multiple	global	full	x	x	x
		x		x	REDcert EU	multiple	Europe (+Ukraine, Belarus)	full		x	
		x		x	Roundtable on Sustainable Biomaterials EU RED (RSB EU RED)	multiple	global	full	x	x	x
		x			Red Tractor Farm Assurance Combinable Crops & Sugar Beet (Red Tractor)	cereals, oilseeds, sugar beet	UK	until first feedstock delivery point		x	
		x		x	Roundtable on Sustainable Palm Oil RED (RSPO)	multiple	global	full	x	x	x
				x	Certification System adressing Indirect Impacts of Biofuel (CIIB)	multiple	EU+	farm gate to first processor		x	x
	Solid biofuels	x		x	Sustainable Biomass Partnership (SBP)	woody biomass	global	from cultivation to energy production	x	x	x
		x		x	DINplus Short rotation coppice sustainably grown according to DIN EN 16214-3	SRC wood	global	cultivation	x	x	x
		x			Nordic Ecolabeling (SWAN)	woody biomass	Denmark, Finland, Iceland, Norway, Sweden		x	x	x
	Biogas	x		x	CHB-Biogasdoneright	multiple	Italy	production process, product		x	x
		x			Green Gas Certification Scheme (GGCS)	multiple	UK	production to use		x	x
	Heat/Power		x		Global Bioenergy Partnership (GBEP)	multiple	global	full	x	x	x
		x			nature made star	multiple	CH	energy production and delivery		x	
		x			OK-Power	multiple	Germany	power production		x	
		x			Grüner-Strom-Label (Green-Power-Label)	multiple	Germany	power production to distribution		x	
		x		x	Green-e	multiple	global	full	x	x	x
Forestry		x		x	Forest Stewardship Council (FSC) CoC	woody biomass	global	full	x	x	x
		x		x	Eco-Certified Composite (ECC) Sustainability Standard	wood fiber	USA	full	x	x	x
Construction		x		x	DGNB System		global	from cradle to grave	x	x	x
		x		x	Green Building Rating System BREEAM		global	design and procurement stage, post construction stage	x	x	x
					Assessment System for Sustainable Building (BNB)		Germany	from cradle to grave	x	x	x
		x			Minergie		Switzerland, Lichtenstein			x	
		x		x	Leadership in Energy and Environmental Design (LEED)		global	full	x	x	x
Food	Fair Trade Certification Systems	x		x	Fairtrade-Label - Fairtrade Labelling Organizations International (FLO)	crops	global	full	x	x	x
		x		x	NATURLAND Fair	multiple	global	full	x	x	x
		x		x	Rapunzel Hand in Hand	multiple	global	full	x	x	x
	Fish Certification Systems	x		x	Marine Stewardship Council (MSC)	fish	global	from fisheries to retailers	x	x	x
	Agricultural Products	x		x	GlobalGAP crops certification	crops	global	pre-farm-gate	x	x	x
		x		x	demeter	multiple	global	cultivation to processing		x	
		x		x	Ecovin	grapes	Germany	cultivation to processing		x	
		x		x	Sustainable Agriculture Network/ Rainforest Alliance Certified (SAN)	crops	global	cultivation	x	x	
		x		x	Roundtable on sustainable palm oil (RSPO)	palm oil	global	cultivation	x	x	
		x		x	UTZ certified	coffee, cacao, tea, hazelnut	global	full	x	x	x
Feed		x		x	GMP+ Feed Responsibility Assurance	soy, fish meal	global	feed production and trade	x	x	x
			x		DLG certificate sustainable agriculture	multiple	Germany	agricultural production		x	
Textiles	Fair Trade Certification Systems	x		x	Fairtrade Textile Standard - Fairtrade Labelling Organizations International (FLO)	certified cotton, other responsible fibres	global	full	x	x	x
		x			NATURLAND Textil (naturleand textile)	natural fibre	global	production		x	
		x			EU Ecolabel - Fabrics	multiple	EU, CH, NOR, ISL, TUR	products	x	x	
		x		x	Textile Exchange Organic 100% content standard	organic fibres	global	product		x	
Chemicals and Plastics		x		x	ISCC PLUS	multiple	global	full	x	x	x
			x		Bioplastic Feedstock alliance	multiple	global	full	x	x	x
Cosmetics		x		x	COSMOS Standard - Cosmetics organic and natural standard	multiple	global	production process, product		x	
Pharmacy		x		x	CRADLE TO CRADLE CERTIFIED PRODUCT STANDARD	multiple	global	production process, product	x	x	x
Materials/ Products			x		INRO Nachhaltigkeitskriterien für die stoffliche Biomassenutzung (sustainability criteria for material use of biomass)	agricultural biomass	global	cultivation		x	x
		x		x	Nature Care Products Standard	multiple	global	products		x	x

Figure 5. Overview of selected existing sustainability certification frameworks.

195

3.2.1. Gaps and Weaknesses in Criteria and Indicator Sets

It is important to mention that our analysis of certification frameworks showed an impressive list of sustainability criteria and indicators included in current sustainability certification of the EU BBE (compare [24]). The criteria and indicators available cover a wide range of sustainability aspects. While some experts stated that the future challenges for the further development of sustainability certification are not about developing more criteria and indicators, but instead the already available criteria and indicators should be used better and more frequently, another expert mentioned that there are still some gaps regarding the specific principles, criteria and indicators currently used in sustainability certification (please see Table A6) We compared the results from the expert interviews to the criteria matrix resulting from our analysis of certification frameworks to identify potential gaps. During the interviews, the work of the S2BIOM consortium was mentioned by one expert. Based on this input, we conducted a review of additional sources to be included in our assessment [32–35]).

Although our analysis and comparison of the questionnaire responses and the criteria assessment showed a wide range of criteria and indicators implemented, the following topics, criteria and indicator seem not to be significantly reflected by certification frameworks so far:

- Land use efficiency
- Tertiary resource efficiency (for the purpose of this paper, we followed the following categorisations: primary biomass resources as composed by plants, secondary biomass resources are related to animals/livestock production, tertiary biomass resources are related to post-consumption, post-production residues/wastes)
- Functionality (output service quality)
- (indirect) land use change GHG emissions
- SO_2 equivalents
- PM10
- Risks for negative impacts on food prices and supply
- Levelised life-cycle cost (excluding subsidies, including CAPEX, OPEX)
- Bio-based content and recyclability/biodegradation

The development of a detailed definition or a consistent conceptual design for these topics, criteria and indicators is outside the scope of this paper; however, to support the discussion and the work in ongoing research activities, the following Table 3 includes proposals for basic definitions of the topics mentioned.

Table 3. Characterization of identified criteria in the context of bio-based products (based on [32–35]).

Topic/Criteria/Indicator	Definition	Nature of the Gap
Land use efficiency	Number of bio-based products (including by- and co-products along the life cycle) per hectare of used area.	Technical criterion, Economic criterion; related to biomass production
Tertiary resource efficiency	Value of the bio-based output divided by the value of the secondary resource. This criterion applies to bio-based products stemming from the conversion of secondary biomass resources such as residues and wastes.	Technical criterion, Economic criterion; Related to by- and co-product use
Functionality (Output service quality)	Economic value of the outputs, compared to the economic value of the heat which could be produced from burning the (dried) primary inputs.	Economic criterion to assess or benchmark added value of a specific production pathway in comparison to alternatives
(indirect) land use change GHG emissions	GHG emissions resulting from carbon stock changes as a direct or indirect effect of feedstock production (e.g., due to the conversion of natural land into cropland caused either as direct land use change or indirect land use change from the production of a bio-based product).	Environmental criterion, related to the conversion of land for biomass production
SO_2 equivalents	Life cycle emissions of SO_2, NO_x, NH_3 and HCl/HF from bio-based product life cycle, expressed in SO_2 equivalents and calculated in accordance to the life cycle emission methodology for GHG. This criterion helps to describe the acidification potential of a bio-based product.	Environmental criterion, related to upstream and downstream emissions throughout the life-cycle of a bio-based product
PM10	Life cycle emissions of PM10 from bio-based product life cycle, calculated in accordance to the life cycle emission methodology for GHG. Supports the quantification of small particle emissions.	Environmental criterion, social criterion, related mostly to the conversion (combustion) of bio-based materials at the end of the life cycle
Risks for negative impacts on food prices and supply	This criterion needs to be fully described and should consider the BEFS methodology [36].	Social criterion, economic criterion, related to increasing competition for land and biogenic resources due to an increasing demand for biomass as a result of a growing BBE
Levelised life-cycle cost	Levelised life-cycle cost, excluding subsidies (excluding subsidies, including CAPEX, OPEX).	Economic criterion, related to the costs associated with the production, utilisation as well as the EOL phase of the bio-based product
Bio-based content and recyclability/biodegradation	The share of a product originating from biomass. Percentage or share of the bio-based products that are biodegradable.	Technical criteria, related to the production of multi-compound materials as well as to EOL scenarios for bio-based materials

It is important to mention that the certification frameworks analysed for this study were developed under specific regulations, for specific markets and applications as well as under consideration of their specific stakeholder perspectives. These influencing factors have led to individual set-ups of topics, principles and criteria included in the various certification activities. Consequently, the uptake and implementation of additional or new criteria and indicators into certification frameworks depend on a number of elements. Among others, the existing legal framework and the requirements regarding sustainability certification in the different sectors of the BBE on EU and member state (MS) level, as well as the availability of appropriate standards and tools to support the implementation, have to be considered. Furthermore, the self-conception (during our analysis we experienced significant differences regarding comprehensiveness of the criteria sets of the different frameworks analysed. While some frameworks work with sets of minimum criteria (e.g., the core sustainability criteria of the RED), others tend to frequently update and expand their criteria and indicator sets) and market positioning of the certification framework, as well as the expectations of the stakeholders involved, are additional elements that influence the possibility to add new criteria to existing certification frameworks.

3.2.2. Harmonisation and Level Playing Field in Criteria Assessment and Operationalisation

The second topic identified during our assessment was specifically mentioned by several of the interviewed experts (please see Table A7). It was directly mentioned by representatives of certification bodies and certification schemes that there is already an overwhelming number of sustainability criteria and indicators available. According to the interview results, the recent challenge and demand is not the development of completely new criteria and indicators, but the (a) adaptation and more precise communication of the existing ones, as well as (b) a harmonisation of the actual operationalisation of the existing criteria by the certification schemes and certification bodies. These statements are slightly contradictory compared to the topic presented in the previous section (see Section 3.2.1). Reason is mainly, that the topics were identified as relevant by experts with different backgrounds. While additional demand for research regarding criteria and indicators was mentioned mainly by experts from scientific institutions, the topic of harmonisation and a better operationalisation was brought up by representatives of certification bodies and certification frameworks.

During our analysis and the interviews, we found that, even though a number of certification frameworks cover the same principles or criteria, the methods or procedures of applying and assessing those criteria in practical audits can differ significantly between the frameworks and even between the certification bodies which conduct auditing processes on behalf of the same certification framework. This can be problematic, especially for criteria whose assessment in an actual auditing process is time or resource intense, or which can be linked to a price benefit for the certified product. An example is the calculation of the GHG mitigation threshold value for liquid biofuels in the EU market. This is a mandatory sustainability criteria under the RED framework. Since in some countries (e.g., in Germany), the outcome of the GHG mitigation threshold assessment is not only relevant for the general acceptance of a biofuel in the market, incentives such as a GHG-related biofuel quota might result in price benefits for additional GHG savings beyond the threshold value of the RED. As a consequence, there is a strong incentive for producers to optimise the GHG footprint of their biofuel. This optimisation process might also involve possibilities which do rather stem from the GHG calculation methodology itself than from an actual optimisation of the process value chain for biofuel production [37–39]. Interestingly, one would expect that especially the wide availability of detailed regulations and rules (e.g., ISO standards or the calculation framework for biofuels as defined in the EU renewable energy directive 2009/28/EC and the related communications) for the calculation of GHG mitigation effects from bio-based materials would help to harmonise the actual calculation procedure. However, even in the highly regulated (with regard to the GHG calculation rules) sector of biofuels, significant differences do exist between the different certification schemes and the certification bodies implementing the respective rules and guidelines. While the general methodology for GHG emission calculation (especially with regards to system boundaries, characterisation values, allocation rules, etc.), as well as comparator values for a determination of mitigation values, are clearly defined, differences in upstream emission factors or definitions of by-products or waste materials can lead to significant differences in results.

Demand for more harmonisation in the actual operationalisation of sustainability criteria and indicators seems to be relevant, mostly in business-to-business markets and sectors with a high degree of regulation (such as the biofuels sector). With the introduction of the RED, the EU commission has created a regulated market for biofuels with a set of mandatory and binding sustainability criteria. Producers of biofuels need to proof that they meet the criteria the RED criteria. Several certification frameworks were developed and recognised by the Commission since the implantation of the RED. These frameworks have implemented the RED criteria into their specific guidelines and auditing processes. However, as described above the frameworks (in this case, the frameworks addressing the biofuels sector) analysed in this study differ not only with regard to the overall comprehensiveness of their criteria and indicator sets, but also with regards to the point how the same criteria are being operationalised and implemented between the different certification frameworks. There are various examples for sustainability criteria that are applied and operationalised differently between the existing frameworks. During our expert interviews, the following criteria were explicitly mentioned in this regard:

- GHG mitigation thresholds or GHG emission calculations,
- the definition and implementation of core labour standards (e.g., based on ILO principles),
- Guarantee of no deforestation after a certain cut-off date,
- Legality of sourcing,
- Land use rights.

During our interviews, especially the representatives of more advanced (basically meaning certification frameworks with more comprehensive criteria sets) certification schemes pointed out the importance of creating a level playing field with regards to the actual operationalisation of the criteria between the existing frameworks. One of the most important reasons for the existing differences in the operationalisation of criteria between certification frameworks can be found in the basic nature of the applied methodologies for criteria assessment. While a huge number of criteria sets, indicators and methods for sustainability assessment are currently available, most of these elements were developed for scientific purposes. For certification practises, they need to be simplified, robust, transparent clear and applicable even if limited data and resources are available. This means that often additional effort for a transfer of existing scientific methodologies into certification practice (often by simplifying the initial methodologies and by making them more robust and useable) is necessary. During the interviews, this was identified as one of the main barriers for the implementation of new scientific methods in the actual practice of sustainability certification. To address the gaps identified under this topic, work on the legal framework and additional guidance/recommendations regarding the technical application of the various sustainability criteria in auditing practice is necessary.

3.2.3. Legislation and Consensus for Minimum Criteria in All BBE Sectors

While the point regarding the harmonisation of the actual operationalisation between certification schemes can be considered a horizontal issue, which is becoming relevant especially in markets with stronger legislations regarding mandatory sustainability criteria (e.g., the EU biofuels market), another important issue mentioned was the lack of a level playing field regarding the general sustainability requirements and consequently, sustainability certification practices across the various sectors of the BBE. During the interviews, it was mentioned several times, especially by experts from policy and industry, that instead of developing new criteria, it might be more important to harmonise the existing criteria and requirements for sustainability certification across the various sectors of the BBE in the EU. One of the experts stated, for example: "it is less important to introduce additional criteria, it is more important to mainstream sustainability requirements to all kind of biomass production. We then might need equivalent sets for production from Agriculture, Forestry, waste management, creation in laboratories and all kinds of technical reactors, etc." (see Appendix C, Table A8 for more details). The point addressed here is different from the topic raised under Section 3.2.2. While the previous topic was basically about harmonising the understanding, interpretation and implementation of one sustainability criterion (from legislation) between different certification frameworks, the point addressed here refers to the general level of sustainability requirements and legislation across the different BBE sectors.

Currently, sustainability certification in the EU BBE is characterised by sectors with and without legally binding sustainability criteria; as a direct consequence, we can observe a number of effects such as:

- Leakage effects (compare Section 3.2.4),
- Missing compatibility between the existing frameworks (e.g., in the sense of meta-stands, i.e., sustainability certification schemes recognising on another) for certification and consequently:
- Missing harmonisation and standardisation activities.

The issue of potential leakage effects is related to topics such as indirect land use change or food security risks. The rationale behind is that due to the different regulations and the bindingness of sustainability criteria between the different sectors of the BBE, pressure, e.g., regarding land resources and hard criteria such as the definition of "no-go-areas", as under the RED it could be shifted from sectors with strong mandatory sustainability requirements in their respective field to sectors with no mandatory sustainability requirements (all sectors other than the sector of biofuels for transportation). Furthermore, these differences in the regulatory framework lead to substantial differences regarding the principles, criteria sets and indicators in certification frameworks between the various sectors of the BBE. As a consequence, compatibility and mutual acceptance between existing frameworks from different BBE sectors are often missing. This can lead to additional burdens and barriers for market actors in the BBE. While to some extent a differentiation of the certification frameworks seems to be desirable, the definition of a consensus for minimum sustainability criteria for all sectors of the BBE would be an important step to reduce negative leakage effects and unnecessary administrative burdens for market actors. Interestingly, this point was brought up especially by experts from industry and policy.

Another important aspect is that a growing BBE, with an increasing cascading use of biomass and bio-based products, might also require an increasing cross-sectoral compatibility and recognition between the different certification frameworks of the various sectors of the bio-based economy. Potential solutions to overcome this barrier are meta-standard-frameworks, which consist of certification frameworks recognising each other. This makes it possible to combine certificates from different frameworks for different parts of the value chain to receive a certification over the complete value chain based on the overarching criteria and indicators of the "meta-standard-framework" (which would be a consensus or an expression of minimum criteria recognised by all certification frameworks recognised under the meta-standard).

3.2.4. Leakage Effects from EU BBE Policies

The introduction of mandatory sustainability requirements in the RED has addressed a number of pressing and highly relevant sustainability issues related to a large-scale rollout of biofuels for the EU transportation sector. As a consequence, some of these pressing sustainability issues such as the conversion of land with high carbon stocks such as forests into cropland being shifted to other sectors of the bio-based economy which are not directly addressed by mandatory sustainability requirements. These leakage effects, which are related to different topics such as indirect land use change, carbon debt or food security risks are still intensively discussed in the EU bioenergy sector [12,40,41].

During the interviews conducted, especially experts from science and policy expressed the opinion that the future success of a BBE will largely depend on the solution of these major sustainability issues (please see Table A9). However, it seems to be widely recognised that a solution to this problem cannot be created solely and isolated out of the biofuel sector [42,43]. There seems to be a risk that, as a consequence of these unsolved issues, the general trust in the development of a sustainable BBE could be tarnished.

An actual solution to problems arising from iLUC and/or a lack of good governance and an unsustainable management of natural land resources such as forests should be based on local solutions (e.g., [42]) and could not come from certification activities, which are focussing on specific sectors and parts of BBE value chains alone. The EU commission is currently organising the negotiations for the development of a RED recast for the 2021–2030 timeframe [15]. Currently available drafts of the document indicate that the sustainability criteria implemented for liquid biofuels will be expanded to electricity and heat production based on biomass. The proposal has two main components: it focuses on limiting the maximum share of food-and-feed-based biofuels by introducing a cap and establishes a sub-target for fuels that are deemed to bring GHG reductions to the transport sector–renewable; electricity included. This is an important first step for the development of a level playing field regarding minimum sustainability criteria for all biomass uses under the EU bio-based economy. Consequently, land use implications pulled by EU BBE activities would be direct land use change implications associated with the responsible sector of the BBE.

Furthermore, recent scientific activities aimed to identify key parameters linked to risks for issues such as iLUC change or food security and to translate these parameters into criteria and indicators suitable for sustainability certification [44–48]. This can be an important step to derive applicable and specific action points to mitigate iLUC risks. While iLUC risks associated with an increasing demand for bio-based materials are often being discussed and quantified using different modelling approaches (e.g., [42,49]), mitigation measures to reduce iLUC effects on market actor level can often not be developed just based on iLUC modelling work [42]. Thus, certification activities could complement the existing research on iLUC assessment. Examples can be found in initiatives aiming at the certification of "low iLUC risk biomass" (e.g., when produced from degraded or abandoned land or from yield increases) [44–48]. Furthermore, the STAR-ProBio project has delivered results regarding the identification of key parameters driving LUC risks from BBE value chains [48]. Future activities will aim at the development of iLUC risk indicators to be used for sustainability certification.

3.2.5. Sustainability Assessment and Certification for New Innovative, Inter-Sectoral Products

The general perception of the BBE is to some extent characterised by an expected high potential for innovation [2,5]. New, innovative products and bio-based resources are expected to be an essential part in the future BBE. Some of these resources and products could even represent interesting links between fossil industry sectors and sectors of the bio-based economy [50,51]. One example could be the utilisation of fossil-based carbon dioxide from power plants to produce algae or power-to-x-products which are then subsequently used in different forms and applications. With regards to the sustainability certification of these products, a number of new challenges and questions (e.g., system boundaries, sustainability criteria, allocation, etc.) for certification activities was brought up during the expert interviews (please see Table A10).

Due to the currently insignificant market relevance of these new products or feedstock, nearly no blueprints for sustainability assessments but also for sustainability certification do exist so far:

- for algae or bacteria production and for
- CO_2 capture (e.g., from air or power plants) and (e.g., for PtX).

3.2.6. EOL in Existing Principles and System Boundaries

The field of sustainability certification in the EU BBE has, in recent years, been driven largely by the developments in the bioenergy sector [52]. Certification in this sector aims mainly to address major issues and topics that are often highlighted in the debate about the general sustainability of bioenergy (e.g., environmental and social issues related to the production or the supply of the biomass used for energy production). Consequently, our analysis of the criteria and indicators currently used showed clear foci on criteria related to sustainable production of biomass as well as on the processing technologies used for the production of energy carriers, materials or other products from biomass. Contrarily, especially EOL scenarios for bio-based products are not adequately reflected so far. With a growing bio-based economy and increasing cascading use effects, this aspect could rapidly gain more importance in the future.

Generally, criteria aiming to address sustainability aspects related to different forms of after-use-phases criteria are only sporadically used so far. Examples which could be found are criteria such as: minimum recycled content in a product, implemented waste management, intended cascading use, etc. (see criteria database [24]). Input regarding this topic received during the expert interviews is described in Table A11.

3.2.7. Traceability of Sustainability and Certificates along the Value Chain

Finally, in addition to the previous points, which mostly address specific aspects related to sustainability criteria for certification, the aspect of traceability of sustainability characteristics (i.e., specific characteristics such as, for example a GHG emission information of products from the different processes of the value chain) and certificate information throughout the complete value chain of a bio-based product was mentioned by several of the experts interviewed (see Table A12).

The point of achieving a consistent, reliable and trustworthy traceability throughout the entire value chain would be an important step to reduce the potential for misuse of certificates, incorrect claims and to increase the overall integrity of sustainability certification. Consequently, pressing sustainability risks such as deforestation or misuse due to false claims (e.g., waste declarations, etc.) could be at least partly addressed by increasing the availability of consistent and complete chains of information.

For this purpose, future activities in the EU BBE should aim to establish instruments to transport sustainability characteristics through the supply chain, ideally independently from the issuing certification framework. This would support the development of meta-standards based on a mutual recognition of different certification frameworks. As a consequence, market actors could use different certification frameworks for different parts of the value chain (e.g., one framework certifying the biomass production process and another certifying the conversion process) under one meta-standard. Potentially database solutions on national (e.g., concepts such as the German NABISY database), or ideally on EU level could be one possible solution for this problem. However, this would require the existence of greater compatibility and established links between the various frameworks as well as general requirements leading to a (possibly mandatory) use of overarching database solutions. A prominent example on the member state level is the Nabisy database, which includes certificate information for all biofuels counted towards the national quota in Germany. The use of this database allows auditors to check and control specific claims made by market actors involved in the chain of custody. According to the currently available drafts for a recast of the RED after the 2021 timeframe, the EU commission recognises the importance of the registries and databases as tools to trace sustainability characteristics in a trustworthy manner. The current proposal for the new RED aims at the development of a European register for biomethane to decrease existing burdens related to international trade and recognition of sustainable biomethane.

3.3. Gaps in Sustainability Standards for Bio-Based Products: General Issues and First Suggestions Derived by Experts Consultation

Our work with standardisation committees and standardisation experts at the European, international and national levels (see also Appendix D) led to valuable inputs, which can be organised into two categories:

- Suggestions related to improving the existing EN 16751 standard (Europe's most prominent sustainability assessment standard for bio-based products),
- Standardisation issues beyond the scope of EN 16751.

Determining which suggestions are related to the horizontal standard EN 16751 described below and which address additional topics was important in this regard. As an example, most of the suggestions we received regarding EN 16751 referred to issues for which a common, horizontal solution for all groups of bio-based products is unlikely. The responses also focussed on LCA, a topic covered by the standard EN 16760. LCA is also important for comparing bio-based products with fossil-derived ones, and this is currently also not covered by the narrow scope of EN 16751. Therefore, we also consider this topic separately.

3.3.1. Suggestions Related to EN 16751

The standard EN 16751 Bio-based products–Sustainability criteria was developed by CEN TC 411, whose main goal is to provide standards for horizontal aspects of bio-based products [53]. The IEC defines a horizontal standard as a standard on fundamental principles, concepts, terminology or technical characteristics, relevant to a number of technical committees (see IEC Guide 108). These standards also have the purpose of avoiding duplication of work and contradictory requirements [54]. According to ANSI, these general, basic standards are to distinguish from vertical, also called application standards.

As the focus of TC 411 is on horizontal standards, it has no intention to present threshold or default values. This is left to specific product standards or political decisions [53]. In this context, we also learned by our interviews that experts distinguish between the system-based approach of horizontal standards and the more performance-based approach in standard setting.

Experts at standardisation committees suggested to develop assessment methods and thresholds for the criteria of EN 16751. Threshold values can be used to check conclusively whether a specific indicator was fulfilled based on measurable and quantifiable parameters. Based on the product-independent focus of this horizontal standard, opportunities to define common evaluation methods and corresponding thresholds are limited. Nevertheless, there are potential exemptions. There are issues that are relevant for the assessment of bio-based products in general, and there are also requirements which can be specified easier than others.

Depending on the specific nature of a criterion and its respective indicator(s), there are areas in which fulfilment requires the definition of threshold values. In other areas, fulfilment can be proven with a simple yes/no condition. Requirements, which can be easily specified by such a simple condition, refer in particular to social and economic criteria, for example to the requirement "no child labour". Therefore, it is important to check the potential to define requirements on a horizontal level whenever possible. Furthermore, suggestions for common yes/no conditions regarding the use of grassland and forests are described at the end of this Section. These are especially relevant for the first life cycle stage of bio-based products.

During our information exchange with the experts interviewed, it was also communicated that economic sustainability is undervalued within EN 16751. In addition to this, our interview series with standardisation experts showed that the adoption of EN 16751 in product-specific standards is not as it should be. According to the understanding of an expert, only a bio-based solvent standard (drafted as EN 16766) and a surfactant equivalent (CEN/TS 17035) contain a requirement to use EN 16751 to show sustainability characteristics in Europe. More standards with a similar requirement would contribute to the promotion of sustainability criteria and stimulate their application. Closer collaboration with the TCs working on vertical standards for bio-based products is suggested.

3.3.2. Standardisation Issues beyond the Scope of EN 16751

This sub-section summarises eight additional suggestions for further standardisation work by standardisation committees and their representatives.

1. Provide assessment methods and thresholds for the criteria of EN 16751, if they cannot be defined via horizontal standardisation

As mentioned briefly before, the development of a certification scheme for sustainable bio-based products requires thresholds based on specified assessment methods. Experts interviewed emphasised that the specification of the assessment methods themselves had also not been carried out yet. Therefore, it is suggested to regard the work on assessment methods and thresholds as a high priority. Considering that the scope of EN 16751 excludes the establishment of "thresholds or limits", additional standards are needed. A specific issue in this regard is also the fact that EN 16751 will not be updated before 2021 due to CEN's review period of 5 years for EN standards. Separate standardisation work might also provide faster results in this regard. A standard providing general assessment methods may be created by the same CEN TC/411, while the TC's scope requires that the specification of thresholds is carried out by other TCs in any case.

2. Provide assessment methods and thresholds for ISO 13065 criteria: if necessary, by additional standards

Specifying assessment methods and thresholds, the sustainability standard ISO 13065, developed by ISO/PC 248, can provide various advantages. The results could also be used as examples and foundation for further standardisation efforts in the field of bio-based products. According to an expert, ISO/PC 248's output should be developed further regarding "criteria that can be evaluated quantitatively and qualitatively and specific levels of sustainability". Nevertheless, he described the challenge of creating levels of sustainability in such an overarching standard that they are suitable for all feedstocks and continents. It was impossible to agree on thresholds in this standard besides providing examples in its annex.

Two solutions are suggested in this regard. The first one refers to the development of very generic indicators which facilitate a basic level of international consistency.

In addition, appropriate coordination between horizontal and vertical standardisation activities is important. Suggesting and assuming the development of additional standards in this regard, an expert added that the creation of a more specific standard based on ISO 13065, e.g., "for a certain fuel from a specific feedstock in certain climatic conditions", could facilitate the determination of such threshold values. Such standards would provide the opportunity to add "a lot more detail and potentially even thresholds".

3. Facilitate a cradle to grave or cradle-to-cradle analysis of bio-based products

EN 16751 has a restricted scope that considers the life cycle stages ("feedstock" and "production" or "cradle" to "gate" only). Advantages of bio-based products further downstream are therefore not recognized. An example, an interviewee illustrates this as follows: "A biodegradable bio-based plastic has an EOL option that emits short cycle carbon only. A petrochemical plastic that is biodegradable contributes to net long cycle carbon emissions as it is decomposed. On the other hand, the energy needed to produce the petrochemical plastic may be much less than an equivalent bio-based plastic. Moreover, if non-renewable energy is used in the production of a bio-based product this may lead to greater (fossil) carbon emissions. This sort of impact (e.g., carbon balance that spans the entire lifecycle) is not currently supported". The interview series led us to suggest the creation of a basic cradle-to-grave standard, which also considers the EOL stage, or, if possible, even a cradle-to-cradle standard, which also considers circular issues of bio-based products. The circular aspect will be also discussed later in this section.

4.   Provide a standard which facilitates comparisons of bio-based and fossil-derived products

As the previous section has partly shown, expert suggestions to specify assessment methods for EN 16751 were linked with comments on the relevance of comparisons with fossil-derived products regarding the three sustainability pillars and entire life cycle considerations. Currently, a comparison between both kinds of products based on EN 16751 is not possible. This standard explicitly excludes fossil content, although, for example, social and economic impacts could be compared relatively easily.

As an example of the advantages of a new standard in this context, an expert mentioned that bio-based solvents can have superior characteristics compared to fossil ones. Further research by the authors identified the solvent Cyrene as a good example for this. It can be directly derived from waste cellulose in two simple steps, therefore having a high stoichiometric biomass utilization efficiency. CyreneTM has demonstrated a similar solvent performance to toxic fossil-derived solvents, whose industrial synthesis involves multiple reaction steps (see [55]).

Several superior characteristics of bio-based products in general have even been highlighted by the European Commission: "(...) higher process efficiency can be obtained (in the production of bio-based products), resulting in a decrease in energy and water consumption, and a reduction of toxic waste. As (bio-based products) are derived from renewable raw materials, (they) can help reduce CO_2 and offer other advantages such as lower toxicity or novel product characteristics (e.g., biodegradable plastic materials)".

An additional specific example for superior characteristics is provided by smart drop-in chemicals. These chemicals are chemically identical to existing ones, but their bio-based pathways provide advantages. Carus et al. 2017 ([56]) uses the term 'smart drop-ins' if at least two of the following superiority criteria apply: the biomass utilization efficiency from feedstock to product is significantly higher compared to other drop-ins, and/or compared to all alternatives, their production requires significantly less energy and/or their time-to-product is shorter due to shorter and less complex production pathways and/or less toxic or harsh chemicals are used or they occur as by-products during their production process.

These examples show the superior characteristics of bio-based products clearly, both in general, and with respect to specific product groups. Therefore, a standard which facilitates demonstrations of these advantages would promote their market up-take or, as an interviewee formulated it, "promote the market and strengthen (the) trade (of bio-based products)". Nevertheless, experts describe that the comparison of upstream environmental impacts is not suitable, given the very different feedstocks. As a solution, we suggest discussions in CEN/TC 411, together with stakeholders of specific bio-based product groups (for example producers and public procurers) in this regard. If facilitating comparisons with fossil-based products by a horizontal standard appears to be difficult, information exchange on the level of product standardisation is important to create synergies where possible.

5.   Consider iLUC and related issues appropriately by standardisation

There is international recognition that the production of bio-based products instead of fossil-based ones can reduce greenhouse gas emissions and contribute to the adaptation to climatic change. However, as bio-based materials are ultimately obtained from land or sea, additional effects require consideration. These effects can moderate environmental performances and the original purpose of sustainability. iLUC has been defined as an unintentional, negative, displacement effect of commodities in the primary sector such as agriculture causing additional land use changes [48].

Screenings and analysis of documents in the Perinorm database showed that iLUC represents a gap on the level of international standardisation. Likewise, experts, for example from the former ISO/PC 248, highlighted the need for action in this regard. The Dutch standard NTA 8080-1:2015 considers low iLUC risk (see also Section 3.2.4). The requirements of this standard are also used for certification based on the standard NCS 8080:2017 and the Better Biomass certification scheme. Nevertheless, NTA 8080:1 describes specific limitations, for example, concerning new understandings and new issues such as "cascading ILUC" and "carbon debt" (see Appendix C). Based on the input of the interview series, we suggest initiating activities to specify iLUC-related requirements on a European level. As mentioned earlier, CENT/TC 383 plans to make changes to the EN 16214-series on sustainability criteria for biomass for energy use to include the revised standards references to the 2015 iLUC Directive modifying both the Fuel Quality Directive (FQD) and the RED. Therefore, this work could be a starting point for standardisation activities for bio-based products, including the determination of assessment measures and thresholds.

6.   Develop standards that provide guidance on social and economic LCA

According to expert opinion (see Appendix C; Tables A1–A4), social LCA (S-LCA) would bring the assessment of social sustainability of bio-based products on a par with environmental sustainability. Considering economic LCA by standardisation was also suggested. In addition to this, an interview of the interview series in the technical committees emphasised the need for a better link between EN 16751 and LCA standards, referring to existing LCA standards and the relation to future ones as well.

7.   Create standards for the circular economy

The European Commission is aware of the importance of the Circular Economy and has developed, for example, the Circular Economy Action Plan [57]. In line with expert suggestions (see Table A5), work on standards in this area has to be regarded as a priority. Appropriate standards should be focused on design aspects of products, promoting products that are designed to be easily refurbished, remanufactured, reused, recycled, biodegraded safely. Specifying the need for action, an interviewee of the general expert interview series referred to the need for standardised methods to measure circularity characteristics. The British standard BS 8001:2017 Framework for implementing the principles of the circular economy in organizations might be used as a starting point in this regard.

8.   Standardise sustainability criteria for bio-based polymers and lubricants

Most suggestions of the experts in both parts of our analysis referred to standardisation issues of bio-based products in general, not to specific product groups. Bio-based polymers and lubricants, for which product-specific standards were suggested, were an exemption in this regard. The bio-based polymer turnover was about €13 billion worldwide in 2016. Nevertheless, they represent only a share of 2% of the global polymer market and a significant increase in their production capacity is forecasted [58]. Likewise, the market of bio-lubricants is growing significantly, from over 630 kilo tons in 2015 to expected 1115 kilo tons by 2024, growing at 6.9% Compound Annual Growth Rate (CAGR) from 2016 to 2024 [59]. These quantities correspond a market size of $2.92 (€2.47 (Exchange rate from 30 November 2017)) billion by 2024. The specific need for action regarding the development of sustainability criteria for both kinds of products was recognised by experts. As mentioned earlier, not only do criteria have to be developed, but assessment methods and thresholds as well.

3.4. Summary of Gaps Identified and Potential Links to Future Research Acticities

The gaps identified and described in Sections 3.1 and 3.2 refer to different sustainability pillars. To summarise the findings and prepare subsequent assessment steps, they were clustered according to a differentiation between general, environmental, social and economic criteria as well as specific ones. This selection might help to structure and prioritise action items for further research work. This information is included in Table 4. Furthermore, it was specified whether the gaps refer to EOL or LCA topics and whether they also address regulatory issues.

Table 4. Structuring the action items for the further work in the context of this report.

Gap and Sustainability Pillar	EOL Topic	LCA Topic	Topics for Work on Regulatory Requirements
General criteria			
Bio-based content and recyclability/biodegradation	x	x	
Environmental criteria			
GHG mitigation thresholds/GHG emission calculations		x	
(indirect) land use change GHG emissions		x	x
SO_2 equivalents		x	
PM10		x	
Guarantee of no deforestation after a certain cut-off date			x
Economic criteria			
Legality of sourcing			x
Land use efficiency			
Secondary resource efficiency	x		
Functionality (Output service quality)		x	
Levelised life-cycle cost		x	
Social criteria			
Core labour standards			x
Risks for negative impacts on food prices and supply			x
Land use rights			x
Specific issues			
Algae or bacteria production	x	x	x
CO_2 capture	x	x	
EOL scenarios (cascading, recycling, etc.).	x	x	
EOL criteria, e.g., minimum recycled content in product, implemented waste management, intended cascade use	x		
Cross compatibility & recognition between the certification systems	x		x

4. Discussion

In recent years, different strategies, policies, certification frameworks, and standards to assess bio-based products have been developed in Europe and worldwide. In addition, the adoption of the European Bioeconomy Strategy in 2012, various additional actions have also shaped the path of the future BBE in Europe.

Nevertheless, need for action has remained, in particular regarding the sustainability assessment of bio-based products. This paper presented first results regarding existing gaps in sustainability certification and standardisation. In total, we analysed approx. 100 certification frameworks (45 in a detailed in-depth analysis), conducted interviews with 20 international experts, and analysed a wide range relevant standards and activities of standardisation committees in the BBE sectors. Information on the sustainability certification landscape and the different schemes was summarised in a database (see [24]), which is publicly available.

Interviews with experts form the starting point for the analyses of research demand regarding certification and standardisation activities. Several topics for future research demand were identified (e.g., to address specific products, such as, for example, bio-based polymers and lubricants) throughout this process.

This analysis revealed an impressive number of existing certification frameworks, criteria, indicators and applicable standards. In particular, experts from certification frameworks and certification bodies stressed the importance of improving the existing work instead of creating completely new criteria or even a completely new certification scheme. Quite contrarily, experts from science, representing a more holistic understanding of sustainability, addressed a number of specific gaps regarding principles, criteria or indicators currently not sufficiently addressed in sustainability certification and standardisation. The assessment of current sustainability certification revealed seven topics with a demand for future research and development (gaps and weaknesses in criteria and indicator sets, harmonisation in criteria assessment and operationalisation, legislation and consensus for minimum criteria in all BBE sectors, leakage effects from EU BBE policies, new innovative, inter-sectoral products, EOL, traceability of sustainability and certificates along the value chain).

In addition to the general standardisation needs identified during expert interviews with standard bodies, five specific gaps regarding general criteria, criteria for the three sustainability pillars, and several specific issues mentioned in Section 3 built the foundation for the analysis of standardisation options and various standardisation recommendations.

Furthermore, specific recommendations related to EN 16751 standardisation activities could be drawn. Of fundamental importance is that the standard is adopted for certification and that certification bodies adjust their schemes appropriately. By the time of our analysis, it was too early to observe changes in this regard. Based on the standard, specified indicators and assessment measures, usable for auditing processes have to be developed. In line with this, thresholds are needed. The work on indicators and assessment methods requires decisions, which solutions can be provided on the level of this horizontal standard and which issues must be addressed by specific product standards. Threshold issues outside the scope of EN 16751 and TC 411 make appropriate coordination with other TCs necessary.

Recommendations to address additional standardisation gaps of the sustainability assessment for the bio-based economy could also be drawn. Comparisons of bio-based and fossil-derived products should be facilitated, as well as analyses, which consider at least the LCA stages, cradle to grave. To provide LCA criteria, methods for social and economic LCA also need to be specified. In line with land use practice in the area of bio-energy, the protection of forests and grassland should be ensured. Regarding iLUC and carbon debt, appropriate measures have to be developed and considered by appropriate standardisation activities. Specifications for specific EOL issues are needed as well.

Based on the horizontal standard EN 16751, suitable product standards have to be created. Specified requirements and thresholds are needed in the relevant sub-areas of the bio-economy, while new product areas also have to be considered appropriately. Regarding specific bio-based products, standardised sustainability criteria and thresholds are needed, for example, for bio-based polymers and lubricants. Bio-gasification is an additional area requiring further exploration regarding standardisation needs. Finally, the standardisation needs of the emerging area circular economy demand specific considerations to make cradle-to-cradle analyses of bio-based products possible.

5. Conclusions

Future research activities in the context of sustainability assessment and certification for bio-based materials can build on a significant amount of existing certification frameworks, criteria sets, tools, and standardisation work. It is an important challenge to adapt and improve the existing building blocks from various sectors of the BBE to be used in a robust, reliable and trustworthy certification approach for the future BBE in the EU.

The currently existing criteria and indicators cover a wide range of sustainability aspects. However, we found that a number of principles and topics have not been adequately reflected so far. The respective criteria listed on the previous page refer to all sustainability pillars of bio-based products.

In addition to the question of additional indicators, criteria and standards, the actual operationalisation, application and implementation in certification practice seems to be carried out very differently between the existing frameworks. In practice, this can lead to differences in quality but also in price differences for the actual certification process. To address these points, support for the practical implementation of tools for sustainability assessment is necessary.

Interestingly, our interviews with experts have shown individual perspectives and expectations regarding the future development of certification and standardisation activities in the EU BBE. Generalising, interviewed experts from science and policy have tend to focus on aspects regarding the development of more holistic and comprehensive criteria and indicator sets (i.e., gaps regarding criteria and indicators) and the elimination of leakage effects (e.g., iLUC effects) from a missing level playing field regarding the sustainability requirements between the different BBE sectors. The latter has also been brought up by experts from industry, since this issue might also create market barriers and distortions hindering the development of meta-standards with mutual recognitions of different certification frameworks across the different BBE sectors. In addition, and quite contrarily to the statements received regarding gaps in criteria sets, experts from certification schemes and certification bodies did not see demand for additional criteria and indicators. Instead, they addressed demand for research regarding assessment tools and guidelines which would allow for a greater harmonisation of the actual implementation of sustainability criteria in certification practices.

Sustainability certification can be considered as one important tool to implement targets regarding a sustainable development from public or private sector and to increase and preserve the general societal acceptance of the BBE. In addition, sustainability assessment tools will allow providing evidence of the claimed better environmental superiority of bio-based products, as requested by policy makers. Currently, some of the major sustainability issues related to advanced sectors of the BBE (e.g., the bioenergy sector) can so far not be directly addressed with the existing sustainability certifications. For example, aspects such as indirect land use change, carbon debt, or food security risks can often not be measured directly. In fact, they have to be modelled [42]. To complement existing modelling activities for the quantification of iLUC risks associated with the development of a BBE, additional research is needed to develop robust criteria and indicators for an iLUC risk assessment during on-farm audits and certification. Secondly, the introduction of (mandatory) sustainability criteria (including criteria related to the protection of land with high carbon content and a high biodiversity value [11]) for all sectors of the BBE would help to create a level playing field and to reduce associated leakage effects.

In general, certification schemes are available for products which are close to market. On the other hand, new products (e.g., from algae) are not appropriately considered by the existing sustainability assessment frameworks.

Finally, bio-based value chains can be long and complex and involve a significant number of market participants, producers, suppliers, and users. For a growing economy, tools, which allow tracing information on sustainability characteristics and certificate parameters, need to be developed. Database solutions can help to solve this problem.

The potential research demand identified with this study covers a wide range of topics to be addressed by different stakeholder and scientific disciplines. The topics identified can be taken up by research consortiums and projects. A number of the topics addressed in this paper will be subject to the research activities of the EU H2020 project STAR-ProBio, which aims to develop tools for sustainability certification.

Author Contributions: Conceptualization: S.M., S.W., L.L. and D.T.; Data curation: D.M., B.S., S.W. and L.L.; Investigation: S.W., D.M., L.L. and S.M. Methodology S.M., D.M., S.W. and L.L.; Supervision, D.T.; Validation, S.W. and B.S.; Writing–original draft, S.M., S.W., D.M., L.L. and B.S.; Writing–review & editing, D.T. and B.S.

Funding: This research was funded by the European Union's Horizon 2020 research and innovation programme under grant agreement No. 727740

Acknowledgments: The contents of the paper are a part of the findings of the project STAR-ProBio. STAR-ProBio has received funding from the European Union's Horizon 2020 research and innovation programme under grant agreement No. 727740. Re-use of information contained in this document for commercial and/or non-commercial purposes is authorised and free of charge, on the conditions of acknowledgement by the re-user of the source of the document, not distortion of the original meaning or message of the document and the non-liability of the STAR-ProBio consortium and/or partners for any consequence stemming from the re-use. The STAR-ProBio consortium does not accept responsibility for the consequences, errors or omissions herein enclosed. This document is subject to updates, revisions and extensions by the STAR-ProBio consortium. Questions and comments should be addressed to: http://www.star-probio.eu/contact-us/."

Conflicts of Interest: The authors declare no conflict of interest.

Appendix A—Information on Analysed Certification Frameworks

Part of the gap assessment for this paper was based on a comprehensive review of existing sustainability certification frameworks. In a stepwise approach, we identified sustainability frameworks relevant for the BBE and selected ~half of the schemes identified to be analysed in an in-depth assessment. Because of the first step, a matrix including the currently available and relevant schemes for the sustainability certification in the EU BBE was produced. The matrix is presented on the following pages. One of the products of our detailed analysis of the currently available certification frameworks is a database, including all criteria and related indicators.

This database is available at: http://www.star-probio.eu/wp-content/uploads/2017/12/Star-ProBio_certification_criteria.xlsx.

Sector		Label	Initiative	Certification scheme	Name	Scope			Criteria		
						Geographic	Feedstock	Supply Chain	Social	Environment	Economic
Bioenergy	Liquid biofuels	x		x	International Sustainability & Carbon Certification (ISCC)	global	multiple	full	x	x	x
		x		x	REDcert EU	Europe (+Ukraine,Belarus)	multiple	full		x	
		x		x	Bonsucro EU	global	Sugar cane	full	x	x	x
		x		x	Roundtable on Sustainable Biomaterials EU RED (RSB EU RED)	global	multiple	full	x	x	
		x		x	Biomass Biofuels voluntary scheme (2BSvs)	global	multiple	full		x	x
		x		x	Red Tractor Farm Assurance Combinable Crops & Sugar Beet (Red Tractor)	UK	cereals, oilseeds, sugar beet	until first feedstock delivery point		x	
		x		x	Scottish Quality Farm Assured Combineable Crops Limited (SQC)	north GP	cereals, oilseeds	until first feedstock delivery point		x	
		x		x	Roundtable on Sustainable Palm Oil RED (RSPO)	global	multiple	full	x	x	x
		x		x	Better Biomass NTA 8080	global	multiple	full	x	x	
		x		x	HVO Renewable Diesel Scheme for Verification of Compliance with the RED sustainability criteria for biofuels	global	multiple	full		x	
		x		x	Gafta Trade Assurance Scheme	global	multiple	farm gate to first processor		x	
		x		x	KZR INiG System	Europe	multiple	full		x	
		x		x	Trade Assurance Scheme for Combinable Crops	UK	multiple	farm gate to first processor		x	
		x		x	Universal Feed Assurance Scheme	UK	multiple	farm gate to first processor		x	
				x	Certification System adressing Indirect Impacts of Biofuel (CIIB)	EU+	multiple	farm gate to first processor		x	

Figure A1. *Cont.*

Category				Scheme	Region	Feedstock	Scope			
Solid biofuels	X			X	ECOFYS - Methodologies for the identification and certification of Low ILUC risk Biofuels	EU+	multiple	farm gate to first processor	X	X	
	X			X	Austrian Agricultural Certification Scheme (AACS)	Austria	multiple	cultivation to production		X	X
	X			X	Sustainable Biomass Partnership (SBP)	global	woody biomass	from cultivation to energy production	X	X	X
	X			X	Green Gold Label (GCL) S1	global	woody biomass	full	X	X	
	X			X	DINplus Short rotation coppice sustainably grown according to DIN EN 16214-3	global	SRC wood	cultivation	X	X	X
	X			X	Nordic Ecolabeling (SWAN)	Denmark, Finland, Iceland, Norway, Sweden global	woody biomass		X	X	X
	X			X	ISCC PLUS	global	multiple	full	X	X	X
Biogas		X		X	CIB - Biogasdoneright	Italy	multiple	production process, product		X	X
		X			Green Gas Certification Scheme (GGCS)	UK	multiple	production to use		X	X
Heat/Power	X		X	X	Global Bioenergy Partnership (GBEP)	global	multiple	full	X	X	X
	X			X	nature made basic	CH	multiple	energy production and delivery		X	
	X			X	nature made star	CH	multiple	energy production and delivery		X	
	X				OK-Power	Germany	multiple	power production		X	
	X			X	Grüner-Strom-Label (Green-Power-Label)	Germany	multiple	power production to distri-bution		X	
	X				Grünes-Gas-Label (Label for sustainable biomethane products)	Germay	multiple	cultivation to distri-bution	X	X	X

Figure A1. *Cont.*

			woody biomass, agricultural products	Biomass production to transportation power production							
Forestry		Laborelec certification system (LBE)	Belgium	multiple			x		x		x
		Austrian Ecolabel	Austria	multiple	x			x	x		
		Council on Sustainable Biomass Production (CSBP)	USA	multiple	conversion of biomass to bioenergy		x		x	x	x
		Green-e The global bioenergy partnership sustainability indicators for bioenergy	global	multiple	full	x		x	x	x	x
	x	Testing framework for sustainable biomass	global	multiple	full	x		x	x	x	x
		Responsible Biomass Certification scheme (RBC scheme)	global	multiple	production and trade of biomass	x	x		x	x	x
	x	Forest Stewardship Council (FSC) CoC	global	woody biomass	full	x	x		x	x	x
	x	Programme for the Endorsement of Forest Certification (PEFC) CoC	global	woody biomass	full	x	x		x	x	
	x	Certified Composite (ECC) Eco-Sustainability Standard	USA	wood fiber	full	x			x	x	x
	x	Sustainable Forestry Initiative (SFI)	USA, Canada	forestland	cultivation	x	x		x	x	x
Con-struction	x	DGNB System			from cradle to grave design and procurement stage, post construction stage (optional)	x			x	x	x
	x	Green Building Rating System BREEAM	global			x			x	x	x
		Assessment System for Sustainable Building (BNB)	Germany		from cradle to grave	x			x	x	x
		Minergie	Switzerland, Lichtenstein							x	x

Figure A1. *Cont.*

Category	Subcategory			Name	Region	Product	Life cycle				
		×		Bewertungssystem Nachhaltiger Wohnungsbau (NaWoh) (Assessment system for sustainable house building)	Germany		from cradle to grave	×		×	×
		×		Bewertungssystem Nachhaltiger Kleinwohnhausbau (BNK) (Assessment system for sustainable family house building)	Germany		from cradle to grave	×		×	×
		×	×	Leadership in Energy and Environmental Design (LEED)	global			×		×	
		×	×	Total Quality Building (TQB)	Austria		planning, con-struction, operation	×		×	×
Food	Fair Trade Certification Systems	×	×	Fairtrade-Label - Fairtrade Labelling Organizations International (FLO)	global	crops	full	×		×	×
		×	×	WFTO Guaranteed System	global	crops	full	×		×	×
		×	×	Fair for Life	global	multiple	full	×		×	×
		×	×	NATURLAND fair	global	multiple	full	×		×	×
		×	×	Rapunzel Hand in Hand	global	multiple	full	×		×	×
	Fish Certification Systems	×	×	Marine Stewardship Council (MSC)	global	fish	from fisheries to retailers	×		×	×
		×	×	Aquaculture Stewardship Council (ASC)	global	fish	full	×		×	
		×		Iceland Responsible Fisheries Management (IRFM) Certification Programme	the fisheries of Icelandic vessels within the Icelandic EEZ;	fish	from fisheries to retailers			×	×
		×	×	Alaska Responsible Fisheries Management (RFM) Certification Program	Alaska	fish	fisheries			×	×
		×	×	GlobalGAP Aquaculture Certification	global	fish	full	×		×	×

Figure A1. *Cont.*

Category			Standard	Region	Crops	Supply chain coverage				
Agricultural Products	×		GlobalGAP crops certification	global	multiple	pre-farm-gate				×
	×		Bioland	Germany	multiple	full	×		×	×
	×		Biopark	Germany	multiple	cultivation to distri-bution	×		×	×
	×		demeter	global	multiple	cultivation to processing			×	
	×		Ecovin	Germany	grapes	cultivation to processing			×	×
	×		FAIR'N'GREEN	Germany	grapes	full	×		×	×
	×		EU organic farming logo	EU	multiple	full	×		×	×
	×		Gäa	Germany, Italy	multiple	full	×		×	×
	×		NATURLAND Lebensmittel (natureland food) Sustainable Agriculture	global	multiple	full	×		×	×
	×		Network/ Rainforest Alliance Certified (SAN)	global	crops	cultivation	×		×	×
	×		IFOAM Standard	global	multiple	cultivation to trade	×		×	×
	×		ISCC PLUS	global	multiple	full	×		×	×
	×		Roundtable on sustainable palm oil (RSPO)	global	palm oil	full	×		×	×
	×	×	ProTerra Standard	global	multiple	full	×		×	×
	×		LEAF Marque Standard	global	multiple	cultivation	×		×	×
	×		UTZ certified	global	cacao, tea, coffee, hazelnut	full	×		×	×
Feed	×		GlobalGAP Compound Feed Manufacturing (CFM)	global	multiple	pre-farm-gate			×	×
	×		Sustainable Feed Standard (SFS)	global	soy	cultivation and trade feed	×		×	×
	×		GMP+ Feed Responsibility Assurance	global	soy, fish meal	production and trade			×	×
	×		DLG certificate sustainable agriculture	Germany	multiple	agricultural production	×		×	×
	×		ISCC PLUS	global	multiple	full	×		×	×

Figure A1. *Cont.*

Sustainability **2018**, *10*, 2455

Category	Standard	Fair Trade Certification Systems	Geography	Coverage	Stage			
Textiles	Fairtrade Textile Standard – Fairtrade Labelling Organizations International (FLO)	×	global	certified cotton, other responsible fibres	full	×	×	×
	Global Organic Textile Standard (GOTS)	×	global	organic fibres	from processing to sale first	×	×	
	NATURTEXTIL zertifiziert BEST (natural textile certified BEST)	×	global	certified organic fibres	processing to sale of organic (100%) fibers and textiles	×	×	
	Better Cotton Initiative (BCI)	×	global	cotton	cultivation	×	×	
	CmiA – Cotton made in Africa	×	global	cotton	cultivation to production	×	×	×
	NATURLAND Textil (natureland textile)	×	global	natural fibre	production	×	×	
	EU Ecolabel – fabrics	×	EU, CH, NOR, ISL, TUR	multiple	products	×	×	
	Goodweave International	×	global	rugs and carpets	production process, product	×	×	
	Textile Exchange Organic 100% content standard	×	global	organic fibres	production product	×	×	
Chemicals and Plastics	ISCC PLUS	×	global	multiple	full	×	×	×
	Bioplastic Feedstock alliance	×	global	multiple	full	×	×	×
	Together for Sustainability initiative	×	global	multiple	full	×	×	×
Pharmacy	COSMOS Standard – Cosmetics organic and natural standard	×	global	multiple	production process, product		×	
Materials/Products	CRADLE TO CRADLE CERTIFIED PRODUCT STANDARD	×	global	multiple	production process, product	×	×	

Figure A1. *Cont.*

			x		INRO Nachhaltigkeitskriterien für die stoffliche Biomassenutzung (sustainablity criteria for material use of biomass)	global	agricultural biomass	cultivation	x	x	x
		x		x	EU Ecolabel - tissue paper	EU, CH,NOR,ISL, TUR	multiple	products		x	
		x		x	Nature Care Products Standard	global	multiple	products		x	x

Figure A1. Overview on certification frameworks in the EU BBE.

Appendix B—Interview Guide/Questionnaire

STAR ProBio

Interview guide on gaps in current sustainability assessment schemes

Introduction

STAR-ProBio is a three-year project (May 2017 - April 2020) and supports the European Commission in the full implementation of European policy initiatives, including the Lead Market Initiative in bio-based products, the industrial policy and the European Bio-economy Strategy. STAR-ProBio does so by developing sustainability assessment tools for bio-based products, and by developing credible cases for bio-based products with the highest actual market penetration and highest potential for the future markets. STAR-ProBio integrates scientific and engineering approaches with social sciences and humanities-based approaches to formulate guidelines for a common framework promoting the development of regulations and standards supporting the adoption of business innovation models in the bio-based products sector. The aim of STAR-ProBio is to fill gaps in the existing framework for sustainability assessment of bio-based products, and improve consumer acceptance for bio-based products by identifying critical sustainability issues in their value chains. This interview is part of the first work package "Screening and analysis of existing sustainability schemes for the bio-economy".

Questions

1. **Your background**

   Please specify your background and stakeholder group in the bio-based economy and the key product- or process-related areas of your knowledge and activity briefly:

   1.1 Product- or process-related background in the bio-based economy?

   1.2 Stakeholder group?

   ☐ Producer, retailer etc.
   ☐ Consumer
   ☐ Certification and standardization bodies
   ☐ Public procurement
   ☐ Scientist
   ☐ Other, please specify

   1.3 Country?

2. **Barriers regarding the acceptance and demand for bio-based products**

   To what extent does (dis)trust of bio-based products affect their acceptance and demand on the market?

3. **Gaps in sustainability assessment**

    3.1 From your point of view, is there a need for new or additional criteria in sustainability assessment schemes to raise consumer trust in bio-based products, for example regarding product-related, environmental, social or economic aspects?

    ☐ Yes   ☐ No

    If yes: Please continue with the remaining questions of block 3 and block 4.
    If no, please continue with block 5.

    3.2 In which topic area is there a need for new or additional criteria?

    ☐ product-related   ☐ economic issues
    ☐ environmental issues   ☐ other, please specify
    ☐ social issues

    3.3 Please explain the kinds of assessment criteria that are needed in each topic area you selected in 3.2.

Topic area of 3.2	Additional assessment criteria that are needed

    3.4 *Optional:* Please explain your considerations using one or more products as examples.

    3.5 How should compliance with the criteria you proposed be measured? What methods should it be based on?

    Criterion:

    Method:

    (please continue with additional criteria, if relevant)

    3.6 Are there areas where the methodology for measuring specific issues in the sustainability assessment of bio-based products needs to be expanded, improved, or adapted?

    ☐ Yes   ☐ No

    If yes: Please specify what is needed.

4. **Standards to overcome current assessment gaps**

4.1 Do you think that current standards from CEN or ISO or national standards provide content that can be integrated into current sustainability assessment schemes to address the missing assessment criteria you described above?

☐ *Yes* → please continue with 4.2   ☐ *No* → please continue with 4.3

☐ *Depends on the criteria that are needed*
   → specific answers are possible below, you can refer to selected criteria by answering question 4.2 and to other criteria by answering question 4.3.

4.3 Do you think that need for additional criteria could be addressed by:

a) *extending current* European standards and legislation?

☐ *Yes*   ☐ *No*

If yes: Please specify the area(s) and your answer:

Need for additional criteria:

Name of the standard(s) and legislation to be expanded to address(es) the need:

Explanation on how this need can be addressed by standard and legislation extension:

b) *new* European standards and legislation?

☐ *Yes*   ☐ *No*

If yes: Please specify the area(s) and your answer:

Need for additional criteria:

Explanation (potential content of new European standards to address this need):

c) If the answer on a) and b) is "no", what other solution(s) would you suggest?

5. Good practice schemes

   5.1 Are you aware of good practice schemes regarding sustainability assessment?

   ☐ Yes   ☐ *No*

   5.2 If yes: Please specify these schemes and describe their specific benefits.

   5.3 Do you think new assessment criteria could enhance these schemes?

   ☐ Yes   ☐ *No*

   5.4 If yes: Please explain your thoughts.

6. Policy gaps

   6.1 Is there a need for policy documents regarding sustainability assessment in new or emerging areas in the bioeconomy?

   ☐ Yes   ☐ *No*

   6.2 If yes: Please specify what is needed.

   a) on the national level in Member States

   b) on a European level

7. Additional comments

   7.1 Do you have additional comments or suggestions for our project?

   ☐ Yes   ☐ No

   7.2 If yes: Please share your thoughts.

Thank you very much for your time!

Appendix C—Identified Gaps and Corresponding Expert Interview Results

Table A1. Selected summary of questionnaire answers with gaps in sustainability standards for bio-based products at standardisation issues beyond the scope of EN 16751 with emphasis on assessment methods and thresholds for criteria of EN 16751, if they cannot be defined via horizontal standardisation.

Identified Gap	Expert Interview Answer
Assessment methods and thresholds for criteria of EN 16751, if they cannot be defined via horizontal standardisation	"Need for sustainability requirements (especially caps and thresholds) for the bio-based economy as a whole (instead of biofuels (bioenergy) only"
	"Additional requirements and criteria should not be associated with economic burdens for operators."
	"The standard EN 16571 will be revised after 5 years automatically The committee will meet again after 4–5 years and decide whether there is a need for revision"
	"Minimum requirement for sustainable procurement ("EU Bio-based"), for specific products (plastics . . .) also minimum bio-based content ("quota") to be raised over time"

Table A2. Selected summary of questionnaire answers with gaps in sustainability standards for bio-based products at standardisation issues beyond the scope of EN 16751 with emphasis on assessment methods and thresholds for ISO 13065 criteria; if necessary, by additional standards.

Identified Gap	Expert Interview Answer
Assessment methods and thresholds for ISO 13065 criteria; if necessary, by additional standards	"ISO may be the best platform to define overarching requirements and guidelines for the definition of specific criteria. ISO 13065 gives an excellent start for bioenergy which should be mainstreamed for other bio-based products."
	"EN 16751:2016 might need an update taking ISO 13065 into account"
	"Need for new EU Standard similar to the RED with a regulatory framework for all supply chains dealing with biomass to ensure no-deforestation in products entering the European market, also for food, feed and chemicals - more obligatory criteria for products containing biomass especially from outside the EU regarding protection of forests, grassland and consideration of indigenous people and their rights and legality - EU may determine best practice certification/ assessment schemes and make specific forms and contents obligatory in such schemes - EU to monitor existing schemes"
	"RSB, ISCC, FSC® (if wood-based), may be considered to be frontrunner schemes, but they also should benchmark themselves against ISO 13065"
	"Standardization of requirements for sustainable biomass and harmonization of certification/certification systems of sustainable biomass (ISO-Standard). The approach of mandatory sustainability criteria for biofuels in EU could be extended to other bio based products/markets. e.g., packaging, food, feed."
	"Increase awareness about the properties of bio-based products utilising current European standards and labelling systems help to specify and communicate the properties of bio-based products in a clear and unambiguous way, thereby contributing to a level of certainty in the market. There will continue to be an ongoing need for new standardisation and labelling (single, unifying, identifiable) to create market certainty for the bio-based sector."

Table A3. Selected summary of questionnaire answers with gaps in sustainability standards for bio-based products at standardisation issues beyond the scope of EN 16751 with emphasis on cradle-to-grave or cradle-to-cradle analyses of bio-based products.

Identified Gap	Expert Interview Answer
Cradle-to-grave or cradle-to-cradle analyses of bio-based products	"Better traceability along entire supply chain" "Demand for more and standardised LCA assessment of greenhouse gas emissions"

Table A4. Selected summary of questionnaire answers with gaps in sustainability standards for bio-based products at standardisation issues beyond the scope of EN 16751 with emphasis on standardisation of iLUC and related issues.

Identified Gap	Expert Interview Answer
Standardisation of iLUC and related issues	Clear demand for "criteria addressing the leakage effects from policies and regulations for the European market to other parts of the world → iLUC and carbon debt"
	Clear demand for "criteria to address risks and leakage effects regarding to land use change, food security, etc."

Table A5. Selected summary of questionnaire answers with gaps in sustainability standards for bio-based products at standardisation issues beyond the scope of EN 16751 with emphasis on standards for the circular economy.

Identified Gap	Expert Interview Answer
Standards for the circular economy	"Consumers need more clarity on "bio-based" and "biodegradation"; more clarity on how to recycle bio-based products"
	Demand for "standardised methods to measure circular issues"

Table A6. Selected summary of questionnaire answers with gaps in current sustainability certification activities with emphasis on gaps & weaknesses in criteria & indicator sets.

Identified Gap	Expert Interview Answer
Gaps & weaknesses in criteria & indicator sets	Build on results from previous projects such as S2Biom
	"No need/demand for new criteria. However, on the level of indicators, additional work is necessary"
	Demand for criteria related to "resource efficiency, bio-based content and recyclability/biodegradation"
	Demand for criteria related to "bio-based content of products; bio-based label such as USDA bio-preferred"

Table A7. Selected summary of questionnaire answers with gaps in current sustainability certification activities with emphasis on harmonisation in criteria assessment and operationalisation.

Identified Gap	Expert Interview Answer
Harmonisation in criteria assessment and operationalisation	Harmonisation is needed for criteria such as: "guarantee of no deforestation after a certain cut-off date, core labour standards, legality of sourcing, land use rights, GHG emissions"
	Main barrier for more harmonisation is "the lack of a level playing field"
	"Existing criteria should be communicated better and be defined more precisely"
	"Criteria are widely available. It is more a question of the actual implementation and acceptance"
	"There seems to be no need for new criteria. More important to make better use of existing criteria and tools"
	Demand for "Better coverage of core social issues during audit. Better risk analysis"

Table A8. Selected summary of questionnaire answers with gaps in current sustainability certification activities with emphasis on legislation & consensus for minimum criteria in all BBE sectors.

Identified Gap	Expert Interview Answer
Legislation & consensus for minimum criteria in all BBE sectors	"It is less important to introduce "additional" criteria, it seems more important to mainstream sustainability requirements to all kind of biomass production. We then might need equivalent sets for production from - agriculture - forestry - waste management - creation in laboratories and all kinds of technical reactors"
	"In contrast to schemes for bio-energy and bio-fuels, which are acknowledged by the Bio-energy Directive, schemes for bio-based products are entirely voluntary. There is no legal requirement to prove that bio-based products are sustainable"
	"Sustainability requirements (especially caps and thresholds) should be introduced for the bio-based economy as a whole (instead of biofuels (bioenergy) only)"
	"New EU Standard similar to the RED with a regulatory framework for all supply chains dealing with biomass to ensure no-deforestation in products entering the European market, also for food, feed and chemicals - more obligatory criteria for products containing biomass especially from outside the EU regarding protection of forests, grassland and consideration of indigenous people and their rights and legality - EU may determine best practice certification/assessment schemes and make specific forms and contents obligatory in such schemes - EU to monitor existing schemes"
	"We need sustainability criteria for ALL types of biomass cultivation and also for food and feed. Equal changes and equal burdens for all of them."

Table A8. *Cont.*

Identified Gap	Expert Interview Answer
	"Generally, all schemes can be enhanced through new criteria that are reasonable to increase sustainable products - obligatory criteria could help to set a baseline for all assessment schemes requirements - new described EU regulations support the certification process of the whole supply chain and its actors - new described EU regulations also give incentives to countries and producers outside the EU to change their habits to still be able to sell their bio-based products - integrity and credibility of schemes must be improved and monitored"
	Need for "Standardization of requirements for sustainable biomass and harmonization of certification/certification systems of sustainable biomass (ISO-Standard). The approach of mandatory sustainability criteria for biofuels in EU could be extended to other bio based products/markets. e.g., packaging, food, feed."
	"Develop minimum consensus for a sustainability criteria set for the bioeconomy"

Table A9. Selected summary of questionnaire answers with gaps in current sustainability certification activities with emphasis on leakage effects from EU BBE policies.

Identified Gap	Expert Interview Answer
Leakage effects from EU BBE policies	Demand for "criteria addressing the leakage effects from policies and regulations for the European market to other parts of the world → iLUC and carbon debt"
	Demand for "bottom up approaches for product certification based on risk based criteria"
	"Leakage effects have to be addressed by both policies and certification"

Table A10. Selected summary of questionnaire answers with gaps in current sustainability certification activities with emphasis on new innovative, inter-sectoral products.

Identified Gap	Expert Interview Answer
New innovative, inter-sectoral products	"For algae or bacteria production there are nearly no blueprints for sustainability assessments"
	No blueprints and frameworks for "CO_2 from air by PtX"

Table A11. Selected summary of questionnaire answers with gaps in current sustainability certification activities with emphasis on End-of-Life (EOL).

Identified Gap	Expert Interview Answer
End-of-Life (EOL)	"Consumers need more clarity on "bio-based" and "biodegradation"; more clarity on how to recycle bio-based products"
	"Capture downstream assessment characteristics (manufacture and so on) right up to the end of life options"
	"Policy considerations for specific minimum requirements (bio-based content, degradability) especially for plastics"

Table A12. Selected summary of questionnaire answers with gaps in current sustainability certification activities with emphasis on traceability of sustainability and certificates along the value chain.

Identified Gap	Expert Interview Answer
Traceability of sustainability and certificates along the value chain	Demand for "better traceability of sustainability information along entire supply chain, performance also compared to non-bio"
	Demand for "Performance indicators; features related to health, origin of biomass (made in.)"
	"Demand for criteria to express the value of local small value chains, closed nutrient cycles, etc."
	"Use of databases transferring sustainability characteristics across supply chain"
	"Use of objective analysis, e.g., remote sensing tool as proof of no deforestation and transfer of information across supply chain with better traceability tools"
	Demand for an "obligatory database to be publicly available for all assessment/certification systems with detailed information about all granted certificates and holders"

Appendix D—Standards Landscape—Selected Examples

Table A13. Standards landscape–Relevant standards in CEN/TR 16208.

Acronym	Title	Sust	Envi	Soci	Econ
ISO 26000	Guidance on social responsibility	x	x	x	x
NTA 8080	Sustainability criteria for biomass for energy purposes	x	x	x	x
ISO 1404X	Environmental management–Life cycle assessment (Series)	-	x	-	-
VDI 4431 ISO 14064	GHG–Part 1: Specification with guidance at the organization level for quantification and reporting of GHG emissions and removals	-	x	x	-
CEN/TR 15932	Plastics–Recommendation for... characterisation of biopolymers & ~plastics	-	x	-	-
ASTM D 7075	Practice for Evaluating and Reporting Environ. Performance of BBP	x	x	-	-
ASTM D6852	Environmental Profile of Materials and Products	-	x	-	-
PAS 2050	Specification for the assessment of the life cycle	-	x	-	-
BP X30-323	General principles for an environ. communication	-	x	-	-
ISO 15380	Lubricants, industrial oils and related products ...	-	x	-	-
SS 155434	Hydraulic fluids–Requirements and test methods	-	x	-	-
SS 155470	Lubricants, industrial oil and related products ...	-	x	-	-
NF U 52-001	Biodegradable materials–Mulching products–Req. & test methods	-	x	-	-

Table A14. Standards landscape–Further examples.

Acronym	Topic	Sust	Envi	Soci	Econ
Selected additional standards based on STAR-ProBio analyses					
EN 16751	Bio-based products–Sustainability criteria	x	x	x	x
EN 16760	Bio-based products–Life Cycle Assessment	x	x	x	-
CEN/TR 16957	Bio-based products–Guidelines for LCI for the EoL phase	x	x	-	-
Standards in related areas					
ISO 13065	Sustainability criteria for bioenergy	x	x	x	x
EN 16214-3 and 4	Sust. criteria for the production of biofuels & bioliquids for energy appli-cations–Principles, criteria, indicators ... - 3: Biodiversity & environmental aspects ... ; 4: Calc. methods of the GHG emission balance using a LCA approach	x	x		
ISO/FDIS 34101-2	Sustainable and traceable cocoa beans–Part 2: Req. For performance (related to econ., social, and environ. aspects)	x	x	x	x
ASTM E 3066a	Standard Practice for Evaluating Relative Sustainability Involving Energy or Chemicals from Biomass	x	x	x	x

References

1. *National Policy Strategy on Bioeconomy: Renewable Resources and Biotechnological Processes as a Basis for Food, Industry and Energy*, 2014th ed.; Federal Ministry of Food and Agriculture: Berlin, Germany, 2014.
2. Delbrück, S. *Future Opportunities and Developments in the Bioeconomy—A Global Expert Survey*; German Bioeconomy Council: Berlin, Germany, 2018.
3. Staffas, L.; Gustavsson, M.; McCormick, K. Strategies and Policies for the Bioeconomy and Bio-Based Economy: An Analysis of Official National Approaches. *Sustainability* **2013**, *5*, 2751–2769. [CrossRef]
4. European Commission. *Bio-Based Economy for Europe: State of Play and Future Potential Europe*; Report on the European Commission's Public on-line Consultation; Publications Office of the European Union: Luxembourg, 2011.
5. Giuntoli, J.; Cristobal, J.; Parisi, C.; Ferrari, E.; Marelli, L.; Torres de Matos, C.; Gomez Barbero, M.; Rodriguez Cerezo, E. *Bioeconomy Report 2016*; JRC Scientific and Policy Report; Publications Office of the European Union: Luxembourg, 2017.
6. Du Pisani, J.A. Sustainable development—Historical roots of the concept. *Environ. Sci.* **2006**, *3*, 83–96. [CrossRef]
7. Wiersum, K.F. 200 years of sustainability in forestry: Lessons from history. *Environ. Manag.* **1995**, *19*, 321–329. [CrossRef]
8. Komives, K.; Jackson, A. Introduction to Voluntary Sustainability Standard Systems. In *Voluntary Standard Systems*; Schmitz-Hoffmann, C., Schmidt, M., Hansmann, B., Palekhov, D., Eds.; Springer: Berlin/Heidelberg, Germany, 2014; pp. 3–19.
9. FSC International. FSC Database. Available online: https://ic.fsc.org/en (accessed on 25 April 2018).
10. Kraxner, F.; Schepaschenko, D.; Fuss, S.; Lunnan, A.; Kindermann, G.; Aoki, K.; Dürauer, M.; Shvidenko, A.; See, L. Mapping certified forests for sustainable management—A global tool for information improvement through participatory and collaborative mapping. *For. Policy Econ.* **2017**, *83*, 10–18. [CrossRef]
11. *Directive (EU) 2015/1513 of the European Parliament and of the Council of 9 September 2015 Amending Directive 98/70/EC Relating to the Quality of Petrol and Diesel Fuels and Amending Directive 2009/28/EC on the Promotion of the Use of Energy from Renewable Sources*; European Union: Brussels, Belgium, 2015.
12. Mitchel, D. *A Note on Rising Food Prices*; The World Bank: Washington, DC, USA, 2008.
13. Searchinger, T.; Heimlich, R.; Houghton, R.A.; Dong, F.; Elobeid, A.; Fabiosa, J.; Tokgoz, S.; Hayes, D.; Yu, T.-H. Use of U.S. croplands for biofuels increases greenhouse gases through emissions from land-use change. *Science* **2008**, *319*, 1238–1240. [CrossRef] [PubMed]
14. Lin, J. Governing Biofuels: A Principal-Agent Analysis of the European Union Biofuels Certification Regime and the Clean Development Mechanism. *J. Environ. Law* **2012**, *24*, 43–73. [CrossRef]
15. *Proposal for a Directive of the European Parliament and of the Council on the Promotion of the Use of Energy from Renewable Sources (Recast). COM/2016/0767 Final/2–2016/0382 (COD)*; European Commission: Brussels, Belgium, 2016.
16. van Dam, J.; Junginger, M.; Faaij, A.; Jürgens, I.; Best, G.; Fritsche, U. Overview of recent developments in sustainable biomass certification. *Biomass Bioenergy* **2008**, *32*, 749–780. [CrossRef]
17. Ladu, L.; Blind, K. Overview of policies, standards and certifications supporting the European bio-based economy. *Curr. Opin. Green Sustain. Chem.* **2017**, *8*, 30–35. [CrossRef]
18. Blind, K.; Petersen, S.S.; Riillo, C.A.F. The impact of standards and regulation on innovation in uncertain markets. *Res. Policy* **2017**, *46*, 249–264. [CrossRef]
19. *M/491 Mandate Addressed to CEN, CENELEC and ETSI for the Development of European Standards and Technical Specifications and/or Technical Reports for Bio-Surfactants and Bio-Solvents in Relation to Bio-Based Product Aspects*; European Commission: Brussels, Belgium, 2011.
20. *M/492 Mandate Addressed to CEN, CENELEC and ETSI for the Development of Horizontal European Standards and other Standardisation Deliverables for Bio-Based Products*; European Commission: Brussels, Belgium, 2011.
21. *Mandate 53/2008 (CEN M/430) M/430 Mandate Addressed to CEN for the Development of European Standards and CEN Workshop Agreements for Bio-Polymers and Bio-Lubricants in Relation to Bio-Based Product Aspects*; European Commission: Brussels, Belgium, 2008.

22. *Directive 2009/30/EC of the European Parliament and of the Council of 23 April 2009 Amending Directive 98/70/EC as Regards the Specification of Petrol, Diesel and Gas-Oil and Introducing a Mechanism to Monitor and Reduce Greenhouse Gas Emissions and Amending Council Directive 1999/32/EC as Regards the Specification of Fuel Used by Inland Waterway Vessels and Repealing Directive 93/12/EEC: FQD*; European Commission: Brussels, Belgium, 2009.

23. Lewandowski, I. Securing a sustainable biomass supply in a growing bioeconomy. *Glob. Food Secur.* **2015**, *6*, 34–42. [CrossRef]

24. Majer, S.; Moosmann, D.; Sumfleth, B. STAR-ProBio Criteria and Indicator Database. 2017. Available online: http://www.star-probio.eu/wp-content/uploads/2017/12/Star-ProBio_certification_criteria.xlsx (accessed on 15 March 2018).

25. Strauss, A.L.; Corbin, J.M. *Grounded Theory: Grundlagen Qualitativer Sozialforschung*; Beltz: Weinheim, Germany, 2010.

26. STAR-ProBio. *STAR-ProBio Deliverable D7.1. Examination of Existing ILUC Approaches and Their Application to Bio-Based Materials*; Unitelma Sapienza University: Rome, Italy, 2018.

27. DIN. *DIN ISO 13065 Nachhaltigkeitskriterien für Bioenergie*; DIN: Berlin, Germany, 2017.

28. D18 Committee. *Guide for Set of Data Elements to Describe a Groundwater Site*; Part One—Additional Identification Descriptors; ASTM International: West Conshohocken, PA, USA, 2004.

29. D18 Committee. *Guide for Using the Seismic Refraction Method for Subsurface Investigation*; ASTM International: West Conshohocken, PA, USA, 2011.

30. D18 Committee. *Guide for Selection of Methods for Estimating Soil Loss by Erosion*; ASTM International: West Conshohocken, PA, USA, 2012.

31. D20 Committee. *Guide for Exposing and Testing Plastics that Degrade in the Environment by a Combination of Oxidation and Biodegradation*; ASTM International: West Conshohocken, PA, USA, 2018.

32. Iriarte, L. *Consistent Cross-Sectoral Sustainability Criteria & Indicators*; European Commission: Brussels, Belgium, 2017.

33. Iriarte, L. *Benchmark and Gap Analysis of Criteria and Indicators (C&I) for Legislation, Regulations and Voluntary Schemes at International Level and in Selected EU Member States: Main Report*; European Commission: Brussels, Belgium, 2015.

34. Buchholz, T.; Luzadis, V.A.; Volk, T.A. Sustainability criteria for bioenergy systems: Results from an expert survey. *J. Clean. Prod.* **2009**, *17*, S86–S98. [CrossRef]

35. Pavlovskaia, E. Sustainability criteria: Their indicators, control, and monitoring (with examples from the biofuel sector). *Environ. Sci. Eur.* **2014**, *26*, 17. [CrossRef] [PubMed]

36. FAO. FAO's BEFS Approach at a Glance. Available online: http://www.fao.org/3/a-h0011e.pdf (accessed on 20 December 2017).

37. Majer, S.; Gröngröft, A.; Drache, C.; Braune, M.; Meisel, K.; Müller-Langer, F.; Naumann, K.; Oehmichen, K. *Technische und Methodische Grundlagen der THG-Bilanzierung von Biodiesel*, 1st ed.; DBFZ: Leipzig, Germany, 2015.

38. Naumann, K.; Oehmichen, K.; Remmele, E.; Thuneke, K.; Schröder, J.; Zeymer, M.; Zech, K.; Müller-Langer, F. *Monitoring Biokraftstoffsektor*, 3rd ed.; DBFZ: Leipzig, Germany, 2016.

39. Oehmichen, K.; Naumann, K.; Drache, C.; Postel, J.; Braune, M.; Gröngröft, A.; Majer, S.; Meisel, K.; Müller-Langer, F. *Technische und Methodische Grundlagen der THG-Bilanzierung von Biomethan*, 1st ed.; DBFZ: Leipzig, Germany, 2015.

40. Bentsen, N.S. Carbon debt and payback time—Lost in the forest? *Renew. Sustain. Energy Rev.* **2017**, *73*, 1211–1217. [CrossRef]

41. Bovari, E.; Lecuyer, O.; Mc Isaac, F. Debt and damages: What are the chances of staying under the 2 °C warming threshold? *Int. Econ.* **2018**. [CrossRef]

42. Finkbeiner, M. Indirect land use change—Help beyond the hype? *Biomass Bioenergy* **2014**, *62*, 218–221. [CrossRef]

43. Laborde, D. *Assessing the Land Use Change Consequences of European Biofuel Policies: Final Report*; IFPRI: Washington, DC, USA, 2011.

44. Mohr, A.; Beuchelt, T.; Schneider, R.; Virchow, D. Food security criteria for voluntary biomass sustainability standards and certifications. *Biomass Bioenergy* **2016**, *89*, 133–145. [CrossRef]

45. Mohr, A.; Beuchelt, T.; Schneider, R.; Virchow, D. *A Rights-Based Food Security Principle for Biomass Sustainability Standards and Certification Systems*; ZEF: Bonn, Germany, 2015.

46. Peters, D. *Methodologies for the Identification and Certification of Low ILUC Risk Biofuels*; Wageningen University & Research: Wageningen, the Netherlands, 2016.

47. Roundtable on Sustainable Biomaterials. *RSB Low iLUC Risk Biomass Criteria and Compliance Indicators*, 4th ed.; Roundtable on Sustainable Biomaterials: Berlin, Germany, 2015.

48. Marazza, D.; Merloni, E.; Compagnoni, L. *Examination of Existing iLUC Approaches and Application to Bio-Based Materials: Star-ProBio Deliverable 7.1*; European Commission: Bologna, Italy, 2018.

49. Ahlgren, S.; Di Lucia, L. Indirect land use changes of biofuel production—A review of modelling efforts and policy developments in the European Union. *Biotechnol. Biofuels* **2014**, *7*, 35. [CrossRef] [PubMed]

50. Scarlat, N.; Dallemand, J.-F.; Monforti-Ferrario, F.; Nita, V. The role of biomass and bioenergy in a future bioeconomy: Policies and facts. *Environ. Dev.* **2015**, *15*, 3–34. [CrossRef]

51. Van Lancker, J.; Wauters, E.; van Huylenbroeck, G. Managing innovation in the bioeconomy: An open innovation perspective. *Biomass Bioenergy* **2016**, *90*, 60–69. [CrossRef]

52. Selbmann, K.; Pforte, L. Evaluation of Ecological Criteria of Biofuel Certification in Germany. *Sustainability* **2016**, *8*, 936. [CrossRef]

53. CEN/TC 411 Bio-Based Products—Biobasedeconomy. Available online: http://www.biobasedeconomy.eu/centc-411-bio-based-products/ (accessed on 29 June 2018).

54. IEC—SMB: Standardization Management Board > Technical Advisory Committees. Available online: http://www.iec.ch/dyn/www/f?p=103:59:0::FSP_ORG_ID,FSP_LANG_ID:3228,25 (accessed on 29 June 2018).

55. Sherwood, J.; de Bruyn, M.; Constantinou, A.; Moity, L.; McElroy, C.R.; Farmer, T.J.; Duncan, T.; Raverty, W.; Hunt, A.J.; Clark, J.H. Dihydrolevoglucosenone (Cyrene) as a bio-based alternative for dipolar aprotic solvents. *Chem. Commun. (Camb. Engl.)* **2014**, *50*, 9650–9652. [CrossRef] [PubMed]

56. Carus, M.; Dammer, L.; Puente, Á.; Raschka, A.; Arendt, O. *Bio-Based Drop-In, Smart Drop-In and Dedicated Chemicals*; Nova-Institut für Politische und Ökologische Innovation GmbH: Hürth, Germany, 2017.

57. *Communication from the Commission to the European Parliament, the Council, the European Economic and Social Committee and the Committee of the Regions. Closing the Loop—An EU Action Plan for the Circular Economy: COM(2015) 614 Final*; European Commission: Brussels, Belgium, 2015.

58. nova-Institut GmbH. Bio-Based Polymers Worldwide: Ongoing Growth Despite Difficult Market Environment—Bio-Based News—The Portal for Bio-Based Economy & Industrial Biotechnology. 2018. Available online: http://news.bio-based.eu/bio-based-polymers-worldwide-ongoing-growth-despite-difficult-market-environment/ (accessed on 27 April 2018).

59. Grand Review Research. Biolubricants Market Size to Reach $2.92 Billion by 2024. Available online: https://www.grandviewresearch.com/press-release/bio-lubricants-market (accessed on 27 April 2018).

![sustainability logo] *sustainability*

MDPI

Article

Exploratory Research of ISO 14001:2015 Transition among Portuguese Organizations

Luis Miguel Fonseca [1,*] and José Pedro Domingues [1,2]

[1]   ISEP—P. Porto, School of Engineering and CIDEM R&D, Porto 4249-015, Portugal
[2]   Department of Production and Systems, University of Minho, Braga 4710-057, Portugal; jpd@isep.ipp.pt
*   Correspondence: lmf@isep.ipp.pt; Tel.: +351-228-340-500

Received: 16 February 2018; Accepted: 10 March 2018; Published: 12 March 2018

Abstract: The purpose of this paper is the assessment of the ISO 14001:2015 transition process among Portuguese ISO 14001 certified organizations, including those that successfully have already achieved ISO 14001:2015 certification. A considerable number of the surveyed companies proceeded with the transition to the ISO 14001:2015 by introducing slight adjustments and were supported by external consultants. Nearly all of the respondent companies (97%) intend to transition until 15th September 2018. The highest ranked reported benefit is the "integrated approach with other management sub-systems" with a well-consolidated perception from the surveyed companies. This is aligned with the ISO 14001:2015 goal of improving the compatibility of management standards supported on the Annex SL. "Alignment with business strategy", "improved top management commitment" and "improved internal and external communication" are also perceived to obtain significant benefits from ISO 14001:2015. The statistical tests carried out (Kruskal–Wallis) confirmed that the perception of some achieved ISO 14001:2015 certification benefits is dependent on the size of the organization. Concerning the motivations to proceed with certification, results suggest that there is not a particular company profile that is compelled to certify their EMS based on a specific type of motivation (Internal or External). Due to ISO 14001:2015 novelty, these exploratory results should be subjected to additional research confirmation.

Keywords: ISO 14001; ISO 14001:2015; Environmental Management Systems; Benefits; Certification

1. Introduction

Traditional approaches aiming at environmental protection often rely on legal frameworks to enforce measures and behaviors in organizations. The 1992 Rio de Janeiro summit on Environment and Development with the approval of the Framework Convention on Climate Change, the Convention on Biodiversity, the Declaration on Forests and the Agenda 21 triggered an increased international emphasis for the development of environmental sustainability and for more environmental friendly products and services. Within this context, the demand for voluntary environmental management systems (EMS) certification has shown a constant rise, which has been depicted by trends of ISO 14001 international standard certification. ISO 14001 has reached 300,000 certificates in 2016 and seems to mimic the same success path as the widely adopted ISO 9001 Quality Management System international standard, which has more than 1,200,000 issued certificates worldwide [1]. ISO 14001 is an International Standard supported on the assumption that better environmental performance can be attained when environmental aspects are systematically identified and managed through pollution prevention, improved environmental performance and compliance with applicable laws [2].

To ensure that ISO 14001 remains updated and relevant for the marketplace by addressing the latest trends and improving compatibility with other management system standards, such as ISO 9001, ISO revised and issued the 2015 ISO 14001 edition [3]. ISO 14001:2015 proposed some new and reinforced approaches, such as the adoption of the Annex SL structure; the requirement demanding organizations to analyze their internal and internal context in order to ascertain the issues and requirements that may impact the EMS; the assignment of specific responsibilities for leaders to promote the EMS; more emphasis on the performance improvement by minimizing the water and energy consumption as well as producing less emission and waste; the introduction of a life cycle perspective that asks for organizations to extend their control and influence to the environmental impacts ascribed to product use and end-of-life treatment or disposal; and the development of an external and internal communications strategy to consider the reporting demanded by regulatory agencies and the expectations of other interested parties [2].

ISO 14001 international standard was first published on 1st September 1996, which established the requirements for EMS, including the organizational structure and the responsibilities; the planning of the activities; the definition of practices, processes and procedures; and the allocation of resources to plan, implement, check and improve the environmental policy [2,4–7].

In the last few decades, many companies aimed to improve the environmental performance of their processes [8] by implementing an EMS. The theoretical literature focusing on the motivations driving organizations to implement and certify an EMS address both external [9–11] and internal [12,13] factors that lead companies to implement such self-regulatory mechanisms.

Additionally, there is a considerable stream of research addressing the benefits [14–24] and the barriers [13,15–18] of ISO 14001 adoption and certification. A systematic review analyzed the content of 48 articles published between January 2012 and April 2017 that addressed the topic of "benefits of EMS adoption and certification". This review provided support for this resulting in fairly positive benefits (average 2.36 in a 1 to 5 Likert type scale) for the certified organizations, although there were some observed variations [20]. The main reported benefits include the ability to manage the environmental aspects more effectively and continually improve the environmental performance; an increased environmental legislation compliance; the prevention of pollution; a lower risk level of penalties and litigation; the improvement of stakeholder satisfaction and employee morale; and the potential to access new markets and new business opportunities with environmentally aware customers. This leads to both lower operating costs and new business opportunities, leveraging the competitive position of organizations [2,13,16,18]. However, there are also some reported negative consequences of EMS implementation and certification, such as the difficulty in measuring the EMS efficiency, increased bureaucracy, higher costs, lack of employee and management EMS support and the generic nature of the ISO 14001 requirements [6,15,16]. Based on an investigation of the benefits and difficulties of adopting ISO 14001:2004 within Italian organizations, Mazzi et al. [17] concluded that the most useful requirements of ISO 14001:2004 are "environmental aspects", "legal and other requirements", "competence, training, and awareness" and "evaluation of compliance". According to this research, the most difficult ISO 14001:2004 requirements to comply with are "legal and other requirements", "competence, training and awareness", "operational control" and "evaluation of compliance". These results highlight that the most difficult requirements match and concur with those considered as the most useful ones.

Boiral, Guillaumie, Heras-Saizarbitoria and Tene [25] concluded that whereas the mainstream literature supports the emergence of positive outcomes after ISO 14001 adoption and certification, the research is often limited in terms of "sample, scope, variables adopted and contextual aspects" and is essentially based on managerial perceptions, which may be influenced by respondent bias. Additionally, according to these authors, positive and negative outcomes of ISO 14001 adoption may co-exist (e.g., less energy and resource consumption costs, but additional bureaucracy and paperwork).

From a theoretical point of view, the adoption of voluntary management systems standards, such as ISO 14001, can be traced back to the following theories: Freeman Stakeholder Theory [26] focusing on the importance of a firms' relationships with critical stakeholders, which may lead to an improved performance by integrating business and societal considerations that subsequently creates value for their stakeholders; Barney [27] Resource Base View Theory considering the unique combination of resources and capabilities (internal) of each firm allowing it to be unique and different with improved performance compared to its competitors; and Meyer and Rowan Institutional Theory [28] addressing the mode of behavior of institutions as a result of societal influence, which explains why organizations converge and become similar (Fonseca et al. [29]; Tuczek, Castka and Wakolbinger [30]). By analyzing the changes introduced in the ISO 9001:2015, Fonseca [31] pointed out that it brings a stronger open systems approach (influence of the environment, dynamic perspective, need for survival) when compared with the 2008 version. Due to the commonalities of both International Standards 2015 edition, this can be generalized to the ISO 14001:2015 when compared to the 2004 version.

As of 15 September 2018, all ISO 14001:2014 certificates will be no longer valid, so ISO 14001 certified organizations should proceed with the transition process until that date. After this time, they should demonstrate their compliance in being successfully audited and certified accordingly to ISO 14001:2015 international standard by a credible and recognized certification body. This is a relevant research issue, both for scholars (since previous research did not address ISO 14001: 2015 edition) and for practitioners (that need to migrate the ISO 14001:2004 EMS to the new edition).

As part of a broader research project to study the phenomenon of the management systems standards certification, an empirical study, supported by an online survey, was designed, developed and made available to the potential respondents, which had the specific purpose of gathering knowledge of the ISO 14001:2015 transition process. The survey was carried out throughout May 2017 and yielded a total of 108 valid responses.

The next section is devoted to Materials and Methods. In Section 3, the authors present, discuss and dissect the empirical results. Section 4 summarizes the soundest results and point out the implications of the research.

2. Materials and Methods

Data was collected by the means of an online survey among ISO 9001 and ISO 14001 Portuguese certified organizations by the leading Portuguese certification body. This approach is consistent with those adopted by other researchers, such as Mazzi et al. [17]. An e-mail was sent to the companies in April 2017, followed by a second call in May 2017. The data were collected anonymously through an automatic online database. The sample comprised a total of 108 Portuguese organizations simultaneously certified according to the clauses of both the ISO 9001 and ISO 14001 standards. The overall response rate was 18%, encompassing 19 organizations already certified against ISO 14001:2015 and the remaining against ISO 14001:2004. This research is included in a broader research project that aims to achieve deeper understanding of the phenomenon of management systems standards certification (e.g., Hypotheses 4 and 11 relate to an ongoing parallel research addressing Quality Management Systems—QMS). The survey was designed to include several sections, which is shown in Table 1. The detail is presented for the sections relevant to this research. IBM Social Sciences Statistical Package (SPSS) v. 22 software was adopted to conduct the statistical tests and calculations (after ordinal to numerical transformation of the Likert scale type of answers). The non-parametric Kruskal–Wallis one-way analysis of variance statistical test was used to determine whether some variables, measured on an ordinal scale, differed based on other variables (namely those related to the characterization of the company). Hence, the following research hypotheses were raised:

Dimension 1—Benefits

Research Statement: The assessment of some benefits derived from the implementation of the ISO 14001:2015 standard differ according to the . . .

Hypothesis 1 (H1). . . . *activity sector where the organization operates.*

Hypothesis 2 (H2). . . . *dimension of the organization (N° of employees).*

Hypothesis 3 (H3). . . . *exposure to international markets.*

Hypothesis 4 (H4). . . . *maturity (years) of the QMS.*

Hypothesis 5 (H5). . . . *maturity (years) of the EMS.*

Hypothesis 6 (H6). . . . *organizational role of the respondent.*

Hypothesis 7 (H7). . . . *experience (years) of the respondent.*

Dimension 2—Motivations

Research Statement: The motivations (internal or external) driving organizations to implement the ISO 14001:2015 standard differ according to the . . .

Hypothesis 8 (H8). . . . *activity sector where the organization operates.*

Hypothesis 9 (H9). . . . *dimension of the organization (N° of employees).*

Hypothesis 10 (H10). . . . *exposure to international markets.*

Hypothesis 11 (H11). . . . *maturity (years) of the QMS.*

Hypothesis 12 (H12). . . . *maturity (years) of the EMS.*

Hypothesis 13 (H13). . . . *organizational role of the respondent.*

Hypothesis 14 (H14). . . . *experience (years) of the respondent.*

Table 1. Survey structure.

Section	Variable(s) ID	Item Assessed	Typology
1—Characterization of the company.	Var_Num_1.1	Activity Sector	Structured (S.O)
	Var_Num_1.2	Dimension (N° of employees)	Structured (S.O.)
	Var_Num_1.3	International market activity (%)	Structured (S.O.)
	Var_Num_3.1.1	Maturity (Years)	Structured (S.O.)
3—Environmental Management System (ISO 14001)	Var_Num_3.1.2		Structured (S.O.)
	Var_Num_3.2.1		Structured (S.O.)
	Var_Num_3.2.2		Structured (M.O.)
	Var_Num_3.2.3	Transitioned to new revision (2015)	Structured (M.O.)
	Var_Num_3.2.4		Structured (M.O.)
	Var_Num_3.2.5		Structured (M.O.)
	Var_Num_3.2.6		Structured (S.O.; L.S.)
	Var_Num_3.3.1		Structured (M.O.)
	Var_Num_3.3.2		Structured (S.O.)
	Var_Num_3.3.2	Not transitioned to new revision (2004)	Structured (S.O.)
	Var_Num_3.3.3		Structured (M.O.)
	Var_Num_3.3.4 (a to d)		Structured (S.O.; L.S.)
	Var_Num_6.1.1		Structured (S.O.)
	Var_Num_6.1.2		Structured (S.O.)
5—Questions relevant to the business processes of the certification entity.	Var_Num_6.2.1	ISO 14001	Structured (S.O.; L.S.)
	Var_Num_6.2.2		Open question.
	Var_Num_6.2.3		Open question.
	Var_Num_6.2.4		Open question.
	Var_Num_6.2.5		Open question.
6—Motivations for certification and impacts	Var_Num_7.1		Structured (S.O.; L.S.)
	Var_Num_7.2 (a to s)		Structured (S.O.; L.S.)
	Var_Num_7.3 (a to h)		Structured (S.O.; L.S.)
8—Respondent characterization.	Var_Num_9.1	Organizational role	Structured.
	Var_Num_9.2	Experience (Years)	Structured.

S.O.—Single option (Solely allows the selection of one option); M.O.—Multiple option (Allows the selection of more than one option); L.S.—Likert scale; Solely the survey items related and pertinent to this research are presented (1, 3, 5, 6 and 8).

3. Results and Discussion

This section is devoted to the presentation and discussion of the results. The first sub-section reports the social-demographical characteristics of the sample. The second sub-section presents the descriptive statistics of the results (mainly summarized as a percentage per item assessed, average and standard deviation) and in the following section, the results of the statistical tests carried out for the validation of the hypotheses are reported and discussed.

3.1. Descriptive Statistics

The typological analysis of the respondent organizations suggests that the sample mimics and resemble the Portuguese reality concerning business demographics. Mainly small and medium enterprises (SMEs) operating at the product and service-oriented activity sectors answered the survey (Figures 1 and 2).

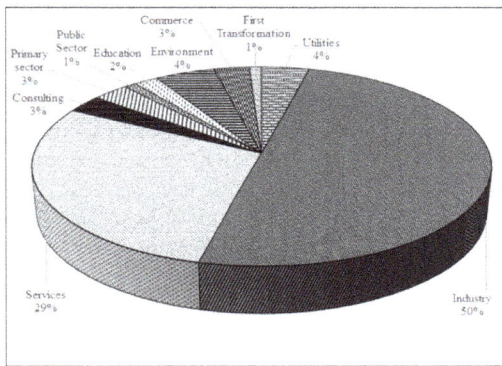

Figure 1. Surveyed organizations—Breakdown by activity sector.

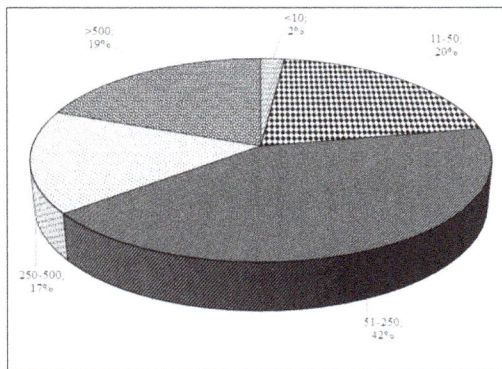

Figure 2. Surveyed organizations—Breakdown by dimension (N° of employees).

A considerable number of the respondent organizations (62%) develop their business activities that are mainly aimed at the Portuguese business market (Figure 3) and held the environmental management system (EMS) certification over more than 4 years (Figure 4).

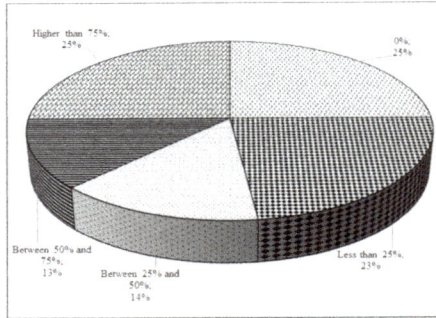

Figure 3. International business orientation.

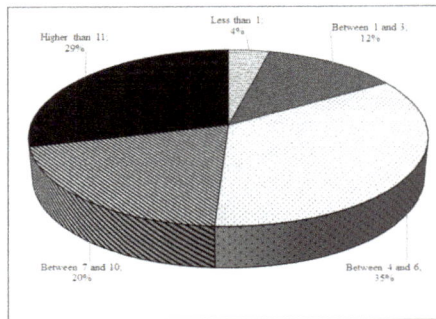

Figure 4. Surveyed organizations—Breakdown by EMS maturity (years).

When the survey was completed online, around three quarters of the respondent organizations had not transitioned to the 2015 revision (Figure 5) due to lack of time and mandatory study and analysis of the novel requirements (Figure 6). However, the majority of the respondent organizations (97%) intend to proceed with the transition before the three-year transition period, with essentially the majority of the organizations expecting to have transitioned by 15th September 2018 (Figure 7). In addition to the major reasons pointed out previously, those organizations that did not proceeded with the transition point out some difficulties in understanding and implementing some requirements and the ascribed costs inherent to the transition process.

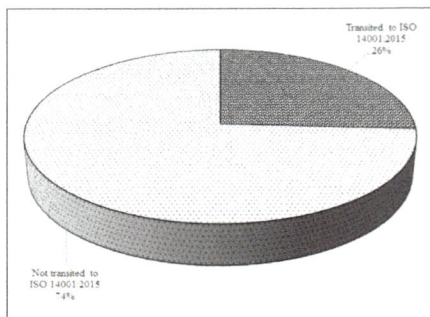

Figure 5. Surveyed organizations- Breakdown by transition to 2015 version.

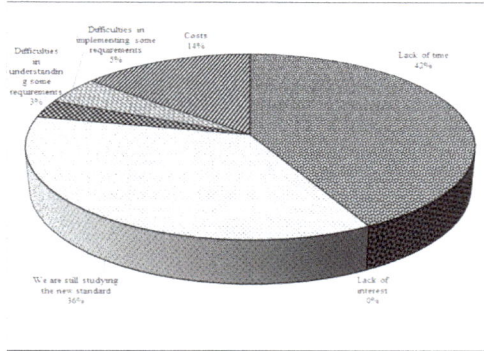

Figure 6. Reasons for not proceeding to the transition.

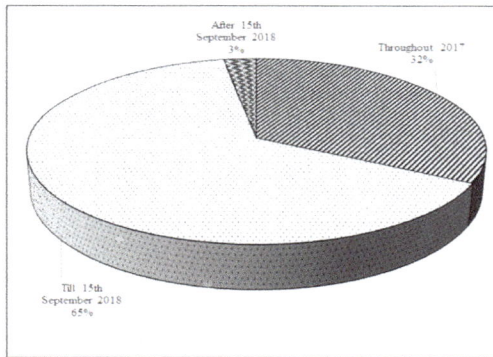

Figure 7. Expected date for transition.

Concerning the characterization of the individual respondent responsible for the submission of the survey, it should be pointed out that these were mainly management system managers (Figure 8) that held more than 11 years of experience (Figure 9).

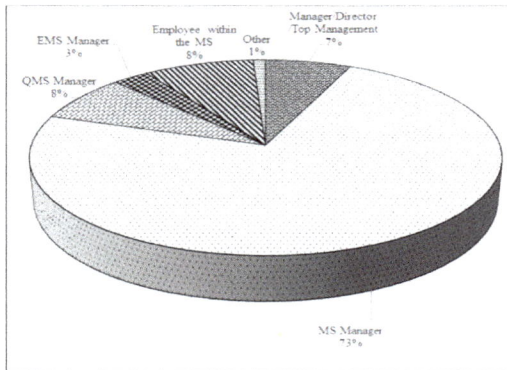

Figure 8. Role of the respondent.

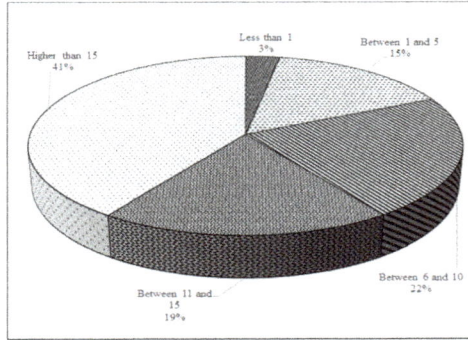

Figure 9. Experience of the respondent (years).

3.2. Summarized Results

The results suggest that no particular resource/strategy was favored over other (Figure 10) and a considerable proportion (around 77%) of the surveyed organizations that proceeded with the transition to the ISO 14001:2015 introduced slight adjustments to the pre-existing EMS (Figure 11). Approximately one-third of the organizations proceeded with the transition and were supported solely on the available internal resources. The remaining organizations reported the support of external consultants, who carried out some training sessions focusing on the ISO 14001 new requirements (Figure 10).

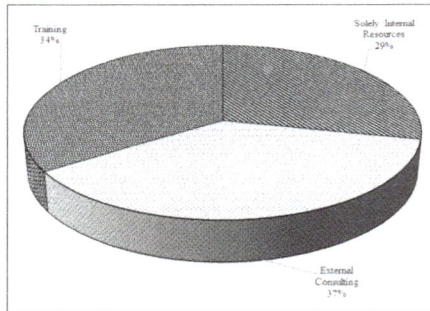

Figure 10. Resources/strategies adopted throughout the ISO 14001:2015 transition.

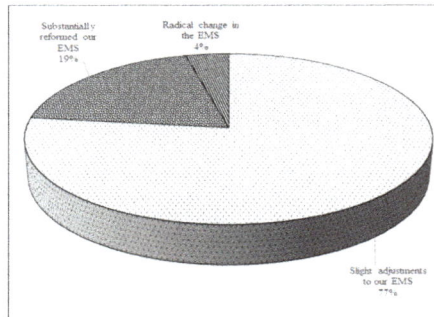

Figure 11. Changes introduced throughout the implementation of ISO 14001:2015 revision.

Concerning the difficulties faced by the organizations (Figure 12) when implementing the new revision requirements, approximately one-quarter of the respondents pointed out the implementation of the life cycle perspective. Other noticeable items pointed out were: "risk management approach", "context–environmental conditions" and "determining results of EMS". Regarding the most useful concepts, nearly one-fifth of the companies found the determination of risks and opportunities, the life cycle perspective and the mapping of the context of the organization the most relevant added values (Figure 13).

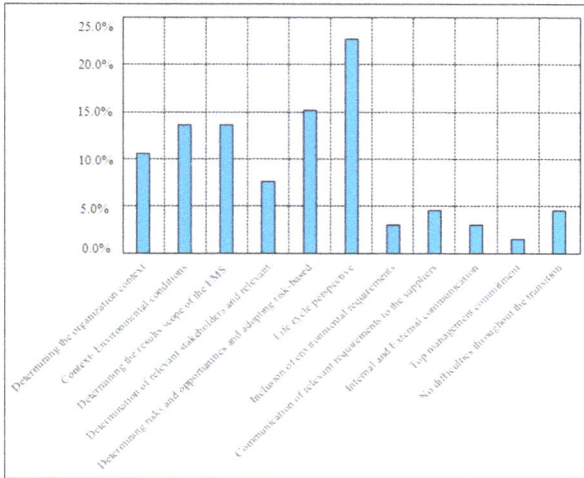

Figure 12. Difficulties faced throughout the transition.

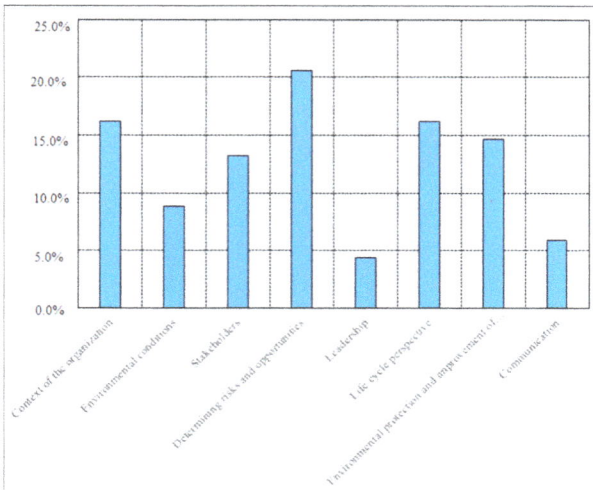

Figure 13. Most useful concepts throughout the transition.

Concurrently, those organizations that did not yet proceeded with the transition expect the implementation of "life cycle perspective" requirement as the one which will present most challenges.

Other difficulties pointed out by organizations include the "determination of risks and opportunities", "mapping of the context of the organization" and the assurance of the "top management involvement" (Figure 14).

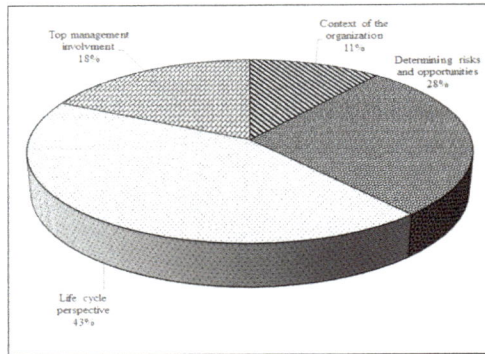

Figure 14. Expected difficulties throughout the transition.

Table 2 presents the items assessed by variable 3.2.6—Benefits from the implementation of the new requirements of the ISO 14001: 2015 edition. Figures 15 and 16 display the summarized results (average and standard deviation). The highest rated benefit pointed out by the surveyed organizations was the "integrated approach with other management sub-systems (Num_Var_3.2.6g)" and the small standard deviation suggests that this is a well consolidated perception from the surveyed organizations. This result is consistent with one of the purposes of the new revision: to improve the compatibility of the standards supported on the Annex SL. Furthermore, one should point out the benefits of "alignment with business strategy (Num_Var_3.2.6a)", "improved top management commitment (Num_Var_3.2.6b)" and "improved internal and external communication (Num_Var_3.2.6c)". The lowest rated benefit perceived by the organizations was "less prescriptive requirements and documentation (Num_Var_3.2.6d)". The benefit of "improved top management commitment (Num_Var_3.2.6b)" presented the greatest standard deviation (Figure 16), suggesting that it is not perceived consistently among the surveyed organizations.

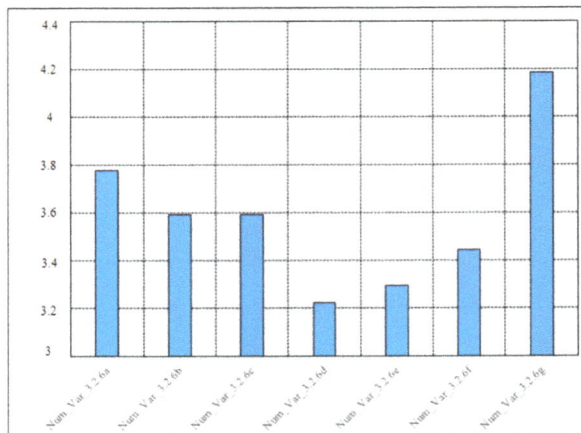

Figure 15. Benefits from the ISO 14001:2015 implementation—Average.

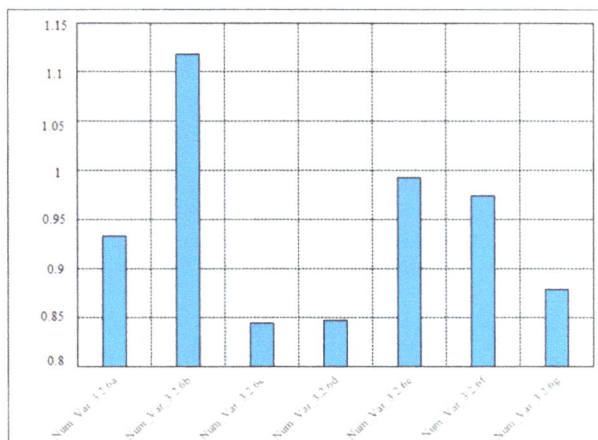

Figure 16. Benefits from the ISO 14001:2015 implementation—Standard deviation.

Table 2. Assessment of the benefits from the implementation of ISO 14001:2015.

Var_ID	Item Assessed	Observations
Num_Var_3.2.6a	Alignment with business strategy.	
Num_Var_3.2.6b	Improved top management commitment.	(1) Not relevant
Num_Var_3.2.6c	Improved internal and external communication.	(2) . . .
Num_Var_3.2.6d	Less prescriptive requirements and documentation.	(3) . . .
Num_Var_3.2.6e	Consumption and cost reduction.	(4) . . .
Num_Var_3.2.6f	Improved environmental performance.	(5) Very relevant
Num_Var_3.2.6g	Integrated approach with other management sub-systems.	

3.3. Statistical Tests

The Kolmogorov–Smirnov and the Shapiro–Wilk statistical tests were used to evaluate the normality of the distribution of the results collected (Table 3) and decide the tests to be used for the research questions. The statistical tests show that the results did not have normal distribution (Sigma ≤ 0.05) and therefore, the Kruskal–Wallis statistical test was used to ascertain and validate the statistical hypotheses formulated.

Table 3. Normality tests (Kolmogorov–Smirnov and Shapiro–Wilk).

	Kolmogorov–Smirnov [a]			Shapiro–Wilk		
	Statistic	df	Sig.	Statistic	df	Sig.
Num_Var_3.3.4a	0.224	75	0.000	0.898	75	0.000
Num_Var_3.3.4b	0.315	75	0.000	0.721	75	0.000
Num_Var_3.3.4c	0.296	75	0.000	0.772	75	0.000
Num_Var_3.3.4d	0.296	75	0.000	0.815	75	0.000
Num_Var_3.2.6a	0.298	27	0.000	0.839	27	0.001
Num_Var_3.2.6b	0.272	27	0.000	0.865	27	0.002
Num_Var_3.2.6c	0.240	27	0.000	0.872	27	0.003
Num_Var_3.2.6d	0.248	27	0.000	0.872	27	0.003
Num_Var_3.2.6e	0.272	27	0.000	0.856	27	0.002
Num_Var_3.2.6f	0.231	27	0.001	0.890	27	0.008
Num_Var_3.2.6g	0.268	27	0.000	0.783	27	0.000
Num_Var_3.2.6h	0.216	27	0.002	0.850	27	0.001
Num_Var_7.1	0.210	87	0.000	0.915	87	0.000

[a] Lilliefors Significance Correction.

Tables 4 and 5 present the results from the Kruskal–Wallis statistical test (Asymptotic Sigma). The results (Table 4) suggest that the benefits of "Num_Var_3.2.6a—alignment with business strategy", "Num_Var_3.2.6b—increased commitment from Top Management" and "Num_Var_3.2.6f—improved environmental performance" are strongly influenced by the size of the organization (N° of employees) ($p < 0.1$). Essentially, the alignment with business strategy seems to be perceived mainly by medium- and large-sized organizations, while the increased commitment from Top Management and improved environmental performance are mainly perceived by small- and medium-sized organizations. Moreover, the perception regarding the benefit of "Num_Var_3.2.6d—less prescriptive requirements and documentation" seems to be dependent on the international exposure of the company to the international market as companies operating mainly in the internal market rate this benefit higher.

Table 4. Benefits from the implementation of ISO 14001:2015 (SPSS Kruskal–Wallis test outputs).

		Tested Variable						
		Var_3.2.6a	Var_3.2.6b	Var_3.2.6c	Var_3.2.6d	Var_3.2.6e	Var_3.2.6f	Var_3.2.6g
	Var_1.1	0.432	0.555	0.727	0.254	0.313	0.375	0.390
	Var_1.2	0.066 *	0.078 *	0.263	0.537	0.264	0.061 *	0.239
	Var_1.3	0.965	0.185	0.584	0.067 *	0.167	0.197	0.922
Grouping Variable	Var_2.1.1	0.336	0.044 **	0.451	0.014 **	0.029 **	0.024 **	0.955
	Var_3.1.1	0.179	0.132	0.306	0.272	0.225	0.218	0.545
	Var_9.1	0.248	0.868	0.865	0.351	0.375	0.386	0.478
	Var_9.2	0.736	0.589	0.339	0.531	0.869	0.782	0.685

* Statistically relevant at $p < 0.1$; ** Statistically relevant at $p < 0.05$.

More robust results were achieved when testing the variable of "Num_Var_7.1—internal or external motivations" to proceed with certification (Table 5). The results suggest that none of the grouping variables impact significantly on the overall results as the differences within the different sub-groups comprising each grouping variable are not statistically relevant. This consistency within the results point out that there is not a peculiar company profile that is compelled to certify their management sub-systems based on a peculiar type of motivation (Internal or External), which is consistent with the generic approach of the meta-standards.

Table 5. Motivations for certification (SPSS Kruskal–Wallis test outputs).

		Tested Variable
		Var_7.1
	Var_1.1	0.203
	Var_1.2	0.962
	Var_1.3	0.507
Grouping Variable	Var_2.1.1	0.299
	Var_3.1.1	0.529
	Var_9.1	0.356
	Var_9.2	0.102

* Statistically relevant at $p < 0.1$; ** Statistically relevant at $p < 0.05$.

Table 6 summarizes the validation of the several research hypotheses raised and pertinent to this research.

Table 6. Validity of the research hypotheses.

Dimension	Research Hypotheses	Validity	Comment
Assessment of ISO 14001:2015 Benefits	H1	× Not confirmed.	activity sector where the organization operates
	H2	√ Confirmed.	dimension of the organization (N° of employees)
	H3	√ Confirmed.	exposure to international markets
	H5	× Not confirmed.	maturity (years) of the EMS
	H6	× Not confirmed.	organizational role of the respondent
	H7	× Not confirmed.	experience (years) of the respondent
Motivations	H8	× Not confirmed.	activity sector where the organization operates
	H9	× Not confirmed.	dimension of the organization (N° of employees)
	H10	× Not confirmed.	exposure to international markets
	H12	× Not confirmed.	maturity (years) of the EMS
	H13	× Not confirmed.	organizational role of the respondent
	H14	× Not confirmed.	experience (years) of the respondent

4. Conclusions

More than 75% of the surveyed organizations proceeded with the transition to the ISO 14001:2015 by introducing slight adjustments to the pre-existing EMS, 37% were supported by external consultants and the remaining 63% proceeded with the transition solely supported on available internal resources. ISO 14001:2015 training was also extensively promoted and carried out. By May 2017, approximately three quarters of the respondent organizations still had not attained ISO 14001:2015 certification, which was reportedly due to lack of time and the mandatory study and analysis of the novel requirements. However, 97% of the organizations intended to transition by 15 September 2018.

The respondents considered that the most useful concepts of ISO 14001:2015 were the "determination of risks and opportunities", the "life cycle perspective" and the "mapping of the context of the organization, which is aligned and supports the intended added value of the new ISO 14001:2015 approaches. Simultaneously, some of these concepts were also reported as the hardest difficulties to overcome by the organizations when implementing or transitioning to the 2015 edition of ISO 14001, which were namely the implementation of the "life cycle perspective", "risk management approach", "context–environmental conditions" and "determining results of EMS". Due to the novelty of ISO 14001:2015 and the fact these requirements were not included in the previous ISO 14001:2014 edition, we cannot discuss these results compared to existing literature. However, these conclusions seem to be aligned and concur with the notion that the requirements usually most difficult to comply with are those also considered the most useful, which was expressed by Mazzi et al. [17] (Italian ISO 14001:2004 certified companies). Similarly, the reported results in the current paper suggest a correspondence between usefulness and difficulty of some EMS requirements of the ISO 14001:2015.

The highest ranked benefit is "integrated approach with other management sub-systems", which seemingly is a well-consolidated perception from the surveyed companies (low score of the standard deviation). This result supports one of the major ISO 14001:2015 objectives: the improvement of the standards compatibility supported on the Annex SL. Furthermore, this is aligned with the conclusions of Domingues et al. on the relevance of Integrated Management Systems [32]. "Alignment with business strategy", "improved top management commitment" and "improved internal and external communication" also achieved high rankings, while the lowest rated benefit perceived by the organizations was "less prescriptive requirements and documentation". The benefit of "improved top management commitment" (due to its high standard deviation) is not perceived consistently by all the surveyed companies.

The results from this research highlight that the perception of some achieved ISO 14001:2015 certification benefits varies according to the organization size. The alignment with business strategy seems to be perceived mainly by medium- and large-sized organizations, whereas the increased commitment from Top Management and improved environmental performance are mainly perceived by small- and medium-sized organizations. Small organizations are usually considered to be more flexible, but less formal and strategic than larger organizations, which might explain these results.

Moreover, medium- and large-sized organizations are more pressured by higher stakeholder scrutiny and therefore, are more accustomed to performance measurement.

Concerning the "internal or external motivations" for proceeding with certification, the results indicate that the differences due to activity sector, organizational dimension, exposure to international markets, the maturity of the EMS and the respondent profile are not statistically relevant. This suggests that there is no specific company profile that is compelled to certify their EMS based on a specific type of motivation (Internal or External), which is consistent with the generic approach of the ISO international meta-standards.

In summary, the results of the statistical analysis highly that in May 2017, only one-quarter of ISO 14001 certified Portuguese organizations had successfully proceeded with the ISO 14001:2015 transition, although 97% of the companies intended to transition by 15th September 2018. The results show that surveyed organizations benefited from the ISO 14001:2015 implementation and the adoption of the new and reinforced requirements, such as determination of risks and opportunities; the mapping of the organizational context and stakeholder identification; and the adoption of the life cycle perspective. Such benefits enhance the environmental performance of the organization and improve the compatibility with other management system standards, such as ISO 9001. According to the statistical tests, the perception of some achieved ISO 14001:2015 certification benefits varies with the size of the organization, whereas the motivation to proceed with certification is independent of organization profile.

These results are valuable for managers and practitioners as they identify the ISO 14001:2015 transition strategies, the most useful requirements and the major obstacles to be overcome as well as pointing out priorities to successfully achieve the transition and maximize the benefits of ISO 14001:2015 adoption. Another interesting conclusion is the relevance of the training made available to the employees (addressing the novel ISO 14001:2015) and the use of external consultants for supporting the transition process. This is consistent with the conclusions of previous research that stressed the importance and criticality of the knowledge of environmental impacts and the need for training and consulting to overcome the difficulties in implementing an EMS [21].

Concerning the limitations of this research, although it addresses a wide range of activity sectors, it is restricted to ISO 14001 certified organizations in Portugal. Additionally, our research suffers from the limitations of the survey methodologies and the potential subjective and biased point of view from the respondents. The analysis of the survey results suggests that it matches (i.e., properly represents) the population, since the distribution of the companies' sample profile is consistent with the population. "Wave analysis" was also adopted to compare the results from late respondents and early respondents [33], with the results showing no significant differences. Thus, this means that there is a minimal possible error from non-respondent bias.

As ISO 14001:2015 implementation is still in the early phase, these exploratory results should be subject to further research confirmation, preferably with a larger sample size and eventually assessing the maturity level attained by the resulting integrated management system [32]. It should be interesting to explore the differences between ISO 14001:2015 and ISO 14001:2004 pros and cons, considering different organizations' sizes, sectors and EMS maturity. The assessment of the environmental outcomes (benefits) of ISO 14001:2015 EMS, its integration with other management systems and the extension of this research to other countries would be also desirable for assessing the possible generalization of these findings.

Acknowledgments: Authors would like to thank all the organizations that kindly answered the questionnaire and to APCER. CIDEM, R&D unit is funded by the FCT—Portuguese Foundation for the Development of Science and Technology, Ministry of Science, Technology, and Higher Education, under the Project UID/EMS/0615/2016. Pedro Domingues benefited from financial support through the FCT post-doc research grant No. SFRH/BPD/103322/2014.

Author Contributions: The manuscript was written by Luis Miguel Fonseca and José Pedro Domingues. Both authors contributed to the study design and the results interpretation, having approved the final manuscript.

Sustainability **2018**, *10*, 781

Conflicts of Interest: The authors declare no conflict of interest.

References

1. International Organization for Standardization. The ISO Survey of Management System Standard Certifications—Executive Summary. 2017. Available online: http://www.iso.org/iso/iso_survey_executive-summary.pdf? (accessed on 3 May 2016).
2. Da Fonseca, L.M.C.M. ISO 14001: 2015: An improved tool for sustainability. *J. Ind. Eng. Manag.* **2015**, *8*, 37–50. [CrossRef]
3. ISO 14001:2015. *Environmental Management System: Requirements with Guidance for Use*; International Organization for Standardization: Geneva, Switzerland, 2015.
4. Abarca, D. Implementing ISO 9000 & ISO 14001 concurrently. *Pollut. Eng.* **1988**, *30*, 46–48.
5. Corbett, C.J.; Kirsch, D.A. International diffusion of ISO 14000 certification. *Prod. Operat. Manag.* **2001**, *10*, 327–342. [CrossRef]
6. Curkovic, S.; Sroufe, R. Using ISO 14001 to promote a sustainable supply chain strategy. *Bus. Strategy Environ.* **2010**, *20*, 71–93. [CrossRef]
7. Darnall, N. Why firms mandate ISO 14001 certification. *Bus. Soc.* **2006**, *45*, 354–382. [CrossRef]
8. Chiarini, A. Sustainable manufacturing-greening processes using specific Lean Production tools: An empirical observation from European motorcycle component manufacturers. *J. Clean. Prod.* **2014**, *85*, 226–233. [CrossRef]
9. Bansal, P.; Roth, K. Why companies go green: A model of ecological responsiveness. *Acad. Manag. J.* **2000**, *43*, 717–736. [CrossRef]
10. Chiarini, A. Strategies for Developing an Environmentally Sustainable Supply Chain: Differences Between Manufacturing and Service Sectors. *Bus. Strat. Environ.* **2014**, *23*, 493–504. [CrossRef]
11. Uchida, T.; Ferraro, P.J. Voluntary development of environmental management systems: Motivations and regulatory implications. *J. Regul. Econ.* **2007**, *32*, 37–65. [CrossRef]
12. King, A.A.; Lenox, M.J.; Terlaak, A.K. The strategic use of decentralized institutions: Exploring certification with the ISO 14001 management standard. *Acad. Manag. J.* **2005**, *48*, 1091–1106. [CrossRef]
13. Heras-Saizarbitoria, I.; Landin, G.A. Do drivers matter for the benefits of ISO 14001? *Int. J. Operat. Prod. Manag.* **2011**, *31*, 192–215. [CrossRef]
14. Poksinska, B.; Dahlgaard, J.; Eklund, J. Implementing ISO 14000 in Sweden: Motives, benefits and comparisons with ISO 9000. *Int. J. Qual. Reliab. Manag.* **2003**, *20*, 585–606. [CrossRef]
15. Boiral, O.; Henri, J.F. Modelling the impact of ISO 14001 on environmental performance: A comparative approach. *J. Environ. Manag.* **2012**, *99*, 84–97. [CrossRef] [PubMed]
16. Tarí, J.J.; Molina-Azorín, J.F.; Heras, I. Benefits of the ISO 9001 and ISO 14001 standards: A literature review. *J. Ind. Eng. Manag.* **2012**, *5*, 297–322. [CrossRef]
17. Mazzi, A.; Toniolo, S.; Mason, M.; Aguiari, F.; Scipioni, A. What are the benefits and difficulties in adopting an environmental management system? *The opinion of Italian organizations. J. Clean. Prod.* **2016**, *139*, 873–885. [CrossRef]
18. Murmura, F.; Liberatore, L.; Bravi, L.; Casolani, N. Evaluation of Italian Companies' Perception about ISO 14001 and Eco'Management and Audit Scheme III: Motivations, Benefits and Barriers. *J. Clean. Prod.* **2018**, *174*, 691–700. [CrossRef]
19. Wang, X.; Lin, H.; Weber, H. Does Adoption of Management Standards Deliver Efficiency Gain in Firms' Pursuit of Sustainability Performance? An Empirical Investigation of Chinese Manufacturing Firms. *Sustainability* **2016**, *8*, 694. [CrossRef]
20. Fonseca, L.M.; Domingues, J.P.; Machado, P.B.; Calderón, M. Management System Certification Benefits: Where Do We Stand? *J. Ind. Eng. Manag.* **2017**, *10*, 476–494. [CrossRef]
21. Ferenhof, H.A.; Vignochi, L.; Selig, P.M.; Lezana, A.G.R.; Campos, L.M.S. Environmental management systems in small and medium-sized enterprises: An analysis and systematic review. *J. Clean. Prod.* **2014**, *74*, 44–53. [CrossRef]
22. Guerrero-Baena, M.D.; Gomez-Limon, J.A.; Fruet, J.D. A multicriteria method for environmental management system selection: An intellectual capital approach. *J. Clean. Prod.* **2015**, *105*, 428–437. [CrossRef]

23. Oliveira, J.A.; Oliveira, O.J.; Ometto, A.R.; Ferraudo, A.S.; Salgado, M.H. Environmental management system ISO14001 factors for promoting the adoption of cleaner production practices. *J. Clean. Prod.* **2016**, *133*, 1384–1394. [CrossRef]

24. Ferreira Rino, C.A.; Salvador, N.N.B. ISO 14001 certification process and reduction of environmental penalties in organizations in Sao Paulo State, Brazil. *J. Clean. Prod.* **2017**, *142*, 3627–3633. [CrossRef]

25. Boiral, O.; Guillaumie, L.; Heras-Saizarbitoria, I.; Tayo Tene, C.V. Adoption and Outcomes of ISO 14001: A Systematic Review. *Int. J. Manag. Rev.* **2017**. [CrossRef]

26. Freeman, R. *Strategic Management: A Strategic Approach*; Pitman: Boston, MA, USA, 1984.

27. Barney, J. Firm resources and sustained competitive advantage. *J. Manag.* **1991**, *17*, 99–120. [CrossRef]

28. Meyer, J.W.; Rowan, B. Institutionalized organizations: Formal structure as myth and ceremony. *Am. J. Sociol.* **1977**, *83*, 340–363. [CrossRef]

29. Fonseca, L.; Ramos, A.; Rosa, A.; Braga, A.C.; Sampaio, P. Stakeholders satisfaction and sustainable success. *Int. J. Ind. Syst. Eng.* **2016**, *24*, 144–157. [CrossRef]

30. Tuczek, F.; Castka, P.; Wakolbinger, T. A review of management theories in the context of quality, environmental and social responsibility voluntary standards. *J. Clean. Prod.* **2018**, *176*, 399–416. [CrossRef]

31. Fonseca, L.M. ISO 9001 quality management systems through the lens of organizational culture. *Qual. Access Success* **2015**, *16*, 54–59.

32. Domingues, J.P.T.; Sampaio, P.; Arezes, P.M. Integrated management systems assessment: A maturity model proposal. *J. Clean. Prod.* **2016**, *124*, 164–174. [CrossRef]

33. Armstrong, J.S.; Overton, T.S. Estimating Nonresponse Bias in Mail Surveys. *J. Mark.* **1977**, *14*, 396–402S. [CrossRef]

MDPI

St. Alban-Anlage 66

4052 Basel

Switzerland

Tel. +41 61 683 77 34

Fax +41 61 302 89 18

www.mdpi.com

Sustainability Editorial Office

E-mail: sustainability@mdpi.com

www.mdpi.com/journal/sustainability

MDPI
St. Alban-Anlage 66
4052 Basel
Switzerland

Tel: +41 61 683 77 34
Fax: +41 61 302 89 18

www.mdpi.com

MDPI

ISBN 978-3-03897-381-2

www.ingramcontent.com/pod-product-compliance
Lightning Source LLC
Chambersburg PA
CBHW051725210326
41597CB00032B/5615